Food Irradiation

FOOD SCIENCE AND TECHNOLOGY

A SERIES OF MONOGRAPHS

A complete list of the books in this series is available from the publisher.

Food Irradiation

WALTER M. URBAIN
Professor Emeritus
Department of Food Science and Human Nutrition
Michigan State University
East Lansing, Michigan

1986

ACADEMIC PRESS, INC.

Harcourt Brace Jovanovich, Publishers
Orlando San Diego New York Austin
Boston London Sydney Tokyo Toronto

ACADEMIC PRESS, INC.
Orlando, Florida 32887

United Kingdom Edition published by
ACADEMIC PRESS INC. (LONDON) LTD.
24–28 Oval Road, London NW1 7DX

LIBRARY OF CONGRESS CATALOGING-IN-PUBLICATION DATA

Urbain, Walter M.
Food irradiation.

Bibliography: p.
Includes index.
1. Radiation preservation of food. I. Title.
[DNLM: 1. Food Irradiation. WA 710 U72f]
TP371.8.U73 1986 664'.0288 85-22815
ISBN 0-12-709370-2 (alk. paper)

PRINTED IN THE UNITED STATES OF AMERICA

86 87 88 89 9 8 7 6 5 4 3 2 1

To my wife, Ruth,
and to our children, Elizabeth and Robert,
and their families

Contents

4. Biological Effects of Ionizing Radiation

5. General Effects of Ionizing Radiation on Foods

6. Meats and Poultry

7. Marine and Freshwater Animal Foods

8. Fruits, Vegetables, and Nuts

9. Cereal Grains, Legumes, Baked Goods, and Dry Food Substances

10. Miscellaneous Foods. Useful Food Modifications

11. Combination Processes

12. Packaging

Preface

Despite an intensive research and development effort in the field of food irradiation for more than 40 years, only a few books covering the general subject have been available. The movement toward commercialization of the process, already underway in several countries, and the recent actions of various governmental agencies to accept irradiated foods as wholesome have stimulated increased interest in the process. This book has been written to provide an overview of food irradiation in all its important aspects and at the same time to give adequate detailed information to assist those who intend to use the process. A strong effort has been made to enable a comfortable understanding of the fundamental aspects of food irradiation as well as its applications.

This task has been made somewhat difficult by the wide range of scientific disciplines that are involved in food irradiation and by the broad utility of the process, not only with respect to what it can do but also with respect to the diversity of foods to which it applies. It is hoped that suitable background material sufficient to orient the reader in the many different areas that are relevant to food irradiation has been provided, but without doing this to an unreasonable degree.

The main objective has been to provide information. The reader will note, however, that at times I have inserted comments that I hope will assist in interpreting certain reported findings or in noting when more information is needed and critical issues have not been resolved.

The establishment of the wholesomeness of irradiated foods has required a great deal of effort over many years. In order to assist in understanding the evidence for wholesomeness, particularly that gained through radiation chemistry, there is provided a brief recounting of the various actions that were taken over the years to get where we are today. It is hoped that some of the mystery that has surrounded the nature of the working of food irradiation as it relates to wholesomeness aspects and to associated concerns will be dissipated by this approach.

The general plan of the book is to start with ionizing radiation, its nature and interaction with matter, and then to consider the biological effects it occasions in living organisms associated with food and including certain foods—mainly raw fruits and vegetables—that also are "alive." The application of these effects in treating foods is covered both broadly and in terms of specific uses for the principal food classes. The final chapter deals with those aspects of food irradiation likely to have major significance in commercial usage, such as consumer attitudes, costs, facilities, safety, etc.

While this writing has been largely a personal effort, no such undertaking can be accomplished without help from others. My interest in food irradiation goes back to about 1936, and I have been fortunate to have had the opportunity to be associated through the years with many of the workers in the field, both in the United States and abroad. Names such as Benjamin Schwartz, Bernard E. Proctor, Arno Brasch, Roland S. Hannan, John Kuprianoff, Maurice Ingram, Clarence Schmidt, Lloyd Brownell, A. W. Anderson, Kenneth MacQueen, D. de Zeeuw, John Hickman, A. Sreenivasan, Harry E. Goresline, K. Vas, Samuel Goldblith, M. J. de Proost, R. S. Kahan, R. M. Ulmann, E. C. Maxie, V. I. Rogachev, R. A. Dennison, Charles Niven, John Liston, D. A. A. Mossel, and A. M. Dollar come to mind among those early workers to whom I acknowledge a personal debt for the various inputs that are reflected in this book.

In the more immediate present, I have sought help in specific ways from contemporaries, and without exception they have gratified me by their responses. Some of these I wish to acknowledge herewith.

First, I name Professor B. S. Schweigert, my former chairman at Michigan State University, now at the University of California, Davis, who encouraged me to undertake the writing of this book and who has helped me in essential ways through many years.

Long years of association with the staff of the U.S. Army Natick Laboratories provided me with much information, especially on radappertization of meats. Of that staff I am especially grateful to Edward S. Josephson, Eugen Wierbicki, Charles Merritt, Jr., Ari Brynjolffson, and John J. Killoran. A similar association with the staff of the U.S. Atomic Energy Commission (now the Department of Energy) likewise has been valuable. From the U.S. Food and Drug Administration staff (at one time as a consultant on food irradiation) I learned many important aspects of government regulation of irradiated foods.

Many colleagues in various American universities have helped in many ways and among these I name W. W. Nawar (University of Massachusetts), James H. Moy (University of Hawaii), Manuel Lagunas-Solar (University of California, Davis), Warren Malchman (Michigan State University), R. B. Maxcy (University of Nebraska), and Jack Schubert (University of Maryland).

For assistance from outside the United States I would first acknowledge the enormous aid provided by the staff of the Food Preservation Section of the International Atomic Energy Agency in Vienna, among whom I name J. G. van Kooij and Paisan Loaharanu. I owe much to F. J. Ley (United Kingdom), A. Tallentire (United Kingdom), J. F. Diehl (Federal Republic of Germany), J. Farkas (Hungary), B. Kalman (Hungary), E. H. Kampelmacher (the Netherlands), L. R. Saint-Lebe (France), K. Umeda (Japan), U. K. Vakil (India), M. Lapidot (Israel), H. J. van der Linde (Republic of South Africa), R. A. Basson (Republic of South Africa), and P. A. Wills (Australia).

Suppliers of irradiation facilities and equipment have been very helpful in providing information, which, as presented in this book, I hope will enable the reader to become familiar with the hardware of food irradiation. While such help was broadly provided, I especially wish to acknowledge the contributions of Atomic Energy of Canada, Ltd.; Radiation Dynamics, Inc.; Radiation Technology, Inc.; Isomedix, Inc.; Far West Technology, Inc.; and Baird Corporation.

Not to be forgotten in any way is the aid provided me on a daily basis for a long time by my dear wife Ruth. Her contributions to this book are evident in many ways, and yet she was helpful in other ways that cannot be seen so readily (patience is an intangible), but which served to get the job done.

WALTER M. URBAIN
Sun City, Arizona

International Symbol for Irradiated Food

In the center is an agricultural product, a food, which is in a closed package denoted by the circle, and which is irradiated by penetrating rays.

CHAPTER 1

Ionizing Radiation

I. RADIATION: DEFINITION AND TYPES

Food irradiation employs an energy form termed ionizing radiation. The particular attributes of ionizing radiation that make it useful for treating foods are several in number. Certain kinds of ionizing radiation have the ability to penetrate into the depth of a food. Through physical effects they interact with the atoms and molecules that make up the food and also those of food contaminants such as bacteria, molds, yeasts, parasites, and insects, causing chemical and biological consequences which can be utilized in beneficial ways. While ionizing radiation frequently is referred to as high-energy radiation, the total quantity of energy needed to secure the beneficial effects with foods is relatively small, and gross changes in a food which could affect its acceptability usually do not occur.

The term *ionizing radiation* includes a number of different kinds of radiation. Not all the radiations classified as ionizing radiation have properties suitable for use in food irradiation, and, as a consequence, restrictions defining what kinds are satisfactory are required.

The propagation of energy through space is termed *radiation* or *radiant energy*. There are two basic types of radiation: (1) electromagnetic and (2) corpuscular.

Electromagnetic radiation consists of self-propagating electric and magnetic disturbances. Conceptually it can be considered to be a wave motion involving oscillating electric and magnetic field vectors. It is characterized by two parameters inversely related to each other:

1. Frequency (symbol ν) is the number of cycles per second and is called hertz (Hz).
2. Wavelength (symbol λ) is the distance between two identical succeeding points on the wave.

Frequency and wavelength are related in accord with the following equation:

$$\lambda = c/\nu$$

in which c is a constant, namely, the velocity of the propagation of the electromagnetic wave, closely equal to 3×10^8 m/sec.

Electromagnetic energy may be considered as behaving as a particle traveling through space in a manner such that the energy is concentrated in bundles called photons. From the quantum theory, the following equation is obtained:

$$E = h\nu = \frac{hc}{\lambda}$$

in which E equals the energy content of the photon, h is Planck's constant, equal to 6.63×10^{-27} erg sec, and ν, c, and λ are the same as given above.

Corpuscular radiation involves particles having mass, which, when in motion, possess kinetic energy. The relation of kinetic energy E with mass and velocity is given by the equation

$$E = \frac{1}{2} \frac{m_0 v^2}{\sqrt{1 - v^2/c^2}}$$

where m_0 is the rest mass of the particle, v its velocity, and c the velocity of propagation of electromagnetic energy. As the equation indicates, E is a function of the square of the particle velocity, and c is a very large number, 3×10^8 m/sec. Unless v also is large, the quantity $\sqrt{1 - v^2/c^2}$ reduces to unity, and E then equals $\frac{1}{2}m_0 v^2$, the term for kinetic energy of classical physics. As v becomes large, however, $\sqrt{1 - v^2/c^2}$ approaches zero. At high particle velocities, therefore, m_0, in effect, increases, as can be seen from the following relationship:

$$m_v = \frac{m_0}{\sqrt{1 - v^2/c^2}}$$

where m_v is the mass for values of v that are large, that is, those approaching c. At very high velocities, therefore, the kinetic energy of the particle becomes very large, due partly to the effect of velocity per se, and partly to the increase of its mass over its rest mass.

II. IONIZATION AND EXCITATION

As the name indicates, ionizing radiation can cause ionization, which is a process in which one or more orbital electrons are removed from an atom. The process is independent of the atom's location, that is, whether it is free or exists as a part of a molecule. Ionization leads to the formation of two or more separate bodies: (1) one or more "free" or unpaired electrons, each of which carries unit negative electric charge, and (2) the residue of the atom, which is positively charged, and is an ion, or more specifically, a cation.

In order to produce ionization, a certain minimum energy content of the radiation is needed. Atoms which have their orbital electrons at minimum energy levels are at their "ground state." These electrons, though still contained within the atom and under the control of the nucleus, can be raised to higher energy levels by absorption of energy. Atoms with electrons above their ground state are "electronically excited." If the electron absorbs sufficient energy, it may leave the atom and become free of control of the nucleus. This is *ionization.* The minimum energies required to free electrons present in the various atomic levels are referred to as *ionization potentials.* Depending upon the atom in question, the ionization potentials of valence electrons vary between about 4 and 20 eV [electron volt (eV) equals 1.6×10^{-12} erg]. Inner electrons can have ionization potentials of many thousand electron volts. If an electron absorbs energy greater than its ionization potential, the excess becomes kinetic energy and enables the electron to travel away from its parent atom.

Reference to Fig. 1.1, the electromagnetic spectrum, shows that energies in the amounts of 4 to 20 eV occur in the region termed ultraviolet and that photons with greater energies are obtained with X and γ rays. Therefore, all these kinds of electromagnetic radiation are capable of causing ionization.

In electronically excited molecules the amount of energy derived from the radiation is less than that needed for ionization. Although most of the excitation energy in molecules is converted to heat and produces various effects of a vibrational, rotational, and translational nature among the

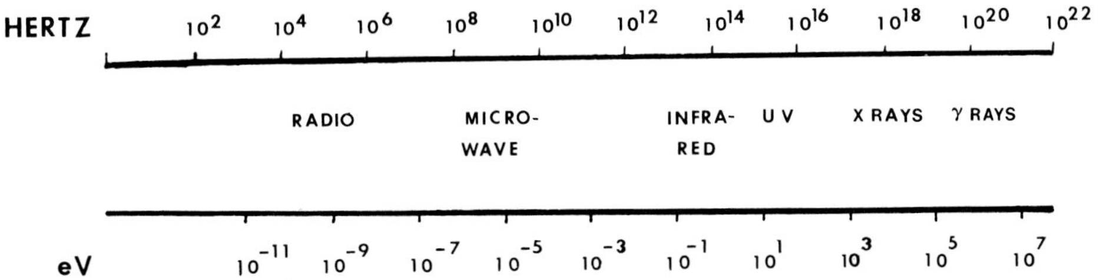

HERTZ 10^{2} 10^{4} 10^{6} 10^{8} 10^{10} 10^{12} 10^{14} 10^{16} 10^{18} 10^{20} 10^{22}

RADIO | MICRO-WAVE | INFRA-RED | UV | X RAYS | γ RAYS

eV 10^{-11} 10^{-9} 10^{-7} 10^{-5} 10^{-3} 10^{-1} 10^{1} 10^{3} 10^{5} 10^{7}

Figure 1.1 Electromagnetic spectrum.

component atoms or groups of atoms, outer electrons of atoms can be displaced to a higher than normal orbit. Electronically excited molecules can undergo chemical change, as is observed in photochemistry. In photochemistry, however, because of its limited energy content, each photon excites only a single molecule. With ionizing radiation, each photon or particle has amounts of energy larger than that needed to produce one excitation or ionization, and by giving to each molecule only a portion of its total energy, it can sequentially excite or ionize many molecules. Unimolecular decomposition and the formation of free radicals and new molecular species, including the results of interaction with other entities present, plus other changes can occur. *It is these chemical changes that are the basis of food irradiation.*

III. RADIATIONS FOR TREATING FOODS

The considerations just reviewed in part have led to the selection of X rays, γ rays, and electron beams for food irradiation. Knowledge of these radiations existed prior to 1900, but it was not until the time of World War II in the 1940s that sources for them became available that offered the possibility of adequate amounts of radiation at costs low enough for commercial use. Although there were a number of isolated early efforts to apply ionizing radiation to foods, it was the availability of large radiation sources that stimulated the development of the food irradiation process in the ensuing decades.

While a number of particles encountered in nucleonics can be given kinetic energy, only one, the electron, has found applicability in food irradiation, largely, but not entirely, because electrons (1) are easily obtained in sufficient quantities, (2) carry an electric charge which enables the use of electrostatic fields to impart controlled amounts of energy, and (3) can be directed and focused magnetically. While each electron acts individually, it is the practice of food irradiation to use very large numbers of electrons, which are made to travel in a given direction and which collectively are referred to as an *electron beam.*

In Table 1.1 are given figures showing the change of electron mass as a function of velocity. In Fig. 1.2 is shown the relationship between velocity and electron energy, expressed in million-electron volts (MeV). In this relationship the effect of change of electron mass is important.

In order to be useful in the treatment of foods, radiation must have the capability of penetrating into the depth of foods. Ultraviolet radiation is essentially absorbed at the surface of materials such as foods and therefore fails to meet this requirement; X rays and γ rays, on the other hand, have capabilities of deep penetration totally adequate for just about any food requirement. Electrons also can be given sufficient energy to penetrate useful

Table 1.1

Increase in Electron Mass as Energy Increases[a]

Energy (MeV)	Velocity (cm/sec × 10^{10})	Ratio[b] m_v/m_0
0.001	0.187	1.002
0.005	0.417	1.010
0.010	0.584	1.020
0.050	1.098	1.098
0.100	1.682	1.196
0.500	2.587	1.979
1.000	2.821	2.957
5.000	2.985	10.79
10.000	2.994	20.58

[a] From "Radiation Preservation of Food." U.S. Army Quartermaster Corps, Washington, D.C., 1957.

[b] Ratio of mass at velocity v (m_v) to rest mass (m_0).

depths in foods. As will be discussed later, effects other than those desired with foods, however, set upper limits for the energy levels of all three types of ionizing radiation when they are used with foods. The necessity to stay below these energy limits does restrict the depth of penetration into foods that is possible, particularly with electron beams.

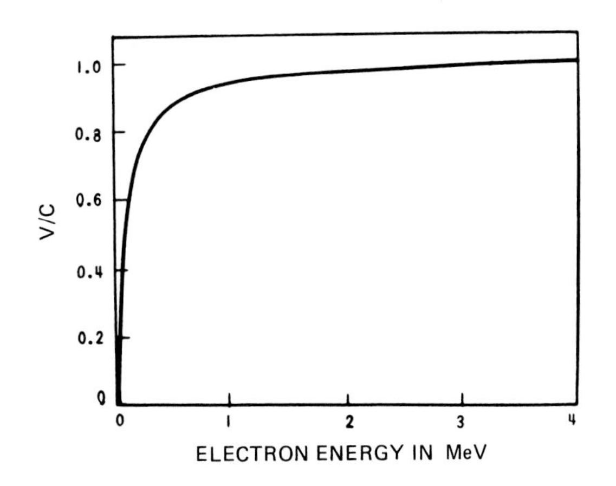

Figure 1.2 Relation between electron energy and velocity. From "Radiation Preservation of Food." U.S. Army Quartermaster Corps, Washington, D.C., 1957.

IV. INTERACTION OF IONIZING RADIATION WITH MATTER

A. General

The interaction of ionizing radiation with matter is complex. What happens depends upon the type of radiation and its energy content, the composition, physical state, and temperature of the absorbing material, plus other factors such as the rate of energy deposition and atmospheric environment. The primary interaction process essentially is the transfer of energy from the incident radiation to the absorber. Only energy that is absorbed is effective in producing changes.

There are three important aspects of the process of interaction, namely, (1) the physical process, (2) the chemical changes which ensue from the physical events, and (3) the biological consequences in the cases of the target materials which include living organisms. An understanding of these aspects of the interaction of ionizing radiation with matter is of good value in applying this energy form to foods.

The physical processes involved in the interaction of ionizing radiations used in food irradiation with matter are covered in this chapter. Succeeding chapters treat the subjects of the chemical changes and biological consequences.

B. Electron Radiation

Charged particles such as electrons of moderate energy interact largely through the electric (coulomb) forces between them and the orbital electrons of the atoms of the absorber. The repulsive force between an incident electron and an orbital electron may cause the latter to move to an orbit of higher energy. This process is termed *excitation.* If sufficient energy is transferred to the orbital electron, it may be ejected from the atom. This process is *ionization.* In both cases some of the kinetic energy of the incident electron can be transferred to the orbital electron. These processes of energy transfer may be repeated many times and ultimately can dissipate all of the kinetic energy of the incident electron. When this happens, the electron is captured by an atom of the absorber having an affinity for electrons (e.g., a cation).

The path of an incident electron in an absorber is not straight. It can change direction as a result of collision with atoms of the absorber. Repeated collisions can cause multiple changes of direction. The consequence is a scattering of electrons in directions different from the direction of the incident electron beam.

Electrons that have been ejected from absorber atoms can acquire sufficient

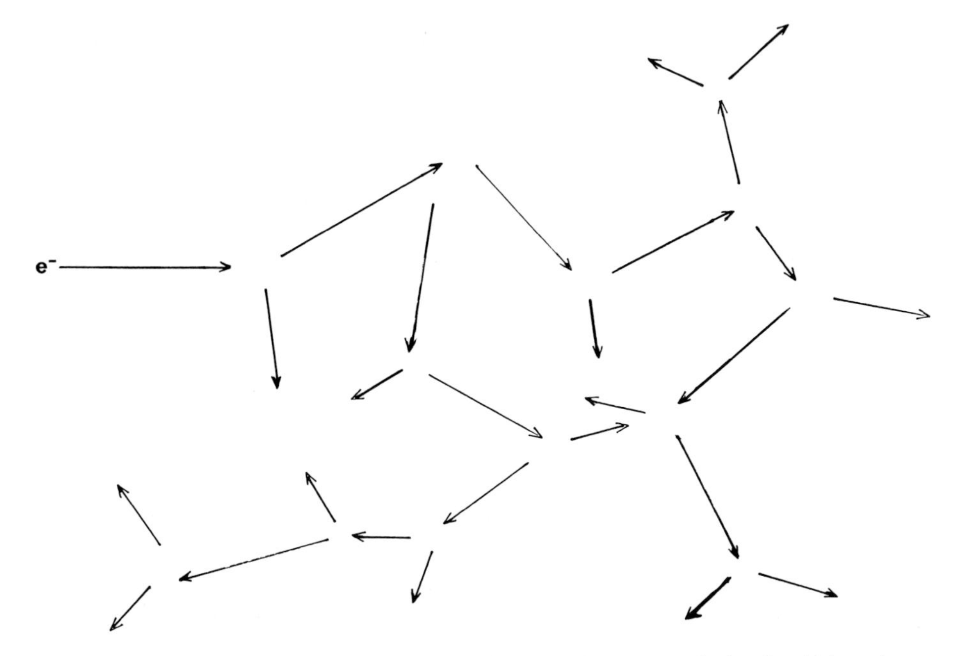

Figure 1.3 Representation of the cascading effect to produce a network of paths which results from the interaction of a single electron with an absorber.

kinetic energy to enable them to undergo the same energy loss process as incident electrons. Each such new electron traverses its own pathway, generally as a spur or branch from the pathway of an incident electron. The result is a network of paths, such as portrayed in Fig. 1.3, except that the actual network is three-dimensional.

In addition to loss of energy through excitation and ionization, electrons can lose energy also through two other processes: *bremstrahlung* (braking radiation) and *Cerenkov* radiation.

Bremstrahlung is electromagnetic radiation of energy comparable with X-radiation (see Fig. 1.1). It results from interaction of a fast-moving electron with the nucleus of an atom and involves the conversion of some of the electron's kinetic energy to electromagnetic radiation. The proportion of the electron's energy converted into bremstrahlung increases with the kinetic energy of the electron and with the atomic numbeı of the absorber atoms. For materials of low atomic numbers (such as foods), bremstrahlung production is appreciable only with electrons having energies above 1 MeV. While the bremstrahlung that may be produced in a food is of insufficient quantity to cause significant chemical change, it may induce radioactivity if the energy level is sufficiently high. This is the principal reason for the need to limit the energy level of electron beams used in food irradiation.

An electron cannot attain a velocity greater than the velocity of electromagnetic radiation. The velocity of the latter, however, is different in different media, and it is possible that an electron traveling in one material may attain a velocity greater than the velocity of electromagnetic radiation in another material. In such a case, the electron upon entering the second medium undergoes an energy loss which usually appears as ultraviolet or visible light and which is called *Cerenkov* radiation. The blue "glow" around ^{60}Co when it is covered with water is a familiar example of Cerenkov radiation. The proportion of energy converted to Cerenkov radiation is very small. Cerenkov radiation has no meaningful role in food irradiation.

C. Electromagnetic Radiation

Unlike electrons which interact with matter largely through coulombic forces between them and orbital electrons, photons of electromagnetic radiation carry no electric charge and, therefore, are not subject to coulombic forces. As a consequence their penetration into an atom is greater than that of electrons. As with electrons, what happens depends upon the photon energy and upon certain parameters of the absorber, such as composition, physical state, temperature, and atmospheric environment.

Three mechanisms of energy transfer exist for X and γ rays: (1) the photoelectric effect, (2) the Compton effect, and (3) pair production.

1. Photoelectric Effect

In the photoelectric effect, a quantum or "bundle" of electromagnetic energy—that is, a photon—gives up *all* of its energy to an atom, from which an electron is ejected. Since the atom loses an electron, ionization occurs and the atom acquires a positive charge. Part of the photon's energy may be used in freeing the electron; the remainder is acquired by the ejected electron and the residual of the atom in the form of kinetic energy. An outer-orbit electron of an atom requires only about 10 eV for ejection. The photoelectric effect, however, is not limited to outer electrons, and for other electrons as much as 500 eV may be needed to eject them. Even 500 eV is greatly less than the energies of the photons of X and γ rays (see Fig. 1.1) and, hence, for these radiations, the photoelectric effect process is easily feasible. The photoelectric effect is portrayed in Fig. 1.4. The ejected electrons operate in the absorber in the manner described earlier for electron radiation. Photoelectric absorption occurs largely with photons of energies below 0.1 MeV in absorbers having low atomic numbers.

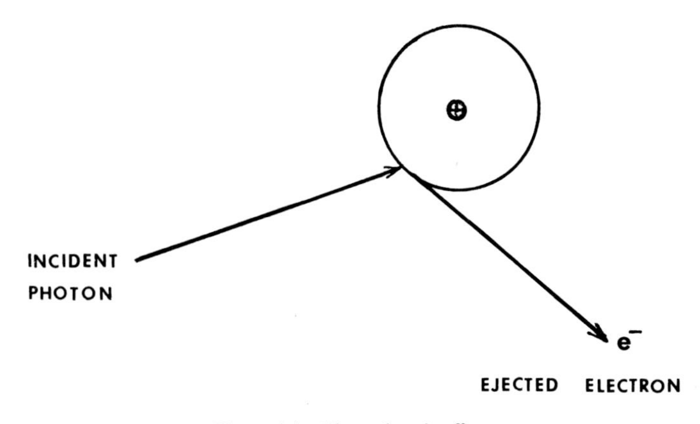

Figure 1.4 Photoelectric effect.

2. Compton Effect

For absorption by water of electromagnetic radiation of energies greater than approximately 0.1 MeV, the photoelectric effect mechanism of energy transfer is replaced by the Compton effect, which is portrayed in Fig. 1.5. Unlike the photoelectric effect, in the Compton effect, only *part* of the photon's energy is transferred to a particular atom. After interaction the incident photon continues through the absorber, but in a changed direction and with less than its original energy. The transferred energy is used in freeing an orbital electron from the atom and also in imparting kinetic energy to the ejected

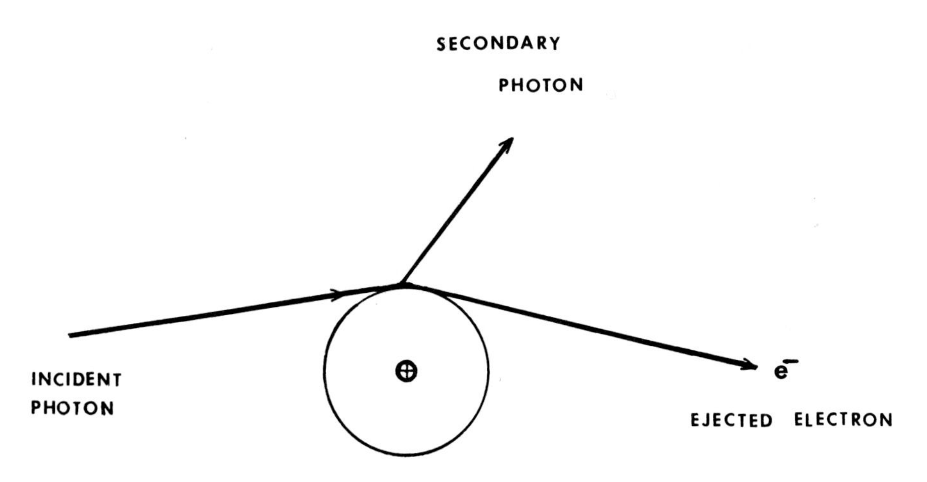

Figure 1.5 Compton effect.

electron and to the remnant atomic ion. The ejected electron, having kinetic energy, itself can cause excitations and ionizations in absorber atoms. The Compton effect accounts for the principal energy transfer over a wide range of energies of both X and γ rays. It is the most important mechanism for electromagnetic energy transfer in food irradiation.

3. Pair Production

The third mechanism of energy transfer for electromagnetic radiation is termed pair production. It involves absorption of a photon and the subsequent formation of one electron and one positron. Unlike the photoelectric and Compton effects, in which electromagnetic radiation is converted largely into kinetic energy, pair production involves conversion of a photon of electromagnetic radiation into matter, two particles: one electron and one positron.

According to Einstein's equation,

$$E = mc^2$$

in which E is energy, m is the rest mass of a body (in this case the electron or positron, equal to 9.1×10^{-28} g), and c is the velocity of electromagnetic radiation *in vacuo* (3×10^8 m/sec). The energy equivalency of an electron or positron, therefore, is 0.51 MeV. Since two particles are formed in pair production, the total energy required is 1.02 MeV. Unless the photon has this energy, pair production cannot occur. If the energy content is exactly 1.02 MeV, an electron and a positron can be formed, but each will have zero kinetic energy. If the photon contains more than 1.02 MeV, the extra energy is divided between the two particles. This permits them to travel through the absorber, producing ionizations and excitations in accordance with their kinetic energy content. At some time the positron will interact with an electron, and in accord with Einstein's equation the masses of both particles are converted into electromagnetic radiation, yielding two photons, each with energy of 0.51 MeV. These photons can interact with the atoms of the absorber in a manner appropriate to their energy content and to the nature of the absorber. Except for atoms of large atomic numbers, the pair production process is an unimportant energy dissipation mechanism for photons of less than 10 MeV. In view of this, its relevancy to food irradiation has to be small.

D. Role of Elemental Composition of Absorber

As has been noted, a critical factor in the dissipation of ionizing radiation by an absorber is its elemental composition. Foods are composed of plant or animal tissue or derivatives thereof. Food contaminants such as bacteria and

Table 1.2
Elemental Composition of Wet Tissue

Element	Atomic no.	%
H	1	10.0
C	6	12.0
N	7	4.0
O	8	73.0
Na	11	0.1
Mg	12	0.04
P	15	0.2
S	16	0.2
Cl	17	0.1
K	19	0.35
Ca	20	0.01

other microorganisms, and insects, have elemental compositions similar to foods. The elemental composition of "wet tissue" is given in Table 1.2.

While different foods have different chemical compositions, they all have a sufficiently common character as to permit the elemental composition of wet tissue as given in Table 1.2 to be considered representative of foods generally. From the standpoint of the physical aspects of the interaction of foods with radiation, the most important feature of their elemental composition is that the elements present are of low atomic number, as indicated by the data of Table 1.2. We generally can exclude from consideration, therefore, interactions relating to elements of large atomic numbers. A few elements of large atomic numbers can be found in foods but usually only in trace amounts. In view of their low concentrations, these also may be ignored generally.

The fact that only elements of low atomic numbers occur in foods makes pair production an unimportant mechanism for energy transfer in food irradiation carried out with X and γ rays. For the same reason the photoelectric effect also is a less important mechanism of energy transfer. The principal mechanism of energy transfer in irradiating foods with X and γ rays is the Compton effect.

The absence of elements of large atomic numbers has importance in avoiding induced radioactivity in foods irradiated with X and γ rays and also with electrons.

E. Quantitative Relationships in Energy Transfer

Usually it is desirable to characterize processes such as energy transfer in a quantitative manner. Several such procedures have been devised for use in connection with irradiation.

1. Specific Ionization

The term specific ionization denotes the number of ions produced per unit of track length (e.g., 1 μm). It does not include energy transfer in producing other effects such as excitations. Usually specific ionization is treated as the average number of ions formed over the distance of the entire energy dissipation track. The use of an average value is necessary because the ion density changes as the particle loses energy.

2. Linear Energy Transfer

Since energy transfer can produce effects in addition to ionization, the term linear energy transfer (LET) has been devised to indicate the average *total* quantity of energy transferred, regardless of what effect is produced in the absorber. LET is the average energy transferred per unit length of the track of the primary particle, and usually is given as kilo-electron volts per micrometer. LET is generally considered to be a better index than specific ionization of the effectiveness of a radiation in producing changes in an absorber.

Because both specific ionization and LET involve not only the nature of the radiation but also the characteristics of the absorber, values of these quantities are specific for a particular radiation interacting with a particular absorber.

3. Stopping Power

Another process requiring quantitative definition is the attenuation of radiation as it passes into the depth of an absorber. In this process corpuscular and electromagnetic ionizing radiations operate in different ways.

For corpuscular radiation the average amount of energy lost by a charged particle, such as an electron, per unit of distance traveled (in the direction of travel) is called *stopping power.* Stopping power is a function of particle energy and charge and of the composition of the absorber.

Because electrons have a relatively low mass ($\frac{1}{1873}$ that of the hydrogen atom), the stopping power of most materials for electrons compared with that for other particles is relatively low. The smallness of the mass causes the electron velocity to be relatively high for a given kinetic energy (see Fig. 1.2). As a consequence, the time period for interaction with the electrons of the absorber is relatively short, and the attainable energy transfer per unit length, relatively smaller.

A charge greater than the unit charge of the electron would increase the coulombic forces and cause a larger energy transfer. The unit charge of the electron is another factor associated with the low stopping power of most materials for electrons.

The number of electrons encountered per unit volume of an absorber by an electron traveling in the absorber affects the stopping power. The more dense the absorber is, the more electrons are contained per unit volume. For this reason, stopping power increases with the density of the absorber.

As the depth of the absorber is penetrated, the energy of the incident electron decreases through transfer to the atoms of the absorber. Below the surface, the stopping power increases as a consequence of this energy loss, but as further depth penetration occurs, the stopping power reaches a maximum and then diminishes as other factors come into play. At low electron velocities, positively charged entities of the absorber act, in effect, to counter the negative charge of the electron and, hence, its ability to interact with orbital electrons of the absorber atoms. In addition, the longer time period between energy transfer events associated with lower velocities of the traveling electrons allows atoms having disturbed orbital electrons sufficient time to readjust their positions in such a way as to reduce the probability for electronic excitation.

4. Absorption Coefficient

The intensity of electromagnetic energy, which can be regarded as a beam of photons, is defined as the energy per unit time crossing unit area perpendicular to the beam. The absorption of electromagnetic energy by a material is in accord with the following equation:

$$I = I_0 e^{-\mu x}$$

in which I_0 is the intensity of the incident radiation, I the intensity after passage through an absorber having a thickness x, μ the linear coefficient of absorption for the absorbing material, which is usually expressed per unit thickness traversed (unit, cm^{-1}), and e the base for natural logarithms. Because x actually is an index of the amount of mass of the absorber traversed by the radiation, an alternate to the linear absorption coefficient, namely, the mass absorption coefficient Υ, is sometimes used. It is defined as the fraction of energy absorbed in 1 g of the absorber for a surface area of 1 cm^2:

$$\Upsilon = \mu/\rho$$

where μ is the linear absorption coefficient and ρ is the density of the absorber, and the units of Υ are square centimeter per gram.

Because of the several pathways by which electromagnetic radiation of energy greater than 10 eV is attenuated by an absorber, one can divide the absorption coefficient into several independent parts: the absorption coefficient for the photoelectric effect, that for the Compton effect, and that for pair production. For a given absorber, each coefficient will vary with the photon

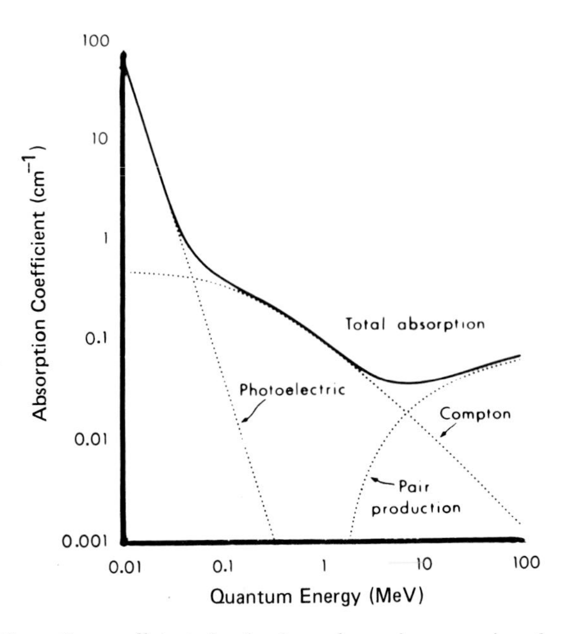

Figure 1.6 Absorption coefficients in aluminum for various energies of γ radiation. The relative contributions of the photoelectric effect, Compton effect, and pair production are indicated. From A. P. Casarett, "Radiation Biology." Prentice-Hall, Englewood Cliffs, New Jersey, 1968. Copyright by U.S. Atomic Energy Commission.

energy. This is shown in Fig. 1.6 for aluminum as the absorber. The relative importance of each mechanism of energy transfer for different photon energies can be determined from information such as given in Fig. 1.6 for aluminum. Absorbers having atoms with atomic numbers similar to aluminum would be expected to show similar relationships. Absorbers containing higher atomic number components, on the other hand, would be markedly different.

5. Absorbed Dose

So far, we have referred to the energy content of an individual particle (electron) or photon. We have spoken also of beams or rays made up of numbers of electrons or photons. Beams or rays carry many multiples of the energy contents of the individual electrons or photons that make them up. The total quantity of energy flowing as a beam or ray per unit time is termed *flux*. When absorption of energy occurs, the mean quantity of energy transferred to a mass of material exposed to the radiation is called the *absorbed dose*, or dose *D*. The SI* unit for *D* is the gray (Gy); it is equal to the absorption of 1 J/kg.

* SI is the abbreviation for the International System of Units. The gray replaces the older unit, the rad, in 1986 (1 Gy = 100 rad).

The energy transfer involved in the absorbed dose occurs over a finite time period as determined by the flux and the radiation–matter interaction. The *dose rate* $\dot{D}$ is the rate of energy transfer per unit time. Dose rate is governed by radiation source characteristics and the manner of its use in the irradiation facility (see Chapter 15, Section III).

While the energy content of an electron or photon enables particular types of interaction with the atoms of an absorber, the effects of ionizing radiation on a food are largely determined by the dose used. For this reason, dose is the most important parameter in food irradiation.

Measurement of dose is discussed in Chapter 15.

6. Spatial Distribution of Energy Transfer

The spatial distribution of energy transferred by electrons to an absorber mass is different from that by X and γ radiation. Each electron of the beam operates as an individual, but, collectively, all the electrons perform in a predictable way.

Each electron gradually loses energy by transfer to the absorber and simultaneously suffers reduction of velocity. As the velocity diminishes, the number of interactions with the atoms of the absorber, per unit distance traveled, increases. At some point the velocity becomes so small that the electron is captured by an appropriate entity of the absorber. When this happens, the limit of penetration of that electron into the absorber (i.e., the range) is reached. With this process one can visualize an electron pathway as displaying initially small energy transfer, but as the electron slows, the density of energy transfer increases. There is, therefore, a nonuniform distribution of energy transfer along the electron's path.

X and γ rays, unlike electrons, do not have a definite range limit in an absorber. Due to interaction with the absorber atoms, the intensity of the ray will diminish in traversing the absorber. Some photons yield their energy to the absorber and disappear, but others pass through unchanged or changed only to a degree.

Primary ionizations occur in accordance with appropriate mechanisms and are distributed along the pathway of the beam, but compared with electron beams, the density of these primary energy-transferring events is relatively low. After the primary events, electrons resulting from the photon interaction with the absorber atoms, and having sufficient kinetic energy, produce excitation and/or ionization in other atoms of the absorber. It is to be noted that X and γ rays interact throughout the absorber mass and in this regard are unlike electron beams which operate only within limited depths of penetration.

It is to be recognized that in the irradiation of foods we are dealing with events that occur at the level of individual atoms. Not all atoms present, however, are involved in receiving energy from the incident radiation. In fact, in view of the total quantity of energy employed in food irradiation, only a very small fraction of those present is involved. Despite the apparent randomness of the radiation–absorber interaction, the collective performance of the individual atoms and molecules is characteristic and predictable. Desired effects can be obtained consistently.

Both corpuscular and electromagnetic radiations act on foods in essentially the same manner—through energetic electrons interacting with the absorber atoms—and can be used interchangeably in many applications. Certain practical considerations, however, can determine which kind of radiation may be preferred.

7. Buildup

Neither electron beams nor X and γ rays accomplish maximum energy transfer at the surface of the absorber. With electron beams some electrons are scattered in directions different from the beam direction and, if they have sufficient kinetic energy in the region near the surface, they may escape from the absorber. Similarly, some secondary electrons are lost. As a consequence of these losses in the region of the absorber near the surface through which the beam enters, energy transfer is less than it would be without this scattering. At some point below the surface, the range of the scattered electrons, both primary and secondary, is inadequate to enable escape from the absorber. This is the region of maximum energy transfer. At depths greater than this, attenuation of the beam intensity reduces the amount of energy available for transfer. This reduction of available energy continues with increasing depth until none remains and the limit of penetration is reached. The net result is that the greatest energy transfer occurs somewhat below the surface of the absorber.

A similar effect occurs with X and γ rays, except that it is less pronounced. While some backscatter does occur, electrons produced by interaction with the absorber travel mostly in the same direction as the incident photon beam, and consequently, most buildup energy is in the forward direction of the beam. As with electron beams, the result is that the maximum energy transfer occurs below the surface of the absorber.

8. Penetration

One necessary property of ionizing radiation used in food irradiation is its ability to penetrate the food. Without this ability, the irradiation of solid foods, at least, would not be possible. The penetration of electron beams and of X and γ rays into matter occurs in different ways.

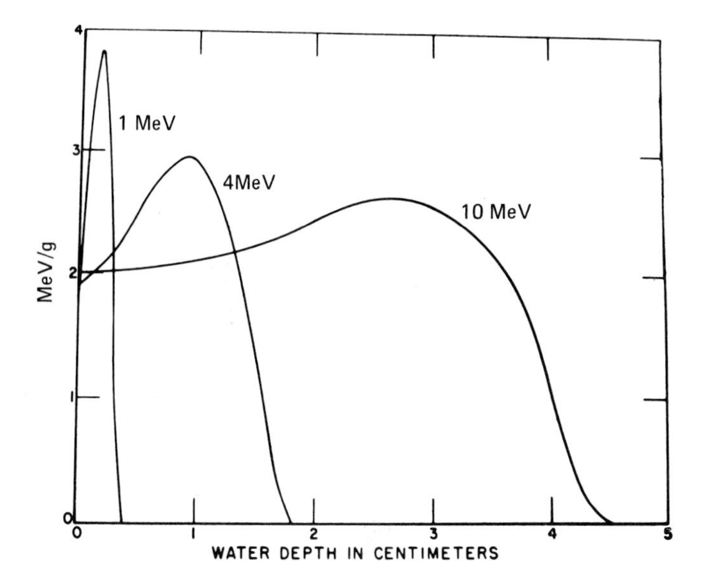

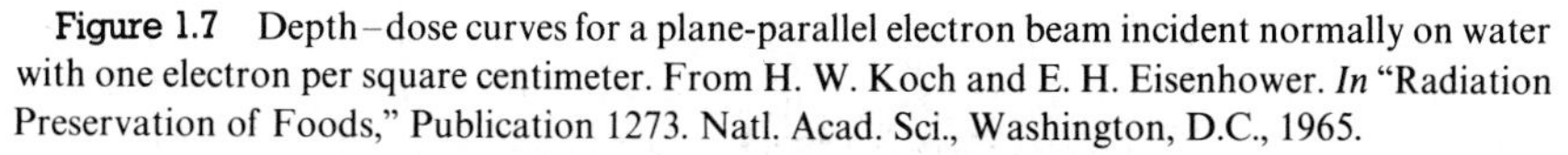
Figure 1.7 Depth–dose curves for a plane-parallel electron beam incident normally on water with one electron per square centimeter. From H. W. Koch and E. H. Eisenhower. *In* "Radiation Preservation of Foods," Publication 1273. Natl. Acad. Sci., Washington, D.C., 1965.

As an electron beam penetrates into an absorber, there is a continuing, although stepwise, loss of energy. The limit of penetration occurs when the electrons constituting the beam collectively have transferred their kinetic energy to the absorber, at least to the degree that energy levels comparable with the thermal energy values of the absorber are all that remain. For water, Fig. 1.7 shows the relationship between depth and energy transfer value (similar to absorbed dose value). The change of this relationship with initial electron energy is indicated by the several curves, each curve being for the indicated energy. From these curves it can be seen that at the energies specified only limited penetration into the water is possible. For consideration of penetrability, water serves as a sufficiently good model for what occurs with foods. These curves indicate that penetration of foods by electron beams may not always be adequate to provide what is needed.

As noted earlier (see Section IV,E,4), X and γ rays transfer energy to an absorber in accordance with the equation

$$I = I_0 e^{-\mu x}$$

Only those photons which interact with the absorber and transfer energy undergo energy loss. The remainder pass through unchanged. The consequence of this process is that there is no true limit of penetration for X and γ

rays. The initial intensity of the ray I_0, nonetheless, will be reduced as the mass of the absorber is penetrated. Although complete absorption is theoretically unattainable, a sufficient reduction of intensity can be regarded as an effective acceptably low level, for whatever objective is sought in performing the irradiation. In the absorption equation above, as x approaches infinity, $e^{-\mu x}$ approaches zero, as also does I. One can choose a value of I obtained from a particular value of x such that I can be regarded effectively equal to zero (or to some other acceptable value) and in this manner obtain the useful penetration limit for the radiation into the absorber.

A useful expression for indicating the penetration of X and γ rays is the *half-thickness*, which is the thickness of the absorber needed to reduce I to half of I_0. The half-thickness of an absorber is given by the equation

$$x_{0.5} = 0.693/\mu$$

The value for μ, the linear coefficient of absorption appropriate to the particular radiation, including its energy level, and to the particular absorber

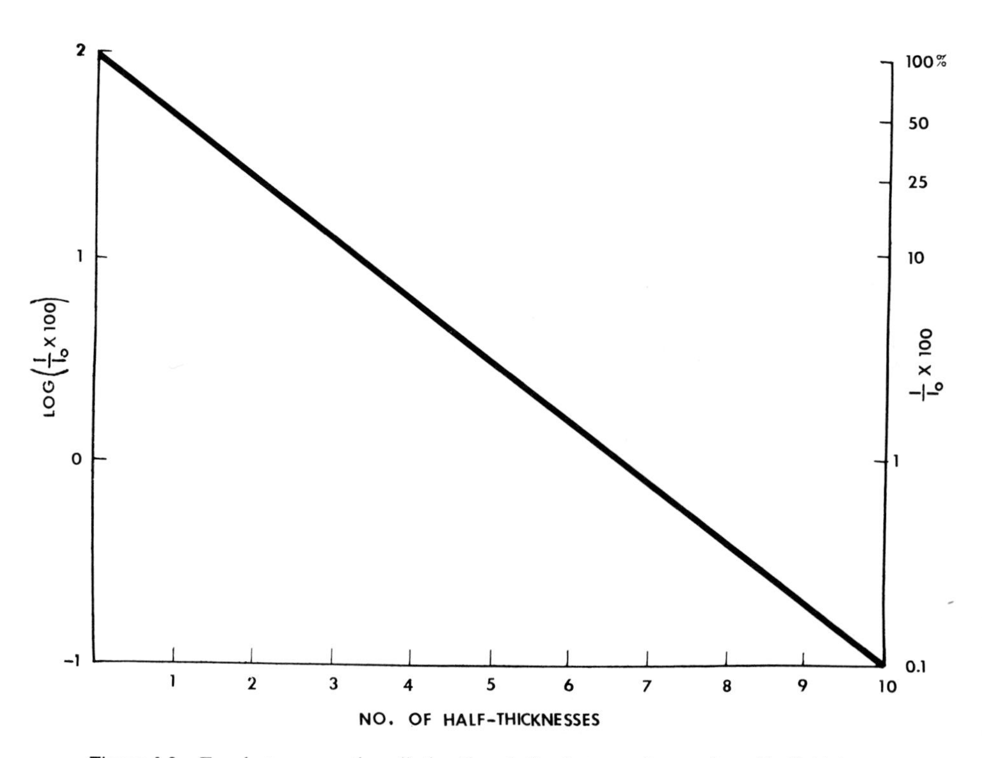

Figure 1.8 For electromagnetic radiation the relation between the number of half-thicknesses of an absorber and the percentage reduction of initial intensity I_0.

Table 1.3
Absorption Coefficients and Half-Thickness Values of γ Rays and X Rays for Water

Energy (MeV)	μ	Half-thickness value (cm)
0.1	0.167	4.14
0.2	0.136	5.10
0.5	0.0968	7.17
0.66[a]	0.084	8.2
1.0	0.0706	9.82
1.17[b]	0.64	10.8
1.33[b]	0.64	10.8
2.0	0.0493	14.05
5.0	0.0301	23.02

[a] ^{137}Cs γ.
[b] ^{60}Co γ.

must be used. In Fig. 1.8, in accordance with the above equation, the half-thickness to reduce the intensity of X and γ rays to desired percentages of I_0 are shown. By using a logarithmic scale for the ordinate, a straight line is obtained. Multiplication of the number of half-thicknesses as required by the appropriate absorption coefficient yields the thickness of the absorber needed to accomplish the desired reduction in I_0. In a practical manner, one can establish the penetration depth by setting a desired value of I.

Table 1.3 gives the linear coefficients of absorption of γ rays and X rays of a number of energy levels for water and the calculated half-thickness value. Such figures are useful in estimating water-shielding requirements and also depth dose quantitation of unit density materials. From these figures it can be seen that 5-MeV γ rays provide adequate penetration for all thicknesses of foods likely to be encountered.

F. Induced Radioactivity

The interactions of ionizing radiation with matter so far covered are all reactions in which the nuclei of the absorber atoms do not participate. Interaction of ionizing radiation with nuclei is possible, however, and, in general because such reactions could induce radioactivity in foods, measures to avoid their occurrence are needed.

Consideration of the significance of radioactivity induced in a food as a consequence of irradiation involves concern for at least four items:

1. What nuclear reactions are possible and what reaction products are formed.
2. The occurrence in foods of those elements which may lead to induced radioactivity.
3. The nature of radionuclides produced. Their half-lives have a dual import. Radionuclides with short half-lives will quickly disappear after irradiation, possibly before consumption of the food can occur. Those with long half-lives generally have low specific activity and therefore contribute relatively small amounts of activity per unit time.
4. The maximum permissible concentration of a radionuclide in a food is related to several factors, including whether it is retained or quickly eliminated from the consumer's body, the nature and the amount of energy transferred to the body, the locus of the deposition and/or incorporation in the body, and the organ(s) involved.

Induced radioactivity primarily is the consequence of excitation of the nucleus of an absorber atom, which then emits a neutron, proton, triton, γ ray, or other secondary radiation. Generally particles are emitted quickly, whereas γ-ray emission can continue over periods of hours or days, or longer. After the initial decay step, the residual nucleus may also be radioactive. Emitted neutrons can activate other nuclei and cause additional radioactivity.

As might be anticipated, the energy of the incident radiation is a principal factor in determining if a reaction with the nucleus is to occur. In order to avoid induced radioactivity in irradiated foods, upper limits are needed for energies of electrons and of X and γ rays. Some forms of corpuscular radiation such as α particles, neutrons, protons, and deuterons can induce radioactivity even at relatively low energy levels and, therefore, are not used.

Another important aspect of induced radioactivity is the elemental composition of the absorber. In this connection both macro- and microconstituents must be considered.

For X and γ rays the principal reactions with atomic nuclei are those in which either a neutron or a proton is ejected. These are referred to as (γ,n) or (γ,p) reactions; (γ,n) reactions lead to a lower isotope of the same element and (γ,p) reactions to a different element of one lower atomic number. In order to initiate these reactions, the incident radiation must be of a minimum or threshold energy, which can differ greatly from one element to another. Of those nuclides likely to occur in foods, only three have threshold energy values for nuclear reactions with X and γ rays of energy levels below 5 MeV. They are ^{2}H at 2.23 MeV, ^{17}O at 4.14 MeV, and ^{13}C at 4.95 MeV. The reactions that they undergo are as follows:

$$^{2}H + \gamma \longrightarrow {}^{1}H + n$$

$$^{17}O + \gamma \longrightarrow {}^{16}O + n$$

$$^{13}C + \gamma \longrightarrow {}^{12}C + n$$

Fortunately the neutron fluxes produced by these reactions are small and do not lead to other nuclear reactions of importance in a food. Fortunate also is the fact that all of the formed isotopes (^{1}H, ^{16}O, and ^{12}C) are stable.

Stable nuclei of some elements can be raised from a stable ground state to a higher and metastable state by direct absorption of electrons or of X and γ rays of energy less than required for particle emission from the nucleus. The spontaneous return to the ground state is accompanied by γ-ray or electron emission. Such nuclear excitations are called isomeric. In all nuclides only 32 isomeric states have half-lives longer than 10 min. Of these only 11 occur in foods, and only in trace amounts. In view of this, isomeric excitation is not a significant mechanism for inducing radioactivity in irradiated foods. It has been estimated that not even one excitation of any nuclide in 1 kg of meat sterilized with electron radiation is likely to occur.

As mentioned earlier (see Section IV,B), a fast-moving electron can interact with the nucleus of an atom to produce electromagnetic radiation called bremstrahlung. If the electron radiation energy is sufficiently great, the resultant bremstrahlung will have sufficient energy to cause nuclear reactions leading to induction of radioactivity. Because foods contain mostly elements of low atomic numbers, bremstrahlung formation by electron radiation is not an effective mechanism of energy transfer. The nuclides ^{2}H, ^{13}C, and ^{17}O, all minor isotopic constituents of foods, have the lowest energy thresholds for bremstrahlung formation. They do not form unstable or radioactive products. The effective threshold energy level for induction of radioactivity in a food by electron radiation through bremstrahlung formation is 10.5 MeV.

The above-described considerations, combined with confirmatory information secured by actual measurement for radioactivity in irradiated foods using the extremely sensitive techniques which radioactivity affords—equivalent to the detection of only a few specific nuclei among approximately 10^{25} nuclei—have led to the establishment of upper energy limits of 5 MeV for electromagnetic radiation and of 10 MeV for electron beams for use in irradiating foods.

The smallness of the quantity of radioactivity that can be induced in foods as a result of using radiations within these limits can be visualized by relating it to radioactivity present in foods due to naturally occurring long-lived radionuclides such as ^{40}K, ^{50}V, ^{87}Rb, ^{115}In, and ^{204}Pb and to the short-lived radionuclides ^{14}C and ^{3}H. In meat this natural activity amounts to about 100 Bq (100 disintegrations/sec). The possible added radioactivity in meat irradiated with what is the largest dose likely to be used in all food irradiation applications, that is, the dose for sterilization (radappertization), is equal to about 10^{-6} Bq, or one disintegration per week. It constitutes about 10^{-7} the natural radioactivity present before irradiation. This small increment of activity is deemed to be radiologically insignificant.

Worth noting is the fact that most irradiated foods are likely to be consumed only after the passage of time after irradiation. This is so not only because most of the uses of food irradiation are for the purpose of preservation, but also in uses such as insect disinfestation time for shipment of the irradiated food may be involved. In these time periods the decay of radioactive components occurs, regardless of their origin and, in commercial practice, irradiated foods contain less radioactivity than nonirradiated fresh foods. The very small increment of radioactivity caused by irradiation in the food is without meaning in this situation.

SOURCES OF ADDITIONAL INFORMATION

Anonymous, "Radionuclides in Foods." Natl. Acad. Sci., Washington, D.C., 1973.

Anonymous, "Training Manual on Food Irradiation Technology and Techniques," 2nd Ed., Tech. Rep. Series No. 114. Int. Atomic Energy Agency, Vienna, 1982.

Becker, R. L., Absence of induced radioactivity in irradiated food. *In* "Recent Advances in Food Irradiation" (P. S. Elias and A. J. Cohen, eds.). Elsevier Biomedical, Amsterdam, 1983.

Brownell, L. E., "Radiation Uses in Industry and Science." U.S. Government Printing Office, Washington, D.C., 1961.

Brynjolfsson, A., and Wang, C. P., Atomic structure. *In* "Preservation of Food by Ionizing Radiation" (E. S. Josephson and M. S. Peterson, eds.). Vol. I. CRC Press, Boca Raton, Florida, 1983.

Charlesby, A., ed. "Radiation Sources." Macmillan, New York, 1964.

Hurst, G. S., and Turner, J. E., "Elementary Radiation Physics." Wiley, New York, 1970.

Wang, C. P., and Brynjolfsson, A., Interactions of charged particles and γ-rays with matter. *In* "Preservation of Food by Ionizing Radiation," (E. S. Josephson and M. S. Peterson, eds.), Vol. I. CRC Press, Boca Raton, Florida, 1983.

Wang, Y., ed. "Handbook of Radioactive Nuclides." CRC Press, Boca Raton, Florida, 1969.

CHAPTER 2

Radiation Chemistry Basics

I. INTRODUCTION

Chapter 1 deals with the transfer of energy from the incident ionizing radiation to the absorber. The second stage of the action of radiation, the subject of this chapter, involves the chemical changes in the molecules of the absorber that result from this energy transfer.

At the beginning of the second or chemical change stage, and as a consequence of having energy greater than normal, absorber atoms either have had orbital electrons moved to higher energy states, or they have lost orbital electrons. Those atoms which have gained energy without losing electrons are termed *electronically excited* or *excited.* Atoms which have lost electrons have become positively charged and are *ions.* These two processes occur without regard to whether the atom exists uncombined or is a constituent of a molecule.

Because they contain abnormal amounts of energy, the ions and the excited atoms, whether they are free or molecular components, are unstable and chemically reactive. They may react among their own kind or with other suitable entities of their environment. The chemisty of these changes is called *radiation chemistry*. Radiation chemistry is a very broad subject which extends into areas not of direct application to food irradiation. Only what applies to food irradiation is covered in this discussion.

The very essence of food irradiation is centered upon the chemical changes in foods and their contaminants caused by irradiation. An understanding of this subject, therefore, is important to the intelligent use of radiation in treating foods. Radiation chemistry also contributes in a very important way to the subject of the safety of irradiated foods for human consumption.

The chemical changes that occur are largely, though not entirely, related to the makeup of the radiation-formed unstable ions or excited molecules. A series of events is likely to take place with each event occurring over a finite time period. The total elapsed time from the transfer of energy from the radiation to the absorber to the establishment of the final stable status is of the order of 10^{-3} sec. As will be noted later, in certain cases, this time period can be extended to hours or days.

The similarity of radiation chemistry to photochemistry has been noted. The principal aspect of similarity is that light in the energy range 2–7 eV can cause electronic excitation in the same way as does ionizing radiation. Subsequent chemical changes are basically similar to those obtained with ionizing radiation, but there are some important differences. The photochemical action is limited to a single plane perpendicular to the direction of the incident light. With ionizing radiation there is a more random distribution of effect at depths in the absorber governed by its composition and certain other aspects of the absorber plus the energy level and kind of ionizing radiation. In addition, with ionizing radiation there is a degree of concentration of events along primary electron tracks and associated branches. Finally, ionizing radiation produces ions as well as electronically excited atoms. Despite these differences some of the knowledge gained through studies of photochemical reactions can be applied to radiation chemistry.

II. PRIMARY CHEMICAL EFFECTS

A. Excitation

While the immediate site of the excitation is in a particular atom, in most cases that atom is part of a molecule. In these cases the molecule is recognized as "excited." Two mechanisms exist for the production of excited molecules by ionizing radiation:

1. Direct:

$$A \longrightarrow A^*$$

2. By neutralization of ions: an excited ion $(A^+)^*$ may be formed at the time of ionization:

$$A \longrightarrow (A^+)^* + e^-$$

Neutralization may occur subsequently and yield the excited molecule A^*:

$$(A^+)^* + e^- \longrightarrow A^*$$

An excited molecule can retain its extra energy for a period of the order of 10^{-8} sec. Loss of the excitation energy may occur by one of a number of pathways:

1. Emission of energy as a photon (fluorescence)
2. Internal conversion to heat
3. Transfer to a neighboring molecule
4. Chemical reaction
   A. Unimolecular:
      a. Rearrangement (isomerization):

         $$A^* \longrightarrow B$$

      b. Dissociation:

         $$A^* \longrightarrow C + D$$

   B. Bimolecular:
      a. Electron transfer:

         $$A^* + E \longrightarrow A^+ + E^-$$

         or

         $$A^* + E \longrightarrow A^- + E^+$$

      b. Abstraction (of hydrogen):

         $$A^* + EH \longrightarrow (AH)\cdot + E\cdot$$

      c. Addition:

         $$A^+ + E \longrightarrow F$$

The second reactant E may be identical with A^* or it may be another entity. The new products B, C, D, and F formed by the above reactions may be stable molecules or they may be reactive entities such as free radicals. [A free radical is an atom or molecule which has one or more unpaired electrons that are available to form a chemical bond. Free radicals are designated by the placement of a dot, as in $(AH)\cdot$ or $E\cdot$.] In cases where free radicals are present, further chemical action is possible.

B. Ionization

Normally an ion reacts with another ion of opposite sign to yield a neutral entity:

$$A^+ + B^- \longrightarrow C$$

$$A^+ + e^- \longrightarrow A$$

Ions may exist as a transient ion–molecule complex, which upon neutralization can yield new compounds:

$$(A \cdot B)^+ + e^- \longrightarrow C + D$$

Ionization may be preceded by other processes similar to those shown for excited molecules. After neutralization the same possibilities exist.

III. SECONDARY CHEMICAL EFFECTS

The net result of the primary chemical effects involving atoms and/or molecules of the absorber is the formation of new compounds and also, possibly, free radicals. Further chemical change, however, still is possible. Here also, what happens is governed by the composition of the absorber and by other factors such as its physical state and temperature. The new molecules formed in the primary chemical effect can react with themselves or with other molecules of the absorber which so far have not been changed. Both the new molecules and the hitherto unchanged molecules of the absorber can interact with the free radicals formed in the primary chemical effect.

Free radicals are common constituents of many systems, including foods, and can be produced by procedures other than irradiation. Irradiation, however, can produce some radicals with greater kinetic or excitational energy than the thermal energy of the molecules of the major portion of the absorber. Such "energetic" radicals react somewhat differently from radicals possessing only energies equivalent to thermal energy.

Although free radicals have been observed to persist in some systems indefinitely, they can be highly reactive through various kinds of reactions including addition, combination, dissociation, rearrangement, disproportionation, and electron transfer. Water content, physical state, temperature, availability of O_2, and other factors affect the stability of free radicals.

The primary chemical effects, as discussed earlier, are the direct consequence of absorber components acquiring energy through interaction with the radiation. While other changes can yet occur, the primary effect, itself, can result in chemical changes in the absorber of a permanent nature. This aspect of radiation-induced chemical change is termed the *direct effect*. Changes resulting from the interaction of primary product with themselves or with other components of the absorber are identified as the result of an *indirect effect* of irradiation. The relative importance of each of these effects in the final total change is related to the composition of the absorber and to certain aspects of the conditions under which the irradiation is carried out and the product held after irradiation.

Direct action of radiation is the consequence of the absorption of energy by particular entities of the absorber, and if this causes chemical changes such that stable end products results, it is clear that this event has no dependency on other factors. Indirect action, on the other hand, occurs through the interaction of the entities formed by direct action with other substances present in the absorber. Factors which can affect this interaction would be expected to play a role in the final outcome. For example, there must be available some means for the reactants to come together. If the system is in the solid state, movement of the reactants is restricted and, consequently, opportunity for reaction is limited. If, however, the system is in the liquid state, movement of the reactants is facilitated and the opportunity for interaction greatly increased. In this way the physical state of the absorber, as determined by temperature during irradiation, is significant in the ultimate chemical change resulting from irradiation.

In a similar manner, a solid substance irradiated in the dry state undergoes chemical change which primarily is the result of direct action. When dissolved in liquid water, it may undergo a greatly different change, largely through indirect action. Control of the factors related to indirect action can be the means for altering the chemical changes in an absorber brought about by irradiation.

IV. *G* VALUE

In order to provide a quantitative basis for radiation-induced chemical changes, a unit, the *G* value, has been devised. The *G* value is the number of absorber molecules changed or of new substances formed per 100 eV of energy deposited. Except for chain reactions, such as occur with certain organic molecules, *G* values seldom exceed 10.

V. AQUEOUS SYSTEMS

Because many foods contain substantial amounts of water, indirect effects are important in food irradiation. The same consideration applies to the biological systems of food contaminants such as those of bacteria.

When pure water is irradiated a number of products are formed in accordance with the following somewhat simplified equation:

$$H_2O \longrightarrow \cdot OH + e^-_{aq} + H\cdot + H_2 + H_2O_2 + H_3O^+$$

Due to recombination effects, continued irradiation of pure water yields a steady-state situation with only the molecular products H_2 and H_2O_2 present, the amounts of which are determined by the dose rate (providing H_2 does not leave the system). That water is not changed by continued irradiation under such conditions is indeed fortunate, since it provides a convenient personnel shield for radionuclide γ sources.

The indicated products obtained when water is irradiated are highly reactive chemically. Electrons and hydrogen atoms are reducing agents. The hydroxyl radical is an oxidizing agent, as is H_2O_2. It is not surprising, therefore, that these products will react with many substances when they are dissolved in water. Due to the strength of the O–H bond they do not, however, react with water molecules present.

The aqueous electron e^-_{aq} is formed when free electrons lose energy down to a level comparable with that of thermal energy of the environment. Surrounding water molecules are bound to the electron by virtue of the electrostatic charge of the electron and the electric dipole moments of the water molecules, as indicated by the following equation:

$$e^- + nH_2O \longrightarrow e^-_{aq}$$

Due to the presence of the bound water molecules, the ability of the electron to move about in the medium is impaired. In the solid state the dielectric properties of water are such that, compared with the liquid state, solvation of the electron is less likely to occur. In the solid state, however, other processes (e.g., trapping and neutralization) limit the ability of the free electron to migrate over relatively long distances. Nonetheless, the solvated electron is a highly reactive substance and can cause important indirect effects of irradiation.

As might be anticipated, by influencing the concentration of radiolytic reactants, the pH of an aqueous system can affect the end result of irradiation. The types of equilibria that are pH dependent are illustrated by the following equations:

Acid solutions

$$e^-_{aq} + H^+ \rightleftharpoons H\cdot$$

$$H\cdot + H^+ \rightleftharpoons {H_2}^+$$

Alkaline solutions

$$(OH)\cdot \rightleftharpoons H^+ + O^-$$

$$(HO_2)\cdot \rightleftharpoons H^+ + {O_2}^-$$

$$H\cdot + OH^- \rightleftharpoons e^-_{aq}$$

The liquid state, by providing a good medium for movement of reactants, aids indirect action of radiation. Freezing the solution limits such movement and consequently reduces indirect action.

Depending upon the solute concentration, aqueous solutions of only one solute show varying response to irradiation. In dilute solutions, that is, at concentrations below 0.1 *M*, indirect action predominates. At greater concentrations direct action may occur and can be significant above 1 *M*. The effect of concentration on the radiation inactivation of L-ascorbic acid in water solution is shown in Fig. 2.1. Noteworthy are the large percentage inactivation at lower concentrations and the progressive lessening as concentration increases. These observations are explained on the basis that, for each of the doses indicated, there is a limited quantity of reactants derived from the radiolysis of the solvent, water. For a given dose, as the concentration of the solute increases, the ratio of reactants to the solute decreases, and, as a consequence, the percentage of inactivated L-ascorbic acid molecules diminishes.

The introduction of a second solute into the solution can bring about competition between the two solutes for the radiolytic products of the water and result in each solute having less opportunity to react than it does when present alone. An example of this effect is shown in Fig. 2.2. The destruction of histidine HCl in water by electron irradiation is reduced by the addition of D-isoascorbic acid. Specific effects of such competitive response vary with the particular solutes involved.

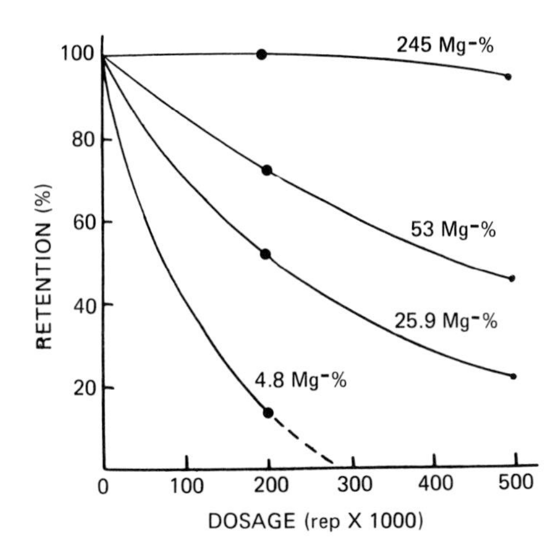

Figure 2.1 Effect of 3-MeV electron radiation on different concentrations of L-ascorbic acid. From B. E. Proctor, *Proc. Int. Conf. Preserv. Foods Ioniz. Radiat., 1959.*

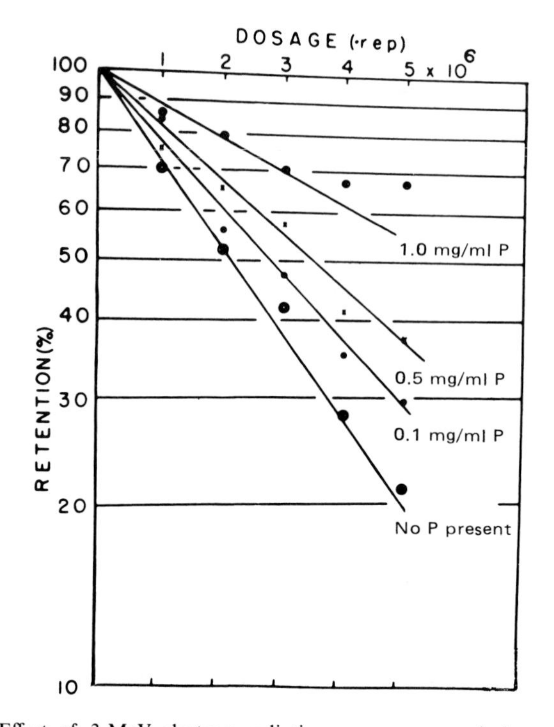

Figure 2.2 Effect of 3-MeV electron radiation on aqueous solutions of histidine HCl (1 mg/ml) in the presence of varying quantities of D-isoascorbic acid (P). From B. E. Proctor, *Proc. Int. Conf. Preserv. Foods Ioniz. Radiat., 1959.*

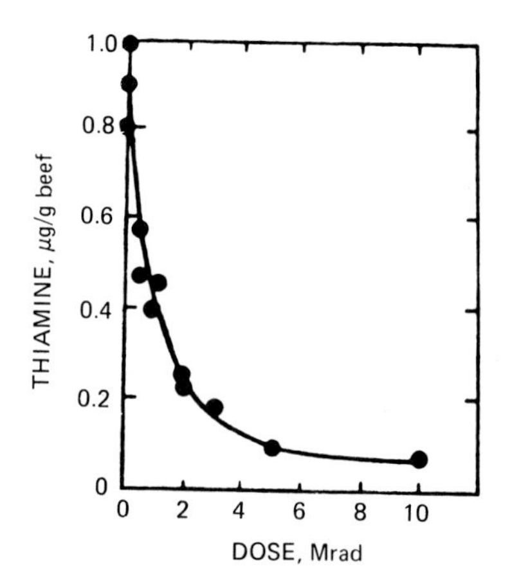

Figure 2.3 Destruction of thiamine in minced beef by irradiation at ambient temperatures. From G. M. Wilson, *J. Sci. Food Agric.* **10,** 295 (1959).

Generally aqueous solutions which contain a number of solutes exhibit only relatively small amounts of reaction and change for each solute. By either direct or indirect action the energy of the dose applied distributes among all the solutes, rather than concentrating on only one or a few of them. As a consequence, there usually are formed many final radiolytic products, but each is in small amount. Since foods generally have a number of component substances, the irradiation of moist foods can be expected to yield numerous radiolytic products, each in low concentration.

Depending upon chemical composition, however, variation from the usual effect can occur. For example, certain substances commonly referred to as *scavengers* react preferentially with free radicals and interfere with the usual reactions. Some molecules, especially large and complex ones, may contain particular groupings of atoms that are more easily attacked by radiation or by radiolytic products than others and, because of this, undergo reactions in preference to other less sensitive parts of a molecule. In this way, for example, the –SH group in cysteine is sensitive to change, especially in oxygenated solutions.

Even when present in complex systems such as a food, some substances have unusual sensitivity for chemical change through interaction with the radiolytic products of water. Figure 2.3 indicates the destruction of thiamine (vitamin B_1) in beef as a function of dose. That this substantial destruction is largely the result of indirect action of radiation associated with the water content of beef can be seen from Fig. 2.4, in which the loss of thiamine is shown to be greatly reduced by irradiation at subfreezing temperatures.

While exceptions do occur, the ordinary effects of irradiation on complex

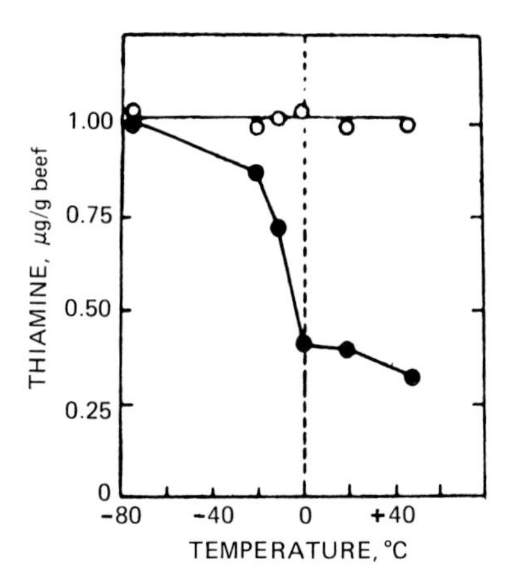

Figure 2.4 Effect of irradiation temperature on thiamine destruction in minced beef. ○, Not irradiated; ●, irradiated (10 kGy). From G. M. Wilson, *J. Sci. Food Agric.* **10,** 295 (1959).

aqueous systems, such as food, which lead to a multiplicity of radiolytic products in small amounts individually is of critical importance in food irradiation in that, overall, foods are little changed chemically by irradiation. This fact will be considered later in connection with (1) doses employed in food irradiation, (2) the highly selective nature of the irradiation process in terms of controlling living organisms that contaminate foods, (3) considerations of toxicological safety of irradiated foods, (4) nutritional value, and (5) sensory qualities of irradiated foods.

VI. ROLE OF IRRADIATION PARAMETERS

A. Dose

As covered in Chapter 5, particular uses of food irradiation require specific doses in order to accomplish desired objectives. Since in every application of food irradiation, the basic mechanism involves chemical change, the dose needed to secure the needed kind and amount of change must be deposited in the food. For this reason, of all the controllable variables in food irradiation, dose is the most important.

Although a radiation-induced chemical change may initiate an action in functioning biological systems (living organisms), the exact relationship between dose and effect often is tied to parameters of the biological system itself. For example, single-cell organisms (e.g., bacteria) may be more resistant to radiation and require larger doses to kill them than do more complex multicellular organisms such as insects. Some biological systems have the capability of repairing the sustained radiation damage. Such systems can require larger doses to secure lethality.

Regardless of the purpose of the treatment, the entire food entity usually is unavoidably pervaded in a gross way by the radiation, and, therefore may undergo radiation-induced changes which are unrelated to the purpose of the treatment. Such changes may not only serve no useful purpose, but may be undesirable from a variety of viewpoints. That such changes occur does not mean that each and every molecule of the irradiated food undergoes chemical change. In fact, as noted later (see Chapter 3), only a small percentage of the total number present is affected. It is clear, however, that as dose is increased more molecules can be affected. In this manner a dose limit for an acceptable amount of change in a food may be reached. In some cases, as discussed in subsequent chapters, the dose needed to obtain the desired effect exceeds the dose permitted by other considerations, and, as a consequence, particular applications of food irradiation may not be practical.

The amounts of radiolytic products formed in a food are not necessarily directly related to dose. Other variables, usually operating on the indirect action of radiation and sometimes controllable, can influence the quantitative relationship between dose and amount of radiolytic products. In order to secure useful applications of food irradiation, it often is necessary to take measures to lessen or avoid undesired changes.

B. Dose Rate

The rate at which radiation can be applied to a food can be varied within wide limits, perhaps by as much as a factor of 10^6. In general, dose rate has not been a very critical parameter in food irradiation. There are, however, some situations where dose rate can affect results. At high dose rates, free radicals can be formed in such high concentrations that recombination rather than reaction with other absorber entities is favored. This reduces the amount of indirect action. Dose rate can also be of significance in circumstances where a concurrent process such as diffusion plays a role in the end result. For example, the subsurface oxygen tension of foods irradiated in the presence of O_2 may be reduced sufficiently at high dose rates to yield an abnormal anoxic condition due to the inability of the diffusion process to keep up with the radiation-caused O_2 depletion. The thus-established anoxic condition may alter the end result of the irradiation.

C. Water Content

In foods and in biological systems most of the indirect action of radiation is associated with their water content. Liquid water provides an effective medium for primary radiolytic products to move and to interact with either other primary products or with other absorber components.

Dry foods, on the other hand, lacking this medium, are less subject to indirect action and tend to display end results mostly attributable to direct action. In dry foods, free radicals formed by irradiation may persist for extended periods. Generally, dry foods undergo less chemical change than do foods containing water.

D. Temperature

The primary effects of irradiation, ionization and electronic excitation, are independent of the irradiation temperature. Subsequent effects, however, may be affected by temperature. Overall, the role of temperature in food irradiation can be important.

Activation energies of chemical reactions vary with temperature, and, as a consequence, yields of reaction products may be altered. Mobility of free radicals and other reactants, as determined by absorber or absorber component viscosity, can be changed with temperature. Sufficiently low temperatures can lead to virtual immobility with consequent greatly reduced capability for interaction. Under such conditions free radicals can persist for relatively long periods without reacting. In some cases the transition from solid to liquid is accompanied by a large change in yield of radiolytic products.

Apart from factors related to the strictly chemical changes, temperature can have important effects in functioning biological systems. For example, temperature changes the metabolic rate of insects, which in some cases, at least, has a bearing upon the response of the organism to radiation.

E. Additives

1. Oxygen

The presence of O_2 during irradiation can alter the chemical changes that follow. Molecular oxygen, having two unpaired electrons, ·(O—O)·, acts as a diradical. It is able to react with other radicals, initially forming peroxy radicals $(RO_2)\cdot$, which can react further. Oxygen also is an oxidant. In some cases irradiation of oxygenated systems yields results that parallel autoxidation. Foods containing lipids are particularly affected by O_2 during irradiation and can develop off- or rancid flavors. From O_2 can be formed ozone (O_3), a very powerful oxidant.

2. Other Substances

Certain substances, used as additives, can alter the chemical changes resulting from irradiation. Usually only small amounts of an additive are considered, and its action probably would be to interact with one or more particular radiolytic products in a preferential way. An additive may also serve to alter an original component of the system so as to prevent production of a particular radiolytic product. Such an additive could be a radical scavenger.

In attempts to solve certain problems encountered in food irradiation, consideration has been given to the use of additives, but so far this approach has been generally unproductive, largely because of limitations imposed by the nature of foods and their uses.

F. Combination of Irradiation with Other Treatments

In some circumstances, irradiation used with other treatments is of value. Very little is known about the effects of such combination treatments upon the radiation chemistry of most sytems, but as discussed in Chapter 11, practical use of them is feasible in certain food irradiation applications.

VII. POSTIRRADIATION EFFECTS

In systems irradiated and held at very low temperature (i.e., cryogenic), free radicals may persist for extended periods of time. When these systems are raised to sufficiently high temperatures postirradiation, the free radicals are able to interact with suitable substances. If they react with other identical free radicals, they may disappear through dimerization or disproportionation. If recombination occurs, no chemical change results. In oxygenated systems peroxides may be reactive for periods after irradiation. Oxygenated aqueous solutions containing H_2O_2 may produce oxidative changes slowly after irradiation.

In functioning biological systems, the biological consequences of chemical changes may not be immediate, but can develop at varying times after irradiation depending upon dose and the specific organism.

VIII. TIMES FOR VARIOUS POSTIRRADIATION EVENTS

It may be useful to know the time periods for the series of events that occur after irradiation. Most events occur in rapid sequence. Some of the last events, however, can occur over hours or days.

The postirradiation events can be classified into three categories as follows:

1. Physical: This initial stage of the interaction of radiation with the absorber involves only energy transfer and no chemical changes.
2. Physical–chemical: This second stage has continued energy transfer events plus systems-associated physical processes such as diffusion and the beginning steps of chemical changes.
3. Chemical: Chemical changes occur in the third and last stage. These changes may involve a sequence of several steps and ultimately lead to an end point in which there is chemical stability in the sense that any additional chemical changes are not associated with irradiation.

These three stages occur in a time sequence which may be approximated as follows:

Physical stage: 0 to 10^{-14} sec
Physical–chemical stage: 10^{-14} to 10^{-10} sec
Chemical stage: 10^{-10} to 10^{-3} sec

The last, the chemical stage, in some cases may be extended for periods of hours or days, depending upon the chemical nature of the system components and other factors such as the system temperature.

As is discussed in Chapter 4, the biological consequences of the chemical changes produced by irradiation in living organisms can be further extended timewise, depending upon the organism and the nature of the changes produced.

SOURCES OF ADDITIONAL INFORMATION

Altman, K. I., Berber, G. B., and Okada, S., "Radiation Biochemistry." Academic Press, New York, 1970.

Bacq, Z. M., and Alexander, P., "Fundamentals of Radiation Biology." Pergamon, Oxford, 1961.

Casarett, A. P., "Radiation Biology." Prentice-Hall, Englewood Cliffs, New Jersey, 1968.

Hart, E. J., Radiation chemistry of aqueous solutions. *Radiat. Res. Rev.* **3**(4), 285 (1972).

O'Donnell, J. H., and Sangster, D. F., "Principles of Radiation Chemistry." Elsevier, New York, 1970.

Spinks, J. W. T., and Woods, R. J., "An Introduction to Radiation Chemistry." Wiley, New York, 1963.

Swallow, A. J., Chemical effects of irradiation. *In* "Radiation Chemistry of Major Food Components," (P. S. Elias and A. J. Cohen, eds.), Elsevier, Amsterdam, 1977.

Taub, I. A., Reaction mechanisms, irradiation parameters and product formation. *In* "Preservation of Food by Ionizing Radiation" (E. S. Josephson and M. S. Peterson, eds.), Vol. II. CRC Press, Boca Raton, Florida, 1983.

CHAPTER 3

Radiation Chemistry of Food Components and of Foods

I. INTRODUCTION

The chemical changes caused by irradiation that are of concern in connection with foods involve (1) changes in the foods themselves and (2) effects on living food contaminants such as bacteria, whose inactivation is an objective of the irradiation. These two topics are treated separately. The radiation chemistry of those food components capable of undergoing change is the subject of this chapter.

Irradiation does not alter the elemental composition of a food and, for this reason, the mineral components need no discussion. (This exclusion assumes that irradiation does not alter the bioavailability of food minerals. This is not necessarily true.)

Carbohydrates, proteins, and lipids are the major food components. Vitamins, while minor constituents of foods in terms of quantities present, are so important nutritionally that they are of major interest. The radiation chemistry of these four groups of food components is covered. Additionally, the radiation chemistry of whole food classes is examined.

II. RADIATION CHEMISTRY OF FOOD COMPONENTS

A. Carbohydrates

There are some foodstuffs that occur as essentially pure and dry carbohydrates (e.g., sugars) or as low-moisture foods having a high carbohydrate content (e.g., grains). For this reason the radiation chemistry of solid-state carbohydrates is of interest.

Physical changes, indicative of chemical changes, that occur with the irradiation of low molecular weight carbohydrates in the solid state include changes in the melting point, optical rotation, and absorption spectra. Irradiation yields various gases, including H_2, CO_2, CH_4, and CO. Generally the amount of gases formed is dose dependent. In addition, irradiation in the solid state yields a number of nongaseous radiolytic products such as formaldehyde, acetaldehyde, acetone, acid derivatives, lactones, glyoxal, malonaldehyde, H_2O_2, and derived sugars. Pure carbohydrates in the

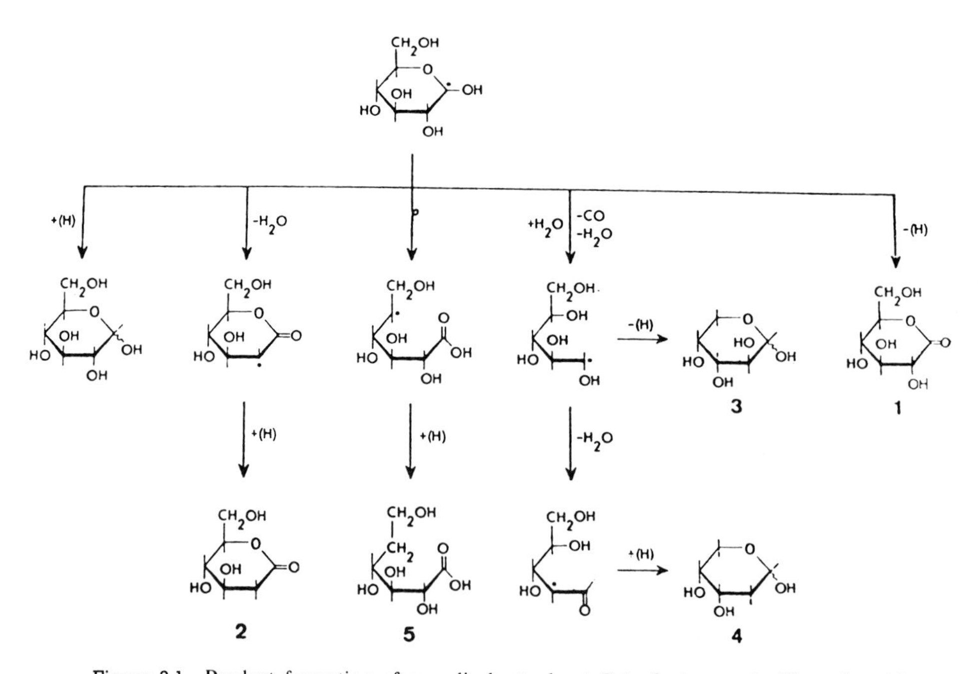

Figure 3.1 Product formation after radical attack at C-1 of glucose. **1,** Gluconic acid; **2,** 2-deoxy-gluconic acid; **3,** arabinose; **4,** 2-deoxy-ribose; **5,** 5-deoxy-gluconic acid. From J. F. Diehl, S. Adam, H. Delincee, and V. Jakubick, *J. Agric. Food Chem.* **26,** 15 (1978).

crystalline state are very sensitive to radiation, sometimes undergoing reactions with high G values, possibly due to chain-type reactions.

In addition to the dry carbohydrate-containing foods, there are others which have sufficient water to require attention to radiation-induced changes in carbohydrates that are related to the action of the radiolytic products of water. Low molecular weight sugars in aqueous solution generally undergo oxidative degradation, partly due to the direct action of the radiation and partly due to interaction with radiolytic products of water, mainly $OH\cdot$ radicals. Oxidation at the ends of the molecule produces acid derivatives. Hydrogen abstraction by $OH\cdot$ radicals can occur with glucose in the absence of O_2 at any of the six possible locations in the molecule. The reactions for the glucosyl radical formed at the C-1 position are indicated in Fig. 3.1. Other similar reactions occur with glucosyl radicals formed at the five other carbon atom locations. Although each occurs in only small amounts, it is clear that many different radiolytic products result.

The presence of O_2 increases the yields of acids and keto acids. Raising the pH increases the amounts of deoxy compounds. The yield of malonaldehyde, a substance of known cytotoxic activity, is small at the normal pH values of most foods, and this is one consideration important in connection with the evaluation of the safety of irradiated foods for human consumption.

The glycosidic bond of disaccharides can be broken by radiation. In O_2-free (N_2O-saturated) solutions of cellobiose, for example, it has been estimated that about one-third of the products formed by $OH\cdot$ radical interaction result from breaking of the glycosidic bond. The presence of O_2 reduces this effect. Irradiation also alters other parts of the cellobiose molecule and forms new 12-carbon compounds. Higher molecular weight derivatives containing 24 carbon atoms also are formed.

Other low molecular weight sugars in aqueous solution undergo similar types of changes, with differences in end products which relate to their initial molecular structure. The very diversity of end products formed produces yields of individual products, which while dose dependent, are small.

From the standpoint of food irradiation, the most important change in polysaccharides caused by irradiation is the breaking of the glycosidic bond. In starches, pectins, and cellulose this results in the formation of smaller carbohydrate units. As a consequence, some foods, especially fruits, undergo softening and loss of texture.

With starches, breakage of the glycosidic bond leads to the formation of dextrins of varying lengths. In addition, other radiolytic products are formed. Generally all starches show similar changes, so much so that information for a

Table 3.1

Radiolytic Products of Maize Starch Irradiated under O_2, Water Content 12–13%[a]

Radiolytic products[b]	Concentration (μg/g per 10 kGy)
Formol	20
Acetaldehyde	40 ($\leq$8 kGy)
Acetone	2.1 (>20 kGy)
Malonaldehyde	2
Glycolaldehyde	9
Glyoxal	3.5
Glyceraldehyde and/or dihydroxyacetone	4.5
Hydroxymethylfurfural	1
Methylglyoxal	<0.25
Diacetyl	<0.1
Acetoin	<0.1
Furfural	<0.4
Formic acid	100
Acetic acid	<1.8
Glyoxylic acid	<0.5
Pyruvic acid	<0.2
Glycolic acid	<0.6
Malic acid	<1.3
Oxalic acid	<1.4
Methyl formate	Traces
Ethyl alcohol	Variable[c]
Methyl alcohol	2.8
Glucose	5.8
Maltose	9.8
Mannose	0.1
Ribose	0.6
Xylose	0.4
Erythrose	1.2
H_2O_2	6.6 (1–4 kGy)

[a] From J. F. Dauphin and L. R. Saint-Lebe, Radiation chemistry of carbohydrates. *In* "Radiation Chemistry of Major Food Components" (P. S. Elias and A. J. Cohen, eds.). Elsevier, Amsterdam, 1977.

[b] Various other acids have been identified at 940 kGy.

[c] Present in control samples.

particular starch may be used to predict changes that occur in other starches. Radiolytic products that have been identified in maize starch are given in Table 3.1. An important consequence of radiation degradation of starches is the reduction of viscosity, as shown in Table 3.2.

Table 3.2

Effect of Irradiation on Viscosity and Degree of Polymerization of Potato Amylose[a]

Dose (kGy)	Intrinsic viscosity ($g^{-1} \times ml$)	Degree of polymerization
0	230	1700
0.5	220	1650
1.0	150	1100
2.0	110	800
5.0	95	700
10.0	80	600
20.0	50	350
50.0	40	300
100.0	35	250

[a] From C. T. Greenwood and C. MacKenzie, *Die Stärke* **15,** 444 (1963).

1. Mixtures of Carbohydrates with Proteins and Lipids

Since carbohydrates occur in many foods along with other substances such as proteins and lipids, the effect of irradiation under these circumstances is of interest.

Irradiation of mixtures of sugars and amino acids leads to polymerization followed by a browning effect. The addition of cysteine or methionine to a glucose solution inhibits formation of carbonyl compounds. Similar results have been obtained with trehalose. The dose dependence of the degradation of trehalose as affected by the presence of each of several amino acids is shown in Fig. 3.2. As shown in Fig. 3.3, proteins also reduce the amount of degradation by irradiation but are not as effective as mixtures of the isolated amino acids of the same composition as they occur in the protein. This difference apparently is related to the effect of the spatial conformation of the acids in the protein which makes them less available for interaction. The protective action of amino acids and proteins is ascribed to interference with the availability of the OH· radical for interaction with the sugar. Supporting this view is the observation that the protective action is only quantitative and not qualitative, since the same radiolytic products are found in mixtures as in pure sugar solutions, but in lesser amounts.

Emulsified lipids have no significant effect on the radiolysis of sugars, as judged from studies on trehalose solutions to which was added homogenized sunflower seed oil.

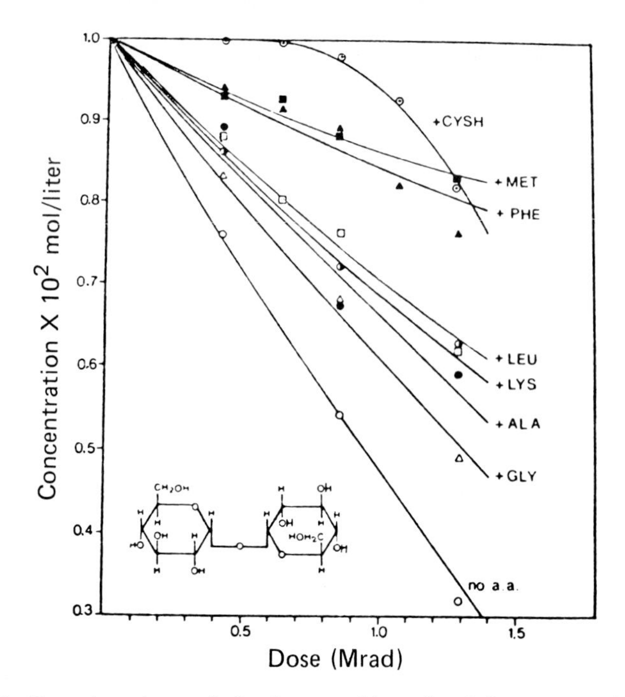

Figure 3.2 Dose dependence of the decomposition of trehalose upon γ irradiation in air-saturated aqueous solution (10^{-2} M) with and without added amino acids (10^{-2} M). From J. F. Diehl, S. Adam, H. Delincee, and V. Jakubick, *J. Agric. Food Chem.* **26,** 15 (1978).

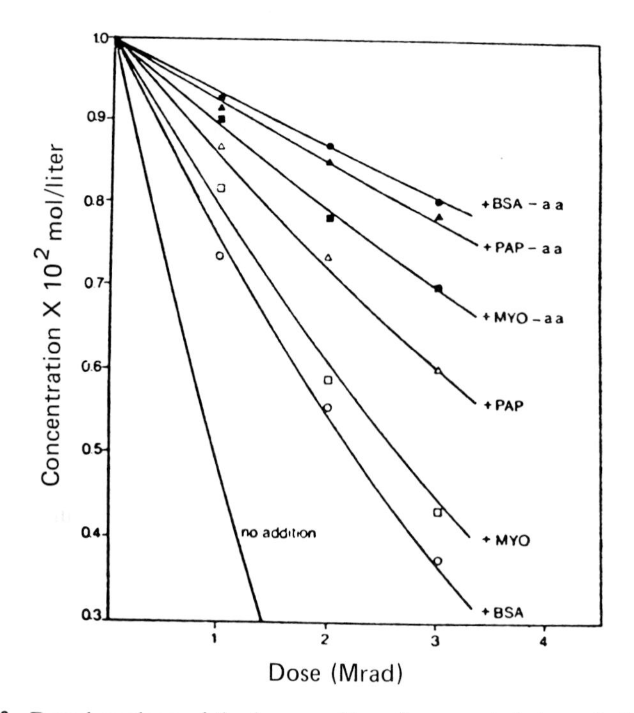

Figure 3.3 Dose dependence of the decomposition of aqueous trehalose solution (10^{-2} M) with and without addition of 0.36% proteins or a mixture of constituent amino acids: BSA, bovine serum albumin; MYO, sperm whale myoglobin; PAP, papain; BSA-aa, amino acid mixture corresponding to the amino acid composition of BSA (PAP-aa and MYO-aa analogous). From J. F. Diehl, S. Adam, H. Delincee, and V. Jakubick, *J. Agric. Food Chem.* **26,** 15 (1978).

2. Comments

The radiation chemistry of carbohydrates is complex and many radiolytic products are possible. The great diversity of these products prevents the formation of large quantities of any single one. In foods which contain other components, there is the additional factor of protection of the carbohydrate from radiolysis as exemplified by the actions of amino acids and proteins.

Perhaps the most significant single effect of irradiation on carbohydrates is the breaking of the glycosidic bond, which can affect not only functional properties of materials such as starches, but also quality aspects of foods, especially fruits, through reduction of firmness and through related texture changes. Except for this effect, generally the radiation-induced changes in carbohydrates are too small to be very important in food irradiation.

B. Proteins and Related Compounds

1. Introduction

With the exception of some scleroproteins, foods as a group contain virtually every kind of protein. Because food irradiation is concerned not only with effects on foods per se, but also with the biological systems of certain living food contaminants, for which proteins provide essential life functions, the radiation chemistry of proteins requires the broadest consideration.

Although foods and living biological systems do not contain isolated amino acids and polypeptides in large amounts, there is important information relative to the radiation chemistry of proteins that can be obtained from an examination of the effects of radiation on these substances. As might be anticipated from the fact that there are about 20 amino acids, each with its own composition and structure, their radiation chemistry can be complex. As with carbohydrates, it is necessary to consider the chemistry of both the dry state and aqueous systems.

2. Amino Acids and Peptides

In the solid (dry) state the simple amino acids, such as glycine, in the absence of O_2 yield NH_3, keto acids, fatty acids, and small amounts of H_2, CO_2, and amines. The sulfur-containing amino acid cysteine yields cystine, H_2, NH_3, H_2S, and an "NH_2-free" fraction. These compounds are attributed to an initial direct reaction of the acid with energetic electrons and subsequent interaction of free radicals and ions to yield stable end products. The latter step is impeded by the limited mobility of the primary active intermediates in the environment of the dry amino acids.

The presence of water makes the indirect action of the radiation more effective, and increases the role of environmental factors. In the absence of O_2, the simple amino acids undergo reductive deamination and decarboxylation, according to the following equations:

Deamination:

$$RCHNH_2COOH + H\cdot \longrightarrow RC\dot{H}COOH + NH_3$$

Decarboxylation:

$$RCHNH_2COOH \longrightarrow CO_2 + RCH_2NH_2$$

Other radiolytic products are formed, the exact nature of which is related to the composition of the amino acid. Glycine, for example, yields in addition to NH_3 and CO_2, H_2, methylamine, acetic acid, and formaldehyde.

If O_2 is present, it removes e_{aq}^- and H· and effectively blocks reductive deamination. Instead, oxidative deamination takes place through interaction with the OH· radical, as indicated by the following equations for glycine:

$$H_2C(NH_2)COOH + OH\cdot \longrightarrow HC(NH_2)COOH + H_2O$$

$$\dot{H}C(NH_2)COOH + O_2 + H_2O \longrightarrow NH_3 + HCOCOOH$$

In the case of aliphatic amino acids, increasing the chain length, by providing additional C–H bonds for interaction with the OH· radical, reduces the amount of oxidative deamination.

Other reactions can take place as a consequence of subsequent interaction of primary products with the radiolytic products of water and with other primary products. Yields of specific products change with dose. At high solute concentrations direct radiation effects become more important.

Other amino acids react differently according to their composition and structure. The presence of a thiol or disulfide group can increase sensitivity to radiation. Among other reactions, oxidation of the sulfur moiety occurs. Cysteine, for example, forms cystine. With aromatic and heterocyclic amino acids hydroxylation of the aromatic ring is the principal reaction.

Peptides undergo main-chain degradation and yield amide-like products, which result from OH· radical attack of the α carbon. With aliphatic amino acids, chain length has little effect on degradation. Aromatic groups reduce amide formation. The principal radiolytic products of peptides are NH_3, fatty acids, keto acids, and amide-like products.

3. Proteins

a. *Introduction*

It can be expected that the radiation chemistry of proteins reflects what occurs with their component isolated amino acids. It also can be anticipated, however, that there are other radiation effects which are related to the large size of protein molecules and to some aspects of their structure.

The absorption of ionizing radiation by a given protein molecule can lead to multiple ionizations and excitations within the molecule. The absorbed energy can transfer from an initial site to another which may be more "sensitive" and where bond breakage may occur. Because of such vulnerable loci, the result of irradiation of a protein is not a fully random process but instead is one that leads to regular patterns of changes.

At least three types of structures or organization can be identified in proteins:

Primary: kind and number of component amino acids and the relationship to each other in chains formed through peptide bonds
Secondary: arrangement of polypeptide chains: open or coiled or folded
Tertiary: spatial configuration in three dimensions of polypeptide chains, when more than one chain occurs in a protein

Because of the secondary and/or tertiary structure, some groups that are responsive to radiation in isolated amino acids may not be sufficiently accessible in proteins to reactants, such as free radicals, that operate through indirect action. As a consequence, reactions that occur with isolated amino acids may not take place when the same acids are components of proteins.

The complexity and size of proteins provide such a large number of loci for interaction through direct and indirect action that the end products are diverse. The amounts formed are dose dependent, but the amount of each is not likely to be large.

b. Dry Proteins

In the absence of water the action of radiation is limited essentially to direct action. The presence of free radicals following irradiation can be demonstrated by electron spin resonance spectra. Radicals formed at low temperatures (e.g., $-196°C$), designated as "primary," are different from those formed at room temperature (secondary radicals). Primary radical formation is greater in proteins having a higher sulfur content. Free radicals ultimately disappear, those containing sulfur lasting longest. Due to limited diffusion at low temperatures, recombination is most likely, whereas at higher temperatures, reaction with other substances is more probable.

The status of a protein before irradiation affects what happens upon irradiation. A heat-denatured protein, for example, yields more free radicals than does its native form. In the heat-denatured protein the secondary and tertiary structures are altered so as to reduce opportunities for the free radicals to recombine and in this way to disappear.

Irradiation can denature native proteins, principally through breaking hydrogen bonds and other linkages involved in secondary and tertiary

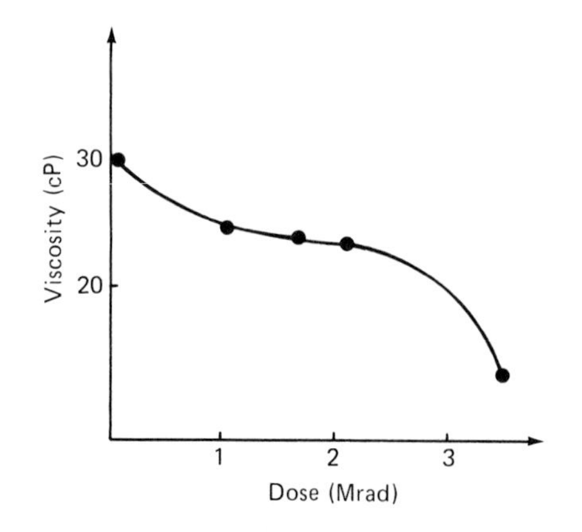

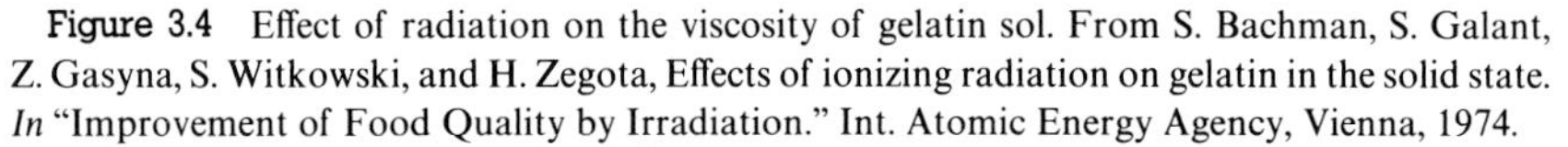
Figure 3.4 Effect of radiation on the viscosity of gelatin sol. From S. Bachman, S. Galant, Z. Gasyna, S. Witkowski, and H. Zegota, Effects of ionizing radiation on gelatin in the solid state. *In* "Improvement of Food Quality by Irradiation." Int. Atomic Energy Agency, Vienna, 1974.

structures. Such denaturation can result in changes in the shape of the molecule and exposure of "sheltered" groups such as disulfide bonds. Breakage of the peptide bond can lead to the formation of smaller units with lower molecular weights. Cross-linking, mainly the result of hydrogen bonding, but also secured by other linkages, can lead to the reverse, namely, aggregation. All of these diverse reactions do not occur with all proteins, but vary with the nature of the protein. Globular proteins, having a "tighter" structure, favor recombination reactions and consequently are more resistant to change. Fibrous proteins, whose structure is more "open," undergo change more easily. Collagen, a fibrous protein, for example, degrades into smaller units when irradiated in the dry state. Some of these changes affect the properties of proteins. In Fig. 3.4 the effect of irradiation of dry gelatin on the viscosity of a gelatin sol is shown. The drop in viscosity with dose is indicative of the formation of smaller units. Reduction of gel strength, explained on the same basis, is shown in Fig. 3.5.

The kind of change caused by irradiation is related to dose. Moderate doses can affect secondary and tertiary structures, the end result varying greatly with the nature of these structures. Higher doses yield detectable changes in the primary structure. Destruction of amino acids occurs. In bovine serum albumin this effect has been observed to be proportional to doses up to 1500 kGy when irradiated in the solid state. As is discussed in sections dealing with food irradiation applications, doses likely to be used in food irradiation will not exceed about 50 kGy. The data of Table 3.3 indicate that the irradiation of gelatin in the dry state with doses up to 35 kGy produces no significant change in amino acid content. Because the principal nutritional

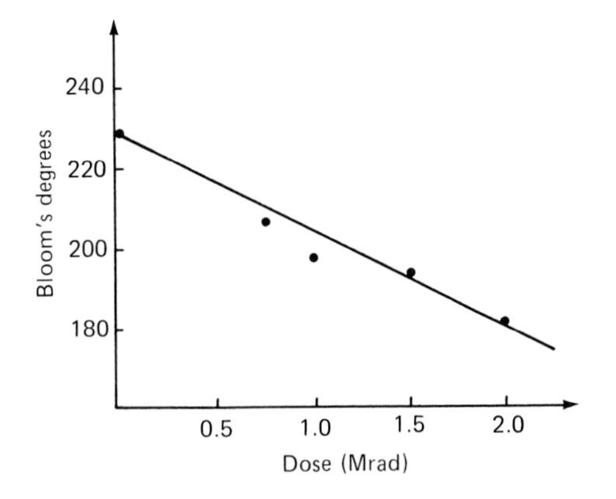

Figure 3.5 Gel strength of irradiated gelatin. From S. Bachman, S. Galant, Z. Gasyna, S. Witkowski, and H. Zegota, Effects of ionizing radiation on gelatin in the solid state. *In* "Improvement of Food Quality by Irradiation." Int. Atomic Energy Agency, Vienna, 1974.

Table 3.3
Amino Acid Content of Dry Gelatin after Irradiation[a,b]

		Dose (kGy)		
Amino acid	Control	10	25	35
Hydroxyproline	11.9	10.60	11.00	13.00
Aspartic acid	5.40	6.00	5.70	5.50
Threonine	1.65	2.12	1.70	1.70
Serine	3.13	3.14	3.15	3.30
Glutamic acid	10.10	10.50	10.30	10.00
Proline	13.55	13.30	13.20	14.40
Glycine	22.30	21.20	21.30	21.50
Alanine	8.60	9.35	9.20	8.60
Valine	2.41	2.53	2.50	2.50
Arginine	8.28	7.75	7.00	7.10
Methionine	0.53	0.90	0.53	0.40
Isoleucine	1.15	1.23	1.16	1.00
Leucine	2.80	2.80	2.74	2.70
Tyrosine	Traces		Traces	
Phenylalanine	2.04	1.90	1.87	1.90
Hydroxylysine	0.90	0.91	0.77	1.10
Lysine	3.45	3.87	3.48	2.70
Histidine	0.77	0.73	0.64	0.60
NH_4^+	1.12	1.01	0.77	0.90

[a] From S. Bachman and H. Zegota, Physicochemical changes in irradiated (gamma ^{60}Co) inulin. *In* "Improvement of Food Quality by Irradiation." Intl. Atomic Energy Agency, Vienna, 1974.

[b] Expressed as percentage of amino acid (grams per gram) in the sample.

value of protein foods is determined by their amino acid content, this observation is important in connection with the irradiation of foods. While very high doses can lead to amino acid destruction, the data on dry gelatin indicate that the doses used in food irradiation lead to insignificant losses. This finding has been confirmed in other ways for protein foods generally.

c. *Wet Proteins (Including Aqueous Solutions)*

Proteins of the category of wet proteins are those that contain more than several percent of water. Proteins found in intact natural biological systems, and in foods derived from them which have not been dried, constitute the majority of proteins of the "wet" category. Also included are aqueous solutions of proteins.

The actual water content may not be the critical parameter. Freezing, for example, makes the water present unavailable for interaction. Likewise, "bound" water may not function as ordinary water.

The presence of water increases the indirect action of radiation through both the presence of active radiolytic products of water and its role of a medium for bringing reactants together. Despite the enhancement of indirect action, direct action of radiation also is important in the overall effect.

Denaturation by irradiation can be demonstrated by changes in the electrophoretic pattern and in the sedimentation coefficient. Change of the latter without change in the molecular weight suggests change of the shape of the molecule due to opening up or unfolding of amino acid chains. Radiation may cause splitting into smaller units. Hemocyanin, for example, through disruption of hydrogen bonds, splits irreversibly into two subunits of equal size. Casein also is split into smaller units, but with increasing dose undergoes aggregation, as evidenced by changes in viscosity.

These varied phenomena of splitting and aggregation are related to disturbances of the secondary and tertiary structures of the protein which expose reactive groups to the action of radiolytic products of water, notably e^-_{aq}, H· and OH·. Aggregation is ascribable to cross-linking among peptide chains.

Changes in the primary structure also occur. These are similar to those found in dry proteins. Deamination, decarboxylation, oxidation of –SH and aromatic groups, and other reactions, occur. In presence of water, however, radiation is less efficient for the reason that part of the incident energy is absorbed by the water.

Of the possible changes in the primary protein structure, from the standpoint of food irradiation, the most important, as was noted earlier, is the destruction of a number of amino acids. Data on the destruction of a number of amino acids of insulin in a 1% aqueous solution are given in Fig. 3.6. It is

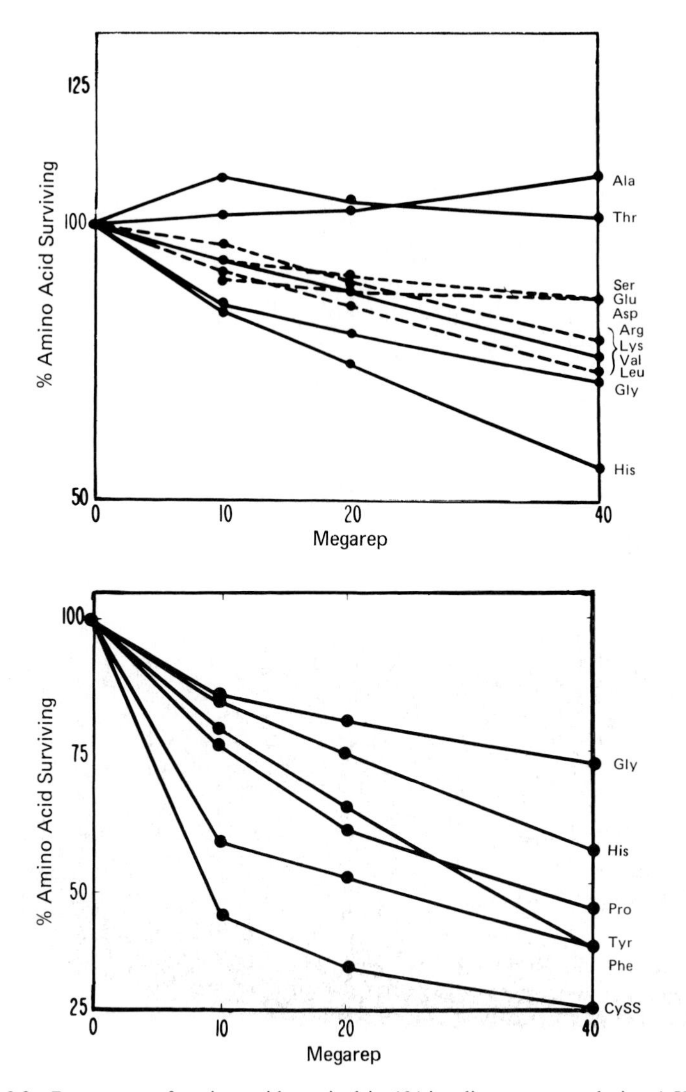

Figure 3.6 Percentage of amino acid survival in 1% insulin aqueous solution (pH 3) as a function of dose. From M. P. Drake, J. W. Giffee, D. J. Johnson, and V. L. Koenig, *J. Am. Chem. Soc.* **79,** 1395 (1957). Reprinted with permission. Copyright (1957) American Chemical Society.

seen that destruction of all amino acids studied occurs, but to varying degrees. Ring-structured amino acids, as a group, tend to be more radiation sensitive.

Two considerations are important with regard to the data of Fig. 3.6: (1) dose and (2) the protein environment.

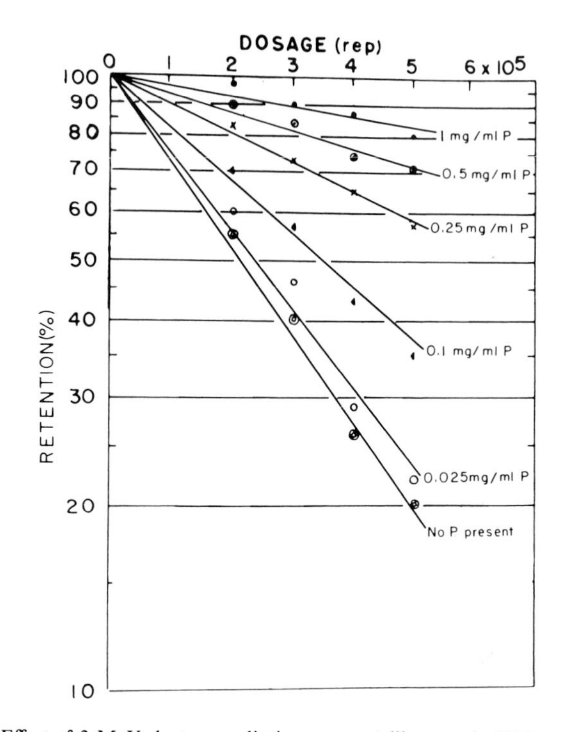

Figure 3.7 Effect of 3-MeV electron radiation on crystalline pepsin (P) in acetate buffer (pH 4.3) and in the presence of varying quantities of sodium D-isoascorbate. From S. A. Goldblith, *Proc. Int. Conf. Preserv. Foods Ioniz. Radiat., 1959.*

Paralleling what was observed with the irradiation of the dry protein bovine serum albumin, the amount of amino acid destruction in insulin is dose dependent. At the doses likely to be used in food irradiation (<50 kGy), in the 1% insulin solution only modest destruction of the more radiation-sensitive amino acids occurs.

The data of Fig. 3.6 are for a simple two-component system, insulin dissolved in water. In such a simple system the solute, insulin, is most vulnerable to attack, especially by indirect action. This simple system is not representative of the more complex systems found in foods. The protective action of a third component is neatly demonstrated by the addition of sodium D-isoascorbate to a pepsin solution, as shown in Fig. 3.7. The full protective effect of a typical complex food system can be seen from the results on the content of 10 amino acids of haddock fillets, before and after irradiation, as shown in Table 3.4.

In considering the use of irradiation for treating foods, the fact that radiation can alter proteins does not necessarily constitute a difficulty. The key

Table 3.4
Effect of Electron Radiation on Selected Amino Acids of Haddock Fillets[a]

Amino acid	Amino acid content[b]	
	Not irradiated	Irradiated[c]
Phenylalanine	3.93	3.63
Tryptophan	1.16	1.08
Methionine	2.99	2.85
Cystine	1.04	1.04
Valine	6.29	6.69
Leucine	8.03	8.25
Histidine	1.85	2.00
Arginine	5.34	5.56
Lysine	9.70	9.29
Threonine	4.87	4.58

[a] From B. E. Proctor and B. S. Bhatia, *Food Technol. (Chicago)* **5**, 357 (1950). Copyright © by Institute of Food Technologists.
[b] Amino acid content expressed as parts of amino acid per 16 parts of nitrogen.
[c] Dose, 53 kGy.

components of proteins from a nutritional standpoint, the amino acids, survive the process very well. Some of the changes in the secondary or tertiary structures of food proteins may interfere with their conventional uses. For example, irradiation impairs the whipping quality of egg white. In general, however, radiation-induced changes are so minor as to be of no particular significance for the ordinary food uses in most cases.

There are certain kinds of proteins that need special consideration in connection with food irradiation. Among these are enzymes, chromoproteins, and the proteins and related compounds such as DNA which play essential roles in biological processes. These all are subject to the radiation chemistry generally applicable to proteins. The effect of radiation on the particular special function of each is treated elsewhere in connection with that function.

So far our attention has been directed to changes in the protein produced by irradiation. Irradiation of proteins also produces other radiolytic products, generally small molecules such as fatty acids, mercaptans, and other sulfur compounds. Although very minor ones, these substances become components of the irradiated foods. Their role in wholesomeness aspects and sensory characteristics of irradiated foods requires consideration and evaluation. These topics are treated separately.

C. Lipids

1. Introduction

Fats generally do not deteriorate through microbial action. Fats are not carriers of insects and parasites. So far no beneficial effect of irradiation in changing a property of a fat has been discovered. In view of these facts the irradiation of lipids is likely to occur only when they are components of intact foods such as meat, and under these circumstances the irradiation of a lipid is largely unintentional and unavoidable.

The effects of radiation upon lipids are of interest almost entirely in terms of their nutritional, toxicological, functional, and sensory characteristics. Unlike the two other major food components, carbohydrates and proteins, lipids exist in foods in a distinct phase essentially totally separate from the aqueous systems that characterize the other two major components. This has significant implications in the radiation chemistry of lipids. Nevertheless, the basic considerations of radiation chemistry apply also to lipids. As with aqueous systems, lipids are subject to both direct and indirect action of radiation. The first stage of interaction with radiation leads to excitation and ionization. The next stage is mainly the production of intermediates, mostly free radicals, which react in various ways, ultimately to form stable end products. As can be anticipated, indirect action of radiation is influenced by environmental and other factors such as whether the lipid is solid or liquid, temperature of irradiation, the presence or absence of O_2, dose, and dose rate. Oxygen plays an unusually important role in that certain lipids readily undergo oxidation, which radiation can accelerate.

2. Fatty Acids

Compared with some carbohydrates and most proteins, lipids are small molecules with a less complex structure. Triglycerides constitute the largest category of lipids found in foods. Apart from glycerol, the constituent subunits are fatty acids. Just as information on the radiation chemistry of amino acids is helpful in understanding changes in proteins, so also is there value in examining the effect of radiation on isolated fatty acids.

In a fatty acid there are parts of the molecule that have an electron deficiency. One such location is the oxygen atom of the carbonyl group. The other occurs in unsaturated fatty acids around the center of unsaturation. Preferential cleavage in the region of the oxygen atom or the double bond occurs as a means of satisfying the electron deficiency. This leads to the formation of particular free radicals as principal intermediates, and ultimately, to particular end products.

The principal end products of fatty acids are CO_2, CO, H_2, hydrocarbons, mainly alkanes, and aldehydes. With unsaturated fatty acids some of the hydrocarbons are unsaturated. Unsaturated acids also form dimers and polymers, the amounts being increased by the presence of O_2.

The formation of these products is the result of cleavage at various locations in the acid to yield free radicals whose existence is terminated by hyrogen uptake or abstraction or by interaction with other radicals.

In the presence of O_2 the free radicals may form hydroperoxides through hydrogen abstraction at carbon atoms adjacent to a double bond, followed by addition of oxygen. The hydroperoxides yield a number of products including aldehydes.

3. Triglycerides

Triglycerides undergo a sequence of events similar to those of fatty acids. Cleavage occurs preferentially at bonds in the vicinity of the carbonyl group (heavy lines, positions a, b, c, d, e) but can occur also at other locations (light lines, positions f_i, f_x) at the carbon–carbon bonds of the fatty acid chain on a

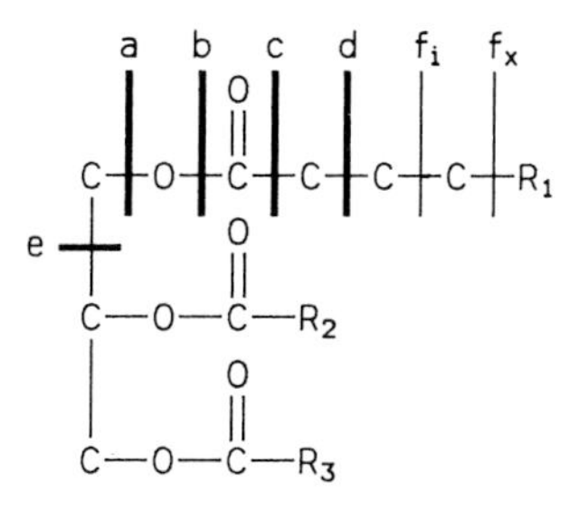

random basis. Some 16 different free radicals have been postulated to be formed by this preferential cleavage of bonds in the vicinity of the carbonyl group, as listed in Table 3.5. These free radicals lead to stable end products by a number of different pathways: abstraction, dissociation, recombination, disproportionation, and radical–molecule interaction. The possible radiolytic products of triglycerides are listed in Table 3.6. If O_2 is present during or after irradiation, radiation accelerates autoxidation, just as with isolated fatty acids. The end results of autoxidation parallel those that occur without radiation.

4. Other Lipids

"Crude fat" is a term used to designate the total lipid content of a food. Triglycerides usually are the principal components of crude fats. Other components are waxes, phospholipids, sterols, hydrocarbons, and pigments.

Table 3.5

Free Radicals Postulated to Be Formed by Irradiation of Triglycerides[a]

$CH_3(CH_2)_xCOO\cdot$ **I**	$H_2C\cdot$ $HCOOC(CH_2)_xCH_3$ $H_2COOC(CH_2)_xCH_3$ **II**	$H_2COOC(CH_2)_xCH_3$ $HC\cdot$ $HCOOC(CH_2)_xCH_3$ **III**
$CH_3(CH_2)_x\dot{C}O$ **IV**	$H_2CO\cdot$ $HCOOC(CH_2)_xCH_3$ $H_2COOC(CH_2)_xCH_3$ **V**	$H_2COOC(CH_2)_xCH_3$ $HCO\cdot$ $H_2COOC(CH_2)_xCH_3$ **VI**
$CH_3(CH_2)_x\cdot$ **VII**	$H_2CO\overset{O}{\overset{\Vert}{C}}\cdot$ $HCOOC(CH_2)_xCH_3$ $H_2COOC(CH_2)_xCH_3$ **VIII**	$H_2COOC(CH_2)_xCH_3$ $HCO\overset{O}{\overset{\Vert}{C}}\cdot$ $H_2COOC(CH_2)_xCH_3$ **IX**
$CH_2(CH_2)_{x-1}\cdot$ **X**	$H_2COOCCH_2\cdot$ $HCOOC(CH_2)_xCH_3$ $H_2COOC(CH_2)_xCH_3$ **XI**	$H_2COOC(CH_2)_xCH_3$ $HCOOCCH_2\cdot$ $H_2COOC(CH_2)_xCH_3$ **XII**
$CH_3(CH_2)_x\overset{O}{\overset{\Vert}{C}}OCH_2\cdot$ **XIII**	$H\dot{C}OOC(CH_2)_xCH_3$ $H_2COOC(CH_2)_xCH_3$ **XIV**	$CH_3(CH_2)_{x-1}\dot{C}HCOOH$ **XV**
$CH_3(CH_2)_y\dot{C}HCH{=}CH(CH_2)_zCOOH$ $\updownarrow$ $CH_3(CH_2)_yCH{=}CH\dot{C}H(CH_2)_zCOOH$ **XVI**		

[a] From W. W. Nawar, *J. Agric. Food Chem.* **26**(1), 21 (1978). Reprinted with permission. Copyright (1978) American Chemical Society.

Only very limited information is available on the radiation chemistry of nontriglyceride components.

The phospholipid, lecithin, yields fatty acids and choline phosphate. DL-α-Dipalmitoylphosphatidylethanolamine, also a phospholipid, with doses of 500 kGy, yields hydrocarbons, alkanals, alkanones, esters, palmitic acid, palmitone, lysophosphatidylethanolamine, and ethanolamine phosphate. The

Table 3.6

Possible Radiolytic Products of Triglycerides[a]

Site of cleavage[b]	Primary products	Recombination products
a	C_n Fatty acid	C_n Fatty acid esters
	Propanediol diesters	Alkanediol diesters
		2-Alkyl-1,3-propanediol diesters
		Butanetriol triester
	Propenediol diesters	
b	C_n Aldehyde	Ketones
		Diketones
		Oxoalkylesters
	Diglycerides	
	Oxo-propanediol diesters	Glyceryl ether diesters
	2-Alkylcyclobutanones (C_n)	Glyceryl ether tetraester
c	C_{n-1} *l*-Alkane	Longer hydrocarbons
	C_{n-1} Alkene	
	Formyl-diglycerides	Triglycerides with shorter fatty acids
d	C_{n-2} Alkane	
	C_{n-2} *l*-Alkene	Hydrocarbons
	Acetyl diglycerides	Triglycerides with shorter or longer fatty acids
e	C_n Fatty acid methyl ester	C_n fatty acid esters
	Ethanediol diester	Alkanediol diesters
		Erythritol tetraester
f_i	C_{n-x} Hydrocarbons	Hydrocarbons
	Triglycerides with shorter fatty acids	Triglycerides with longer fatty acids

[a] From H. Delincee, Recent advances in radiation chemistry of lipids. *In* "Recent Advances in Food Irradiation" (P. S. Elias and A. J. Cohen, eds.). Elsevier, Amsterdam, 1983.

[b] See triglyceride general formula in text (Section II, C, 3). $i = 1, 2, \ldots, n - 3$; x = any carbon number from 3 up to $n - 1$.

major volatile radiolytic products of cholesterol are hydrocarbons derived from the side chain. Other radiolytic products of cholesterol also have been identified, such as cholestane-3β:5α:6β triol. The sterols, vitamins D_2 and D_3, are resistant to radiation; in petroleum ether solution they yield hydrocarbons.

5. Identified Radiolytic Products

Table 3.7 lists radiolytic products identified in a chloroform extract of beef irradiated with the extraordinarily high dose of 500 kGy. The high dose employed permits identification of radiolytic products of high molecular weight and low concentration. The data of Table 3.7 generally fit the picture provided by radiation chemistry concepts.

Table 3.7

Radiolytic Compounds Identified in Chloroform Extract from Beef[a]

Alkanes
- Heptane
- Octane
- Nonane
- Decane
- Undecane
- Dodecane
- Tridecane
- Tetradecane
- Pentadecane
- Hexadecane
- Heptadecane

Alkenes
- Nonene
- Decene
- Undecene
- Dodecene
- Tridecene
- Tetradecene
- Pentadecene
- Hexadecene
- Heptadecene

Alkadienes
- Decadiene
- Dodecadiene
- Tridecadiene
- Tetradecadiene
- Pentadecadiene
- Hexadecadiene
- Heptadecadiene

Alkynes
- Decyne
- Undecyne
- Dodecyne

Aldehydes
- Hexanal
- Nonanal
- Undecanal
- Dodecanal
- Tetradecanal
- Pentadecanal
- Hexadecanal
- Octadecanal

Esters
- Me dodecanoate[b]
- Me tetradecanoate
- Me pentadecanote
- Me hexadecenoate
- Me heptadecanoate
- Me octadecanoate
- Me octadecenoate
- Et tetradecenoate[b]
- Et pentadecanoate
- Et hexadecenoate
- Et octadecanoate
- Pr hexadecanoate[b]
- Pr octadecanoate

Alcohols
- Hexanol
- Decaonol
- Undecanol
- Tridedanol
- Tetradecanol
- Hexadecanol
- Octadecenol

Fatty acids
- Heptanoic acid
- Octanoic acid
- Nonanoic acid
- Decanoic acid
- Tetradecenoic acid
- Pentadecanoic acid
- Hexadecanoic acid
- Heptadecanoic acid
- Octadecanoic acid

Lactones
- γ-Palmitolactone
- δ-Palmitolactone
- γ-Sterolactone
- δ-Sterolactone
- γ-Oleolactone
- δ-Oleolactone

Ketones
- 2-Pentadecanone
- 2-Heptadecanone
- Butyl tridecenyl ketone
- Palmitone
- 16-Tritriaconta 24-enone

Diol esters
- 2-Hydroxypropyl hexadecanoate
- 1,2-Tetradecanoyl propanediol diesters
- Hexadecanoyl, tetradecanoyl 1,2-propanediol diesters
- Tetradecanoyl, hexadecanoyl 1,3-propanediol diesters
- 1,2-Hexadecanoyl propanediol diesters
- 1,3-Hexadecanoyl propanediol diesters
- Tetradecanoyl, octadecenoyl 1,2-propanediol diesters
- Hexadecanoyl, octadecanoyl 1,2-propanediol diesters
- Glyceryl 1-tetradecanoate, 2 octadecanoate or isomers
- 1,3-Palmitin
- 1,2-Octadecenoyl propanediol diesters

Miscellaneous
- Cholesterol
- Dehydrocholesterol
- Cholestatriene
- Cholestadiene 7-one

[a] From M. Vadji and W. W. Nawar, *J. Am. Oil Chem. Soc.* **56** (6), 611 (1979).
[b] Me, Methyl; Et, ethyl; Pr, propyl.

6. Changes in Physical and Chemical Properties

Only insignificant changes in the physical properties of lipids, such as melting point and viscosity, occur with doses below 50 kGy. At higher doses changes which alter physical properties occur. The same statement applies to chemical properties as measured by acid number, peroxide values, iodine value, and so on.

7. Significance of Acceleration of Autoxidative Process by Irradiation

The role of radiation in the autoxidative process primarily is the production of free radicals. These radicals in the presence of O_2, either during irradiation or later, cause autoxidation of the lipid through formation of hydroperoxides, aldehydes, ketones, and other compounds typical of oxidized fats. Oxidation of antioxidants, including vitamin E, also can occur.

While some exceptions have been reported, because of the autoxidation role of radiation, the irradiation of foods containing lipids as major components is best done is the absence of O_2. Holding in the absence of O_2 postirradiation is likewise indicated.

D. Vitamins

1. Introduction

Concern for the fate of vitamins when foods are irradiated is almost entirely centered on maintenance of their biological function as essential nutrients. For this reason most of the information available on the effects of radiation on vitamins relates only to the degree of survival after irradiation. In cases in which there is a vitamin loss, there is often not a great deal of information on the identity of the resultant radiolytic end products.

Part of the problem in determining how irradiation alters a vitamin is that the smallness of the quantities of vitamins present in foods makes analyses for the radiolytic products difficult. While studies of model systems in place of foods can overcome this difficulty, the effects of radiation on vitamins observed in such simple systems may not be indicative of what occurs in the more complex food systems. As it happens, vitamins do survive irradiation of foods quite well in all but a few cases, and for this reason lack of knowledge of the radiation chemistry or their degradation is not especially serious.

Vitamins usually are grouped as to their solubility in water or lipids. This grouping has a value in connection with indirect effects of radiation.

2. Water-Soluble Vitamins

a. Vitamin C (Ascorbic Acid)

In a simple aqueous solution vitamin C is easily destroyed by irradiation. Dehydroascorbic acid, diketogluconic acid, plus other acids and compounds have been identified as radiolytic products. At neutral pH ascorbic acid is attacked by e_{aq}^- and $OH\cdot$.

In contrast, with doses up to 5 kGy, only small losses of ascorbic acid have been observed in fruits and in vegetables such as potatoes, onions, and tomatoes.

b. Vitamins of the B Group

Among the B vitamins, thiamine (vitamin B_1) appears to be the most radiation labile. This lability may be due to the presence in the thiamine molecule of hetero double bonds (e.g., —C=O and —C=N). In water solution dihydrothiamine is formed. Evidence that the attack of the thiamine molecule involves e_{aq}^- and $OH\cdot$ is the observation that the destruction of B_1 is greatly reduced in the presence of N_2O or O_2 (e_{aq}^- scavengers) or glucose ($OH\cdot$ scavenger).

Substantial loss of B_1 occurs in foods, such as meats, when they are irradiated. Irradiation in the frozen state greatly improves retention (Fig. 2.4).

Riboflavin (vitamin B_2) also contains a number of hetero double bonds and for this reason, in simple aqueous solutions undergoes change by interaction with active radiolytic products of water. Probably because it can bind protein, riboflavin is fairly resistant to radiation in many foods. Low-temperature irradiation improves retention.

Niacin (nicotinic acid), pyridoxine, vitamin B_6, and cobalamin (vitamin B_{12}) also undergo degradation when irradiated in water solution. In foods they are, however, only moderately affected by irradiation.

Choline, pantothenic acid, biotin, and folacin (pteroylglutamic acid) all are fairly resistant to radiation when in foods.

3. Fat-Soluble Vitamins

a. Vitamin A

Food sources of vitamin A include the actual vitamin A (retinol) and the provitamin A carotenoids (primarily β-carotene). While vitamin A and the provitamin carotenoids are essentially equivalent nutritionally, their radiation chemistry is not necessarily the same.

In lipid solvents destruction of all forms occurs and involves both direct and indirect action of radiation. The amount of degradation varies substantially with the lipid solvent used and with minor compositional differences, such as whether vitamin A is present as the acetate or alcohol. Based on spectrophotometric analysis, β-carotene undergoes cis–trans isomerization, progressive saturation of the polyene chain, and formation of α isomers. By complexing carotenoids, proteins can exert a protective action and reduce degradation. A similar protective action of certain carbohydrates has been observed.

In foods, radiation-induced losses of vitamin A and provitamin carotenoids occur to varying degrees. In fats and milk losses are high. Carotene destruction can be reduced by the addition of ascorbic acid or α-tocopherol. In vegetables there is little loss of vitamin A and carotenoids. Vitamin A losses generally increase when irradiated foods are held in air.

b. Vitamin D

Vitamin D exists in at least two forms: D_2 (calciferol) and D_3 (irradiated 7-dehydrocholesterol). These two forms differ in the composition of the hydrocarbon side chain. Both are steroids and contain particular groups that ordinarily are vulnerable to attack by radiation. Studies of D_3 in iso-octane solution indicate it is not especially radiation sensitive. Hydrocarbons have been identified as radiolytic products. The irradiation of foods generally produces little loss of vitamin D.

c. Vitamin E

Vitamin E (α-tocopherol) ordinarily is easily oxidized, particularly by products of the oxidation of unsaturated fats. The irradiation of lipid-containing foods in the presence of O_2, or their storage in air causes large losses of vitamin E. Such losses are avoided by exclusion of O_2.

d. Vitamin K

There are a number of related compounds which have vitamin K activity. All are napthaquinones with varying hydrocarbon side chains. The sensitivity to radiation is dependent upon the particular compound and the medium in which it occurs. Of the various forms, vitamin K_3 is most radiation sensitive. In beef, doses of about 30 kGy result in loss of all activity. In plant foods, on the other hand, irradiation causes little loss.

III. RADIATION CHEMISTRY OF FOODS

A. Introduction

The chemical changes in foods caused by irradiation are very much the same as those that occur in the isolated major food components. The sum total of what is formed in a given food reflects its initial composition, in that products formed by both direct and indirect action originate with the individual different food components. In the complex system that is present in most foods there is little opportunity for concentration of the action of radiation on only one or even several loci. As a consequence the formation of a number of different radiolytic products is likely. The amount of each is related to the amount of radiation absorbed, or dose.

In food irradiation, dose is determined by what is required to secure the objective of the treatment, and varies widely accordingly. Dose specification is discussed in later chapters in connection with the various applications of food irradiation. The maximum dose likely to be employed is about 50 kGy.

An estimate of the maximum number of chemical bonds that can be broken by a dose of 50 kGy can provide an indication of the quantity of radiolytic products formed. If it is assumed that every bond broken results in one new radiolytic product, that the absorption of 25 eV breaks one bond, and if the triglyceride tripalmitin (molecular weight 806) is taken as a typical food component, then 50 kGy performing with 100% efficiency in bond-breaking action (the actual efficiency probably is at most ~10%) can break 1.25×10^{19} bonds per gram of tripalmitin and form that many new molecules. This action at most involves only about 0.01% of the molecules of tripalmitin present and leaves the remainder unaffected by the irradiation. This estimate undoubtedly is not precise due to the simplifying assumptions involved in securing it, but it does clearly point to the conclusion that 50 kGy is capable of altering only a small fraction of the molecules present in an absorber. This fact, plus the large number of radiolytic products formed as a result of the complexity of the system in a food, add up to the result that the amount of any given radiolytic product is very small compared with the quantity of molecules which undergo no change.

Not all foods have been examined to determine the chemical changes that occur as a result of irradiation. There is, however, sufficient information available on a number of foods to enable a good understanding of the basic pathways of change, especially when the general concepts discussed above are kept in mind. The major findings on the radiation chemistry of the principal food groups are indicated in the following sections.

B. Meats and Poultry

Based on work done so far, doses ranging between 1 and 50 kGy have been considered in the irradiation of meats and poultry. That the amount of radiolytic products formed in these foods is dose dependent can be seen from the data of Fig. 3.8. These data not only establish the relationship between dose and amount of radiolytic products, but also point to the small quantities

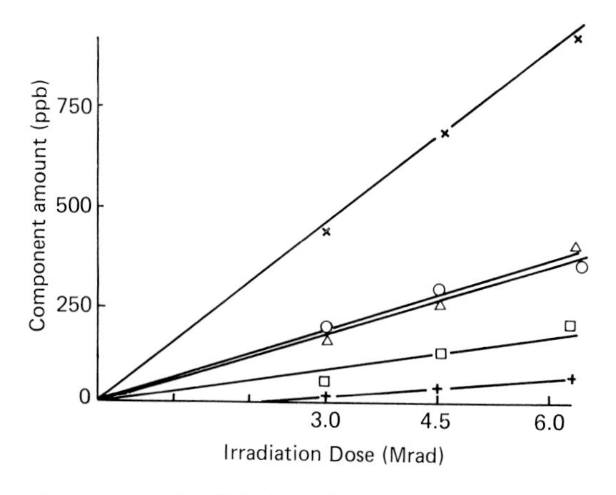

Figure 3.8 Relative amounts of radiolytic products produced in beef as a function of dose at −185°C. ○, Sulfur compounds; △, carbonyls; □, hydrocarbons; +, alkyl benzenes; ×, total. From C. Merritt, Jr., P. Angelini, E. Wierbicki, and G. W. Shults, *J. Agric. Food Chem.* **23,** 1037 (1975). Reprinted with permission. Copyright (1975) American Chemical Society.

formed. The amounts are so small as to require the use of the most sensitive analytical techniques for their detection. For this reason most of the determinations of radiolytic products in these foods have been done with large doses of the order of 50 kGy. To a considerable degree at least, this is valid, since, as the data of Fig. 3.8 show for the dose range covered, the same classes of radiolytic products are formed at all doses.

As shown in Table 3.8, some 65 volatile compounds have been identified in irradiated uncooked beef (dose 56 kGy). The amounts of the individual compounds range from 1 to 706 μg per kilogram (1–706 ppb). The total amount of all substances identified is 9892 μg per kilogram (9892 ppb). Ninety percent of the weight of these radiolytic products is accounted for by hydrocarbons. Of the 65 compounds, 70% are hydrocarbons. As can be anticipated, the origin of the various radiolytic products is in the different major components of the beef. Table 3.9 indicates the meat components associated with the various classes of volatile compounds identified in meat. The major radiolytic products, the hydrocarbons, originate in the lipids or lipoproteins. Aromatic and sulfur compounds derive from protein.

While analytical techniques available have dictated concentration of identification effort on the volatiles formed as a result of irradiation, it is known that less volatile products also are formed such as longer chain hydrocarbons and aldehydes, diglycerides, and diol esters, and that they are more abundant than the volatiles.

It is probable that most of the volatile radiolytic products identified in beef

Table 3.8

Volatile Compounds Identified in Irradiated Beef[a,b]

	Amount present (ppb)								
	Cobalt		Electron		Thermal		Frozen		
Compound	Cooked	Uncooked	Cooked	Uncooked	Cooked	Uncooked	Cooked	Uncooked	Raw
Alkanes									
Ethane	0[c]	172	0	179	0	0	0	0	0
Propane	60	164	65	173	0	0	0	0	0
Butane	125	208	127	221	0	0	0	0	0
Pentane	170	205	180	203	4	8	2	1	1
Hexane	207	209	248	217	67	125	7	6	1
Heptane	281	417	298	438	62	102	8	10	1
Octane	284	348	302	367	0	47	0	0	0
Nonane	125	266	146	—	0	0	0	0	0
Decane	175	362	184	—	0	0	0	0	1
Undecane	217	176	203	—	—	0	—	0	—
Dodecane	326	207	286	—	—	0	—	0	—
Tridecane	—[c]	321	—	—	—	0	—	0	—
Tetradecane	—	313	—	—	—	0	—	0	—
Pentadecane	—	696	—	—	—	0	—	0	—
Hexadecane	—	221	—	—	—	0	—	0	—
Heptadecane	—	394	—	—	—	0	—	0	—
Alkenes									
Ethene	0	28	0	28	0	0	0	0	0
Butene	12	32	13	33	0	0	0	0	0
Pentene	2	36	2	38	3	4	3	0	0
Hexene	2	34	2	35	22	31	1	1	4

Heptene	46	111	45	116	12	31	2	1	0
Octene	22	95	20	97	0	0	0	0	0
Nonene	33	59	48	61	0	0	0	0	0
Decene	101	126	116	—	—	0	—	0	—
Undecene	82	78	93	—	—	0	—	0	—
Dodecene	171	156	162	—	—	0	—	0	—
Tridecene	—	178	—	—	—	0	—	0	—
Tetradecene	—	488	—	—	—	0	—	0	—
Pentadecene	—	121	—	—	—	0	—	0	—
Hexadecene	—	156	—	—	—	0	—	0	—
Heptadecene	—	618	—	—	—	0	—	0	—
Isoalkanes									
2-Methyl propane	27	45	28	47	0	9	0	0	0
2-Methyl butane	4	19	5	22	0	143	0	0	1
2-Methyl pentane	10	33	11	34	1	28	1	0	0
2-Methyl heptane	11	29	12	24	27	47	4	4	0
Isoalkenes									
2-Methyl propene	2	37	2	39	0	4	0	0	0
Alkynes									
Decyne	25	23	23	—	—	0	—	0	—
Undecyne	9	4	11	—	—	0	—	0	—
Dienes									
Tetradecadiene	—	98	—	—	—	0	—	0	—
Pentadecadiene	—	73	—	—	—	0	—	0	—
Hexadecadiene	—	706	—	—	—	0	—	0	—
Heptadecadiene	—	16	—	—	—	0	—	0	—

(continued)

Table 3.8 (*Continued*)

Compound	Amount present (ppb)								
	Cobalt		Electron		Thermal		Frozen		
	Cooked	Uncooked	Cooked	Uncooked	Cooked	Uncooked	Cooked	Uncooked	Raw
Aromatic hydrocarbons									
Benzene	15	18	14	19	2	0	3	0	0
Toluene	50	65	50	66	48	73	3	6	0
Xylene	4	1	3	1	7	7	1	1	0
Alcohols									
Methanol	16	20	15	19	23	40	41	0	0
Ethanol	76	122	73	124	9	15	18	0	0
Ketones									
Acetone	108	137	106	140	65	120	3	4	1
2-Butanone	71	88	72	90	5	10	5	0	0
Aldehydes									
2-Methyl pentanal	11	10	10	11	0	0	0	0	0
Undecanal	—	76	—	—	—	0	—	0	—
Dodecanal	—	63	—	—	—	0	—	0	—
Tetradecanal	—	54	—	—	—	0	—	0	—
Pentadecanal	—	46	—	—	—	0	—	0	—

Hexadecanal	—	127	—	—	—	0	—	0	—
Octadecanal	—	30	—	—	—	0	—	0	—
Hexadecenal	—	33	—	—	—	0	—	0	—
Octadecenal	—	398	—	—	—	0	—	0	—
Sulfur compounds									
Carbonyl sulfide	2	2	2	2	22	75	0	0	0
Hydrogen sulfide	2	2	1	1	1	5	0	0	0
Ethyl mercaptan	7	9	7	11	0	0	0	0	0
Dimethyl sulfide	4	5	4	6	0	0	0	0	0
Dimethyl disulfide	3	4	3	4	7	13	1	1	0
Chloro compounds									
Tetrachloroethylene	9	8	9	8	9	11	10	12	4
Miscellaneous									
Acetonitrile	3	1	3	1	21	57	6	3	0

[a] From "Evaluation of the Health Aspects of Certain Compounds Found in Irradiated Beef." Fed. Am. Soc. Exp. Biol., Bethesda, Maryland, 1977.

[b] Note: Irradiation of beef was with dose of 56 kGy at $-30°C \pm 10°$. Thermally sterilized beef was processed to $F_0 = 5.8$. Frozen control beef was stored at -20 to $-18°C$. Irradiated beef was heated to 68 to 75°C for enzyme inactivation prior to irradiation. Cooking was done by heating while covered 5- to 6-cm thick layers for 15 min in convection oven at 204°C. An additional 2 min heating was done with the cover removed. Packaging was in either metal cans or flexible pouches.

[c] 0, Level below detectable limits; —, not determined.

Table 3.9

Volatile Compounds Isolated from Meat Substances[a]

Protein	Lipid	Lipoprotein
Methyl mercaptan	C_1–C_{12} *n*-Alkanes	C_1–C_{14} *n*-Alkanes
Ethyl mercaptan	C_2–C_{15} *n*-Alkenes	C_2–C_{14} *n*-Alkenes
Dimethyl disulfide	C_4–C_8 Iso-alkanes	Dimethyl sulfide
Benzene	Acetone	Acetone
Toluene	Methyl acetate	
Ethylbenzene		
Methane		
Carbonyl sulfide		
Hydrogen sulfide		

[a] From C. Merritt, Jr., P. Angelini, and D. J. McAdoo, Volatile compounds induced by irradiation in basic food substances. *In* "Radiation Preservation of Foods," Advances in Chemistry Series 65. Reprinted with permission. Copyright (1967) Americal Chemical Society.

(Table 3.8) are formed through indirect action of radiation. Evidence for this is seen in Fig. 3.9. At a dose of 45 kGy the amount of the radiolytic products increases with temperatures. Were these compounds the result simply of direct action, there would not be increased yield with increasing temperature.

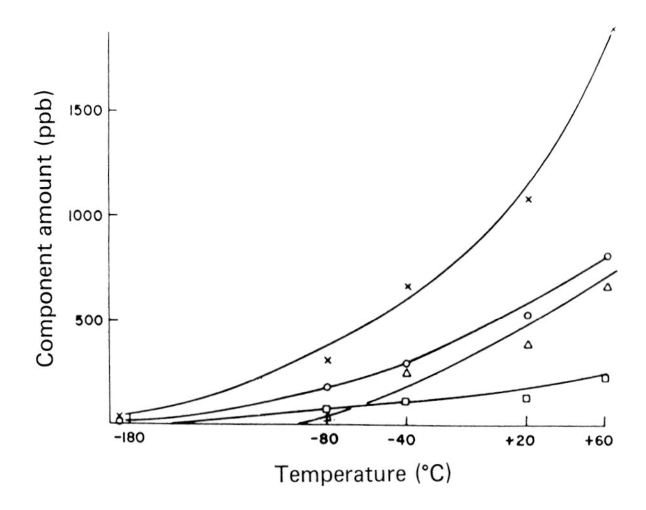

Figure 3.9 Relative amounts of radiolytic products produced with a dose of 45 kGy (4.5 Mrad) in beef as a function of temperature. Symbols as in Fig. 3.8. From C. Merritt, Jr., P. Angelini, E. Wierbicki, and G. W. Shults, *J. Agric. Food Chem.* **23,** 1037 (1975). Reprinted with permission. Copyright (1975) American Chemical Society.

Table 3.10
Amino Acids in Beef Irradiated at 60 kGy[a]

	Percentage of total amino acids			
		Irradiation temperature		
Amino acid	Not irradiated	20°C	−20°C	−196°C
Lysine	15.6	16.6	14.9	16.8
Histidine	1.3	1.2	1.0	1.1
Aspartic acid	11.5	11.4	11.7	11.4
Threonine	0.4	0.4	0.5	0.4
Serine	2.4	2.4	2.2	2.3
Glutamic acid	19.5	18.3	19.1	19.0
Proline	4.7	4.8	4.6	4.9
Glycine	9.4	9.1	9.4	9.4
Alanine	11.1	11.0	11.1	11.1
Cystine	1.1	1.2	0.9	0.7
Valine	2.6	2.4	2.2	2.4
Methionine	2.8	2.7	2.8	2.9
Isoleucine	0.8	0.9	0.7	0.8
Leucine	8.7	8.7	8.6	8.8
Tyrosine	3.1	4.0	3.9	3.8
Phenylalanine	4.0	3.8	4.2	4.2
Alloisoleucine	0.8	1.1	1.1	1.2

[a] From F. L. Kauffman and J. W. Harlan, "Effect of low temperature irradiation on chemical and sensory characteristics of beef steaks," Technical Report 69-64-FL. U.S. Army Natick Laboratories, Natick, Massachusetts, 1969.

Table 3.10 shows the amino acid content of beef before and after irradiation at 60 kGy. These figures indicate virtually no change as a result of irradiation and are in line with the small amount of radiolytic products that have been detected.

The radiolytic products from all meats are essentially the same. This conclusion can be made from the information in Table 3.11. That this is so is in accord with the concept that the radiolytic products formed reflect the composition of the food. In the case of foods of similar composition, such as the various meats, the radiolytic products are similar.

Apart from producing radiolytic products of the kind just described, irradiation does cause other changes in meats. The solubility of collagen, the protein that is the major component of connective tissue, is increased by irradiation. The increase in solubility is attributed to scission of the peptide chain to yield units of lower molecular weight. Figure 3.10 shows the increase in solubility of collagen when beef muscle is irradiated at 0°C with increasing

Table 3.11

Compounds Identified in Various Irradiated Meats and Associated Substances[a, b]

Compound	Beef[c]	Veal[c]	Mutton[c]	Lamb[c]	Pork[c]	Chicken[d]	Haddock[e]
n-Alkanes							
Methane	S	S	S	S	S	S	—
Ethane	M	M	M	M	M	T	—
Propane	M	M	M	M	M	—	—
Butane	M	M	M	M	M	—	S
Pentane	M	M	M	M	M	S	M
Hexane	M	M	M	M	M	S	S
Heptane	M	M	M	M	M	S	S
Octane	M	M	M	M	M	S	T
Nonane	M	M	M	M	M	T	—
Decane	M	M	M	M	M	—	—
Undecane	M	M	M	M	M	—	—
Dodecane	M	M	M	M	M	—	—
Tridecane	M	M	M	M	M	—	—
Tetradecane	M	M	M	M	M	—	—
Pentadecane	L	L	L	L	L	—	—
Hexadecane	M	M	M	M	M	—	—
Heptadecane	L	L	L	L	L	—	—
l-Alkenes							
Ethene	S	S	S	S	S	—	—
Propene	S	S	S	S	S	—	—
Butene	S	S	S	S	S	—	T
Pentene	S	S	S	S	S	T	T
Hexene	S	S	S	S	S	T	T
Heptene	S	S	S	S	S	S	T
Octene	S	S	S	S	S	T	—

Nonene	S	S	S	S	S	—	—
Decene	S	S	S	S	S	—	—
Undecene	S	S	S	S	S	—	—
Dodecene	M	M	M	M	M	—	—
Tridecene	S	S	S	S	S	—	—
Tetradecene	L	L	L	L	L	—	—
Pentadecene	M	M	M	M	M	—	—
Hexadecene	L	L	L	L	L	—	—
Heptadecene	L	L	L	L	L	—	—
l-Alkynes							
Decyne	T	—	T	—	T	—	—
Undecyne	T	—	—	—	—	—	—
Dienes							
Tridecadiene	T	—	—	—	T	—	—
Tetradecadiene	S	—	—	—	M	—	—
Pentadecadiene	S	—	—	—	S	—	—
Hexadecadiene	S	—	—	—	L	—	—
Heptadecadiene	M	—	—	—	L	—	—
Aromatic hydrocarbons							
Benzene	T	T	T	T	T	T	S
Toluene	T	T	T	T	T	S	S
Ethylbenzene	T	T	T	T	T	T	—
Propylbenzene	T	T	T	T	T	—	—
Alcohols							
Ethanol	M	M	M	M	T	S	S
Propanol	S	S	S	M	S	—	—
Butanol	S	S	T	S	T	—	—
Pentanol	S	S	S	S	—	—	—
Hexanol	T	—	—	—	—	—	—

(*continued*)

Table 3.11 (*Continued*)

Compound	Beef[c]	Veal[c]	Mutton[c]	Lamb[c]	Pork[c]	Chicken[d]	Haddock[e]
Ketones							
Acetone	S	T	S	M	M	T	S
Butanone	T	S	S	M	M	—	S
Acetone	—	T	T	T	T	—	—
Aldehydes							
Butanal	—	—	—	—	—	—	S
Pentanal	S	—	—	—	—	—	T
Hexanal	S	T	T	T	S	—	T
Heptanal	T	—	—	—	—	—	—
Octanal	T	—	—	—	—	—	—
Esters							
Ethyl acetate	T	T	T	T	T	—	—
Sulfur compounds							
Methanethial	—	—	—	—	—	T	—
Ethanethial	—	—	—	—	—	—	T
2-Thiapropane	T	T	T	T	T	—	M
2-Thiabutane	—	—	—	—	T	—	—
2,3-Dithiabutane	S	S	T	T	S	—	S
3,4-Dithiahexane	T	T	S	T	T	—	—
3-Thiaheptane	T	T	T	—	T	—	—
Methional	T	—	T	—	T	—	—

[a] From C. Merritt, Jr., *Radiat. Res. Rev.* **3,** 353 (1972).
[b] L, Large (>1 ppm); M, moderate (>0.1 ppm); S, small (>0.01 ppm); T, trace (<0.01 ppm). Quantities estimated from gas chromatographic data.
[c] 60 kGy.
[d] 5 kGy.
[e] 56 kGy.

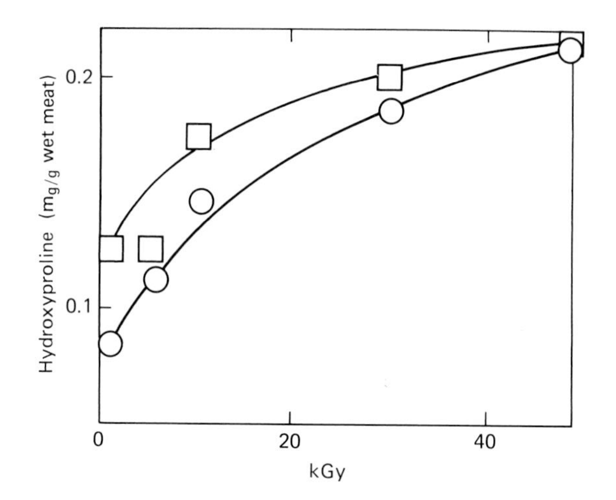

Figure 3.10 Effect of irradiation on the extractability of collagen from meat (as indicated by changes in amounts of soluble hydroxyproline). ○, Irradiated raw; □, irradiated after blanching. From A. J. Bailey and D. N. Rhodes, *J. Sci. Food Agric.* **15,** 504 (1964).

dose. Twenty-two percent of the collagen becomes soluble in water when the dose is 50 kGy.

Both fresh and cured meats can undergo color changes when irradiated. Oxidation of the pigments occurs in the presence of O_2 and forms a brown color. Cooked fresh meats in the absence of O_2 turn from brown or gray to pink. This color change is attributed to the formation of globin myohemochromogen.

One chemical change in meats of importance is the development of an unpleasant flavor. This flavor, or one very similar, occurs in many animal foods. This is discussed in Chapter 6.

Irradiation, even with doses of the order of 50 kGy, does not fully inactivate endogenous proteolytic enzymes in meats. Apparently the low concentration of these proteins and the complexity of the meat system provide protection against radiation inactivation.

C. Marine and Freshwater Animal Foods

While there is less information available on the radiation chemistry of marine and freshwater animal foods, what is known indicates that the effects of irradiation on these foods parallel those on the red meats. There are, however, certain differences which reflect differences in composition. Only certain fish, for example, contain appreciable amounts of lipids, and in fish of

Table 3.12

Compounds Identified in Fresh Nonirradiated and Irradiated Haddock Fillets[a]

Nonirradiated	Irradiation dose (kGy) 1.3	2.6	6.5	13	65
Acetaldehyde	Acetone	Benzene	Benzene	Acetone	Acetone
Acetone	Benzene	Dimethyl disulfide	Dimethyl disulfide	Benzene	Benzene
Butene	Butene	Dimethyl sulfide	Dimethyl sulfide	*n*-Butane	*n*-butane
Diethylether	Dimethyl disulfide	Methyl mercaptan	Dimethyl trisulfide	Butene	2-Butane thiol
Dimethyl disulfide	Dimethyl sulfide		Methyl mercaptan	Dimethyl disulfide	Butene
Dimethyl sulfide	Hydrogen sulfide		Methyl thioacetate	Dimethyl sulfide	Carbon disulfide
Toluene	Methyl mercaptan		Toluene	*n*-Heptane	Carbonyl sulfide
	Toluene			1-Heptyne	Dimethyl disulfide
				Hydrogen sulfide	Dimethyl sulfide
				2-Methyl 1-Butene	2,3-Dithiohexane
				Methyl ethyl ketone	Ethyl alcohol
				Methyl mercaptan	*n*-Heptane
				2-Methyl propene	1-heptene
				n-Octane	1-Heptyne
				1-Octyne	1-Hexene
				n-Pentane	Hydrogen sulfide
				1-Pentene	Isopentane
				2-Thiobutane	2-Methyl 1-Butanal
				Toluene	Methyl ethyl ketone
					Methyl heptane
					2-Methyl pentane
					2-Methyl propanol
					n-Octane
					1-Octyne
					n-Pentane
					1-Pentene
					2-Thiobutane
					Toluene

[a] From L. J. Ronsivalli, V. G. Ampola, F. J. King, and J. A. Holston, "Study of Irradiated-Pasteurized Fishery Products," TID 24256. U.S. Atomic Energy

low lipid content, hydrocarbons are formed in smaller numbers than are found in meats. Table 3.12 lists the compounds identified in nonirradiated and irradiated fillets of haddock (a fish of low oil content). Table 3.4 lists the amino acid composition of haddock fillets irradiated with a dose of approximately 53 kGy. As is expected, there is no significant change in the amino acid composition. Irradiation causes an increase in the amount of free amino acids. The mechanism for this has not been identified. It could be the direct consequence of radiation or the result of proteolytic enzyme action, or an artifact resulting from the breakdown of tissue structure facilitating the extraction that is part of the analytical procedure.

In those seafood products containing appreciable amounts of lipids, changes in the unsaturated components have been observed. These changes are indicative of oxidation.

Irradiation can cause pigment changes. The irradiation of frozen tuna converts the oxidized pigment ferrimyoglobin to oxymyoglobin. Irradiation blackens raw lobster and shrimp due to release of enzymes and subsequent tyrosine formation.

D. Eggs

That irradiation causes the formation of radiolytic products in eggs can be concluded from observations of radiation-induced off-flavors. There is, however, no specific information on the nature of these and other radiolytic products in eggs. Irradiation causes a decrease in the viscosity of liquid egg white, as shown in Table 3.13. This suggests either denaturation of the albumen or degradation into smaller units.

Table 3.13
Effect of Irradiation on the Viscosity of Liquid Egg White[a]

Dose (kGy)	Viscosity (centipoises)
0.0	5.26
4.3	3.03
5.8	2.91
7.2	2.84
8.6	2.81

[a] From H. R. Ball and F. A. Gardner, *Poult. Sci.* **47**(5), 1481 (1968).

E. Dairy Products

Dairy products generally develop off-flavors at low doses. Acetaldehyde and dimethyl sulfide have been identified as radiolytic products. An increase in the sulfhydnyl content of raw skim milk and skim milk powder has been observed.

The volatile compounds formed in anhydrous butter fat immediately upon irradiation are comparable with those formed in other foods. Of all the butter fat radiolytic products, CO_2 is produced in greatest amounts. Of the other radiolytic products, aliphatic hydrocarbons are present in greater amounts and number. Acids, alcohols, aldehydes, ketones, and esters are formed in smaller quantities. The presence or absence of O_2 has little effect. This is explained by the relatively fast action of radiation in forming free radicals and their subsequent interaction. Oxygen, whose availability is generally limited by its solubility in the fat, does not affect the fast radiation-induced changes. If O_2 is in fact available adequately, hydroperoxides form and result in oxidative changes.

Irradiation of cottage cheese whey produces various changes in the proteins present. Both fragmentation and aggregation have been observed.

F. Cereal Grains and Legumes

All three of the major food components—protein, lipid, and carbohydrate—are present in the cereal grains and legumes as they are harvested. Generally these foods have a low moisture content as they are handled in storage and distribution. Manufacturing practices in some cases may fractionate the harvested product in order to derive other products of particular composition. It is to be expected that the radiation chemistry of these foods reflects the composition of the particular product that is irradiated.

The protein content of these foods is significant in their nutritional value, and in the case of certain ones, such as wheat and rice, also relates to their functional characteristics. Even with high doses (≤ 30 kGy), irradiation does not affect the total protein content. That the amino acid contents of the proteins of wheat and the legume red gram are essentially unchanged can be seen from the data of Tables 3.14 and 3.15. Some increases in certain of the free amino acids occur. In wheat small increases of isoleucine, tyrosine, valine, and alanine have been observed with a dose of 10 kGy. In red gram similar increases occur with isoleucine, tyrosine, valine, glutamic acid, lysine, and phenylalanine.

Both fragmentation and aggregation of gluten of wheat occur; these

Table 3.14
Effect of Irradiation with a Dose of 10 kGy on Amino Acid Composition of Wheat[a]

	Grams per 16 g of nitrogen	
Amino acid	Control	Irradiated
Aspartic acid	4.62	4.49
Threonine	2.97	2.88
Serine	5.32	5.41
Glutamic acid	31.60	32.30
Proline	11.07	11.00
Glycine	3.57	3.42
Alanine	3.03	2.71
1/2 Cystine	1.02	1.04
Valine	4.55	4.48
Methionine	1.73	1.68
Isoleucine	4.10	4.08
Leucine	6.83	7.08
Tyrosine	3.25	3.20
Phenylalanine	4.74	4.82
Lysine	2.42	2.38
Histidine	1.91	1.93
Arginine	4.59	4.39
Tryptophan	1.02	0.95

[a] From H. Srinivas, H. V. Ananthswamy, U. K. Vakil, and A. Sreenivasan, *J. Food Sci.* **37,** 715 (1972). Copyright © by Institute of Food Technologists.

changes affect solubility. Similar effects occur with other wheat proteins. The moisture content of the wheat at the time of irradiation affects the course of these changes. At 13.4% moisture, for example, gluten undergoes changes which indicate fragmentation; whereas, at 25.8% moisture the changes suggest aggregation. These differences have been related to predominance of the direct effects of radiation at low moisture contents and of indirect effects at higher levels.

In general the enzymes present in cereal grains are unaffected by irradiation. The data of Table 3.16 demonstrate this for amylase and protease activities. An increase in maltose values occurs, as shown in Table 3.17. This increase is not attributed to increased amylase activity, but to radiation effects on the starch of the wheat which make it more susceptible to enzymatic hydrolysis. The trypsin inhibitor of red gram is not inactivated by doses as high as 30 kGy.

Carbohydrates also are major components of cereal grains. In cereal grains

Table 3.15

Effect of Irradiation with a Dose of 30 kGy on Amino Acid Composition of Red Gram[a]

Amino acid	Grams per 16 g of nitrogen	
	Control	Irradiated
Aspartic acid	7.73	7.63
Threonine	3.36	3.13
Serine	4.36	4.12
Glutamic acid	23.71	23.47
Proline	3.54	2.78
Glycine	3.02	2.95
Alanine	3.88	3.91
Cysteine	Trace	Trace
Valine	3.55	3.61
Methionine	0.79	0.81
Isoleucine	3.14	3.02
Leucine	6.03	5.15
Tyrosine	1.45	1.79
Phenylalanine	8.79	8.74
Lysine	6.29	6.84
Histidine	3.22	3.65
Arginine	4.87	5.46
Tryptophan	1.25	1.20

[a] From S. P. Nene, U. K. Vakil, and A. Sreenivasan, *J. Food Sci.* **40,** 815 (1975). Copyright © by Institute of Food Technologists.

Table 3.16

Effect of Irradiation of Wheat Grains on α-Amylase and Protease Activities[a]

Irradiation dosage (kGy)	α-Amylase activity (SKB units/g)	Protease activity (Hb units)	Maltose values (maltose/10 g)
0	0.027	12.4	153
0.46	0.028	12.4	156
0.93	0.026	12.4	179
4.65	0.026	12.2	216
9.3	0.024	12.1	244
27.9	0.020	11.9	315

[a] From P. Linko and M. Milner, *Cereal Chem.* **37,** 223 (1960).

Table 3.17
Effect of Radiation on Diastatic Activity of Wheat at Three Moisture Levels[a]

	Maltose value[b]		
Dose (Gy)	10% Moisture	20% Moisture	30% Moisture
0	152	197	232
200	172	207	236
400	190	211	356
600	196	247	368
2000	211	261	368

[a] From H. N. Ananthaswamy, V. K. Vakil, and A. Sreenivasan, *J. Food Sci.* **36,** 792 (1970). Copyright © by Institute of Food Technologists.
[b] Expressed as milligrams of maltose per 10 g starch per hour at 30°C.

starch constitutes about 90% of the carbohydrate present. Other carbohydrates include dextrins and sugars. Legumes have lower carbohydrate contents.

Large doses of radiation result in changes in wheat carbohydrates. The starch content is reduced and that of reducing sugars increased. Irradiation increases the amounts of soluble pentosans, although the total quantity of pentosans is unchanged. The moisture content of the grain at the time of irradiation affects what happens to the carbohydrate. The data of Table 3.17 demonstrate this. Higher maltose values are obtained with increasing moisture content of the wheat. This indicates greater radiation-induced depolymerization of the starch component at the higher moisture levels. Indications of similar depolymerization of starch components in rice and in the legumes red gram and navy beans have been obtained.

At the very high dose of 100 kGy small decreases in triglycerides, galactolipids, and phospholipids occur. Simultaneously there are increases in the content of free sterols and mono- and diglycerides. With doses in the range 0.1–1.75 kGy no effects were observed. Peroxide values of wheat flour, however, increase with dose, as shown in Table 3.18. A "tallowy" odor of irradiated wheat has been observed and this undoubtedly is associated with the increase in peroxide values.

Evidence of the formation of free radicals by irradiation in rice and wheat has been secured through the use of data obtained with electron spin resonance measurements. With low moisture contents and high doses (20 kGy), radicals persisted for many days, as can be seen from the curve in Fig. 3.11. Higher moisture levels and the availability of O_2 accelerate the disappearance of the free radicals. At doses below 1 kGy the free-radical content of rice and wheat is very low.

Table 3.18

Peroxide Values of Stored Whole-Meal Flour from Irradiated Soft Winter Wheat[a, b]

Storage (days)	Irradiation dose (kGy)				
	0	1	10	50	100
0	1.08	1.30	1.41	1.70	2.09
25	1.49	1.37	1.26	1.48	1.48
124	1.94	1.61	1.64	1.39	1.11
185	3.12	2.69	3.02	1.31	1.07

[a] From K. H. Tipples and F. W. Norris, *Cereal Chem.* **42,** 437 (1972).
[b] Values expressed as micromoles per gram.

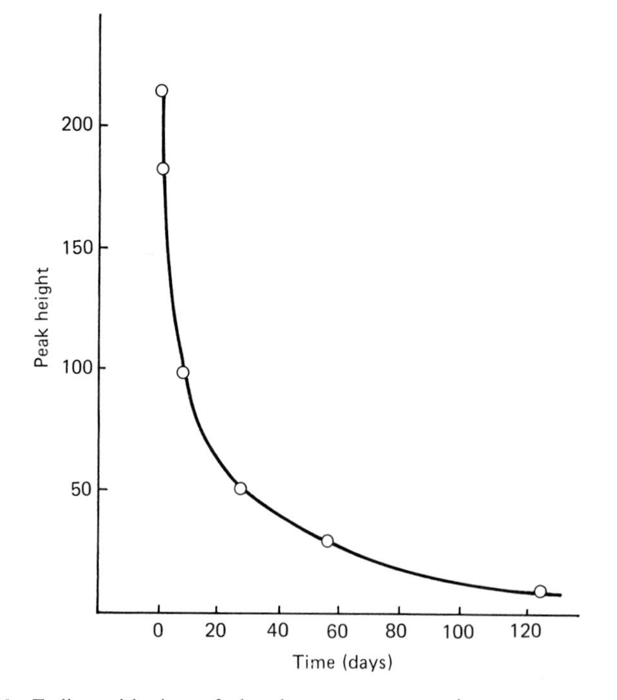

Figure 3.11 Fading with time of the electron paramagnetic resonance spectrum of 4% moisture flour irradiated with 20 kGy. Peak height is expressed in arbitrary units. From C. C. Lee, *Cereal Chem.* **39,** 147 (1962).

G. Fruits and Vegetables

The great majority of applications of irradiation to fruits and vegetables is concerned with fresh and raw products. This circumstance distinguishes these kinds of usage of irradiation from the irradiation of other foods in that the fresh fruits and vegetables are living organisms with active metabolic processes, whose course may be influenced by irradiation and whose consequences can cause additional postirradiation chemical changes in the foods. The immediate effects of irradiation, therefore, are only a part of the total result of irradiation of these living foods.

Fruits and vegetables are characterized generally as having carbohydrates as their major component, other than water. As a group, they have only small amounts of protein and lipids. They are also characterized as having textural properties as very important quality attributes, particularly in the fresh state.

Texture of fruits and vegetables is related to the turgor of the living cells which are part of their tissues and to certain structural other tissues, including materials that produce intracellular cohesion, which provide the rigidity characteristic of the particular food. Turgor is the pressure of cell contents on the cell wall. Anything that affects this pressure alters texture. Reduction of the amount of cell contents, for example, reduces pressure causing a lessening of turgor and softening of the food. Changes in the cell wall can affect its permeability and in this way affect turgor. Rupture of the cell wall can release the cell contents and cause texture changes.

The cell walls contain cellulose, hemicellulose, and lignins. Pectic substances also are present in the area of the cell wall. In very starchy foods such as green bananas, the starch present can be an additional factor in their firmness.

Because of the importance of texture in the quality of fruits and vegetables and because their texture is so closely linked to carbohydrate components, changes in the carbohydrates caused by irradiation are of paramount importance.

Of the immediate changes in fruits and vegetables caused by irradiation, softening—that is, loss of texture—is the principal limiting factor in using the process. For most fruits and vegetables it limits the dose that may be used.

Softening of fruits and vegetables is primarily related to radiation degradation of the carbohydrates associated with normal texture, namely cellulose, pectin, and starch. Such degradation can affect texture in at least two ways: (1) by weakening the rigid structural tissues or (2) by altering cell walls to reduce turgor. A third possibility is by affecting processes of endogenous enzymes, either by releasing the enzymes from their normal locations into the plant tissue generally to where they can attack carbohydrates, or by altering the carbohydrate substrates so as to make them more susceptible to enzyme action.

At relatively low doses (2–3 kGy), change of texture appears to be associated with release of calcium from the calcium–pectin association present in fruits and vegetables.

Other immediate radiation-induced chemical changes in fruits and vegetables include reduction of ascorbic acid content and loss of –SH groups. In potatoes a transient rise in reducing and nonreducing sugars and in free amino acids occurs. In strawberries a decrease in acidity results from irradiation. This causes an apparent sweeter than normal taste.

It may be noted that, in general, studies of the irradiation of fruits and vegetables have been limited largely to relatively low doses (below ~3 kGy). Most applications of irradiation to these foods require only low doses. Additionally, however, limitations of dose imposed by the softening effect of irradiation on these foods has deterred investigation of the effects of higher doses. Except for the important effect on carbohydrate components associated with firmness, it is likely that chemical changes caused by the indicated low doses are small.

Since most of the irradiated fruits and vegetables are likely to be marketed and consumed some time after irradiation, changes in composition related to the metabolic processes of these living foods can be important in terms of quality and other factors. This area is covered in Chapters 4 and 8.

H. General Comments on the Radiation Chemistry of Foods

It is clear that there are many gaps in the present knowledge of the radiation chemistry of foods. Only a few foods have been studied to determine the radiolytic products that are formed. Nonetheless, what is known in a specific way regarding foods plus the general knowledge of radiation chemistry enables a sufficient understanding of the changes irradiation produces in foods to be of great value in a number of important areas such as the appraisal of the wholesomeness of irradiated foods, the technology involved in food irradiation, and the biological aspects of the process. In a sense, the radiation chemistry knowledge unifies the many separate areas of food irradiation.

A final comment on the radiation chemistry of foods is best made by providing a direct quotation of one whose contributions to the subject have been outstanding, Charles Merritt, Jr. (1972): "The conclusion arising from these comparisons is simple. No volatile compounds produced in foods by irradiation have been found that are not similarly found qualitatively and quantitatively in other products resulting naturally." That this is so, places food irradiation as regards the chemical changes it causes in foods in the domain of previous experience and knowledge regarding foods.

SOURCES OF ADDITIONAL INFORMATION

Adam. S., Recent developments in radiation chemistry of carbohydrates. *In* "Recent Advances in Food Irradiation" (P.S. Elias and A. J. Cohen, eds.). Elsevier Biomedical, Amsterdam, 1983.

Anonymous, "Evaluation of the Health Aspects of Certain Compounds Found in Irradiated Beef," Report 1977, Supplements I and II. Fed. Am. Soc. Exp. Biol., Bethesda, Maryland, 1979.

Basson, R. A., Advances in radiation chemistry of food and food components—An overview. *In* "Recent Advances in Food Irradiation" (P. S. Elias and A. J. Cohen, eds.). Elsevier Biomedical, Amsterdam, 1983.

Basson, R. A., Recent advances in radiation chemistry of vitamins. *In* "Recent Advances in Food Irradiation" (P. S. Elias and A. J. Cohen, eds.). Elsevier Biomedical, Amsterdam, 1983.

Beyers, M., Drijer, D., Holzapel, C. W., Niemand, J. G., Pretorius, I., and van der Linde, H. J., Chemical consequences of irradiation of subtropical fruits. *In* "Recent Advances in Food Irradiation" (P. S. Elias and A. J. Cohen, eds.). Elsevier Biomedical, Amsterdam, 1983.

Dauphin, J. F., and Saint-Lebe, L. R., Radiation chemistry of carbohydrates. *In* "Radiation Chemistry of Major Food Components" (P. S. Elias and A. J. Cohen, eds.). Elsevier, Amsterdam, 1977.

Delincee, H., Recent advances in radiation chemistry of lipids. *In* "Recent Advances in Food Irradiation" (P. S. Elias and A. J. Cohen, eds.). Elsevier Biomedical, Amsterdam, 1983.

Delincee, H., Recent advances in radiation chemistry of proteins. *In* "Recent Advances in Food Irradiation" (P. S. Elias and A. J. Cohen, eds.). Elsevier Biomedical, Amsterdam, 1983.

Diehl, J. F., Radiolytic effects in foods. *In* "Preservation of Food by Ionizing Radiation" (E. S. Josephson and M. S. Peterson, eds.), Vol. I. CRC Press, Boca Raton, Flordia, 1982.

Diehl, J. F., Adam, S., Delincee, H., and Jakubick, V., Radiolysis of carbohydrates and of carbohydrate-containing foodstuffs. *J. Agric. Food Chem.* **26,** 15 (1978).

Handel, A. P., and Nawar, W. W., Radiolysis of saturated phospholipids. *Radiat. Res.* **86,** 437 (1981).

Handel, A. P., and Nawar, W. W., Radiolytic compounds from mono-, di-, and triacylglycerols. *Radiat. Res.* **86,** 428 (1981).

King, F. J., Mendelsohn, J. M., Gadbois, D. F., and Bernsteinas, J. B., Some chemical changes in irradiated seafoods. *Radiat. Res. Rev.* **3**(4), 399 (1972).

Merritt, C., Jr., Qualitative and quantitative aspects of trace volatile components in irradiated foods and food substances. *Radiat. Res. Rev.* **3**(4), 353 (1972).

Merritt, C., Jr., and Angelini, P., Chemical changes associated with flavor in irradiated meat. *J. Agric. Food Chem.* **23,** 1037 (1975).

Merritt, C., Jr., and Taub, I. A., Commonality and predictability of radiolytic products in irradiated meats. *In* "Recent Advances in Food Irradiation" (P. S. Elias and A. J. Cohen, eds.), Elsevier Biomedical, Amsterdam, 1983.

Merritt, C., Jr., Angelini, P., and Graham, R. A., Effect of radiation parameters on the formation of radiolysis products in meat and meat substances. *J. Agric. Food Chem.* **26,** 29 (1978).

Merritt, C., Jr., Vadji, M., and Angelini, P., A quantitative comparison of the yields of radiolysis products in various meats and their relationship to precursors. *J. Am. Oil Chem. Soc.* **62**(4), 708 (1985).

Nawar, W. W., Radiolytic changes in fats. *Radiat. Res. Rev.* **3**(4), 327 (1972).

Nawar, W. W., Radiation chemistry of lipids. *In* "Radiation Chemistry of Major Food Components" (P. S. Elias and A. J. Cohen, eds.). Elsevier, Amsterdam, 1977.

Nawar, W. W., Reaction mechanisms in the radiolysis of fats: A review. *J. Agric. Food Chem.* **26,** 21 (1978).

Nawar, W. W., Radiolysis of nonaqueous components in foods. *In* "Preservation of Food by Ionizing Radiation" (E. S. Josephson and M. S. Peterson, eds.), Vol. II. CRC Press, Boca Raton, Florida, 1983.

Simic, M. G., Radiation chemistry of water-soluble food components. *In* "Preservation of Food by Ionizing Radiation" (E. S. Josephson and M. S. Peterson, eds.), Vol. II. CRC Press, Boca Raton, Florida, 1983.

Tobback, P. P., Radiation chemistry of vitamins. *In* "Radiation Chemistry of Major Food Components" (P. S. Elias and A. J. Cohen, eds.). Elsevier, Amsterdam, 1977.

Urbain, W. M., Radiation chemistry of proteins. *In* "Radiation Chemistry of Major Food Components" (P. S. Elias and A. J. Cohen, eds.). Elsevier, Amsterdam, 1977.

Vadji, M., and Nawar, W. W., Comparison of the radiolytic compounds from saturated and unsaturated triglycerides and fatty acids. *J. Am. Oil Chem. Soc.* **55**(12), 849 (1978).

Vas, K., Estimated radiation chemical changes in irradiated food. *In* "Recent Advances in Food Irradiation" (P. S. Elias and A. J. Cohen, eds.). Elsevier Biomedical, Amsterdam, 1983.

CHAPTER 4

Biological Effects of Ionizing Radiation

I. GENERAL

In most cases the purpose in irradiating foods is to affect a living biological system. This system is either in organisms such as spoilage bacteria, which are present in the food as contaminants or, in cases where the objective is to control a process such as ripening or senescence of raw fruits and vegetables, it is a part of the food itself. In seeking to affect these functioning biological systems, we unavoidably also irradiate the food itself, but this is largely incidental to our purpose. An exception to this targeting on biological systems is the group of applications in which radiation is used to alter by chemical changes the usual characteristics of foods.

When ionizing radiation acts upon a biological system, it does so at the cellular level. The chemical changes resulting from the absorption of radiation that are of interest are those that interfere with the complex set of activities that occur in cells. Not every change necessarily does this. The fact, however, that radiation is lethal to all forms of life suggests that a common mechanism may be involved.

Cells have been considered to be the unit of life. They are highly organized bodies having a continuous enclosing membrane which defines the cell's limits. Materials pass through this membrane in a controlled manner in accordance with the cell functions. The cell's interior is filled with cytoplasm, an unspecialized protoplasm, a viscous translucent material. Inside the cell there usually is a continuous membrane network. Within this network are a number of inclusions, which may be different in different kinds of cells. Each inclusion performs a particular function in the activity of the cell. Cells also contain enzymes which participate in the cell activities.

Some organisms, such as bacteria, have only one cell. Others have a great many. In multicellular organisms there may be a number of different kinds of cells, each kind performing a particular function, as, for example, germ cells, whose role is in the reproduction of the organism. Cells generally reproduce by division (mitosis).

The absorption of ionizing radiation causes chemical changes in the cell components, which can have consequences for the cell's activities. It cannot be claimed at this time, however, that there is a full understanding of the events and the mechanisms that are involved in action of radiation on cells. While there is evidence for several particular effects of radiation on cells, it is not known how significant some are. The following have been considered possibilities: (1) alteration of the cell membrane, which affects its function of transferring materials critical to cell activity, (2) effects on enzymes, (3) effects on synthetic processes, particularly the synthesis of DNA and RNA, (4) effects on energy metabolism through reduction of phosphorylation, and (5) compositional changes in the DNA of a cell, which affect normal cell functions, including reproduction. Of these changes, the changes in the cell DNA are considered very important. The size of the DNA molecule and its complexity make it especially vulnerable to change by radiation.

As a consequence of the radiation-induced changes, cell death may occur. Generally cell death is associated with inability to reproduce. Although a number of causes of cell death have been identified, it is possible that more than one mechanism is involved. In some cases (usually with relatively small amounts of radiation) damage may not cause cell death, but, by altering the general material of the cell, it will result in progeny that are different from the original cell; that is, mutants will result. Some cells have an innate capability of repairing cell damage and in this way they are more radiation resistant.

Radiation effects on cells can be due to both direct and indirect action. In view of this, environmental factors can change the effects of radiation.

Although the changes caused by radiation are at the cellular level, the consequences of these changes vary with the organism in which they exist. Some effects may not be discernible immediately, because the response of the organism may require passage through one or more of its life cycle phases. In this time-final-effect relationship, dose has an important role. This delay of

Table 4.1
Approximate Lethal Doses for Various Organisms

Organism	Dose (Gy)
Mammals	5–10
Insects	10–1000
Vegetative bacteria	500–10,000
Sporulating bacteria	10,000–50,000
Viruses	10,000–200,000

effect has significance in some applications of food irradiation.

Although some effects occur with sublethal doses, the use of sufficient radiation always leads to cell death, and often, as a consequence, to death of the organism of which it is a component. As shown on Table 4.1, the lethal dose varies with the organism. It is to be noted that the lethal dose decreases with increasing complexity and size of the organism.

Since our interest in the irradiation of foods is generally directed to the inactivation of organisms that contaminate the food, it is fortunate that only a small amount of energy in the form of ionizing radiation is required for this inactivation. While the food mass receives this lethal energy, it is generally insufficient to cause significant alteration of the food. In this sense food irradiation is a selective process, since it does have an effective impact on the targeted contaminants without having a comparable impact on the food. In terms of what happens to the food, irradiation can be described as a "gentle" process.

II. EFFECTS OF RADIATION ON ORGANISMS THAT CONTAMINATE FOODS

A. General

Concern for organisms that contaminate foods arises from two kinds of possible consequences of their presence: (1) food spoilage and (2) disease or illness following consumption. The following kinds of organisms are associated with foods:

Viruses	Molds
Bacteria	Parasites
Yeasts	Insects

It is recognized that each type differs greatly from the others. Viruses are essentially nucleoproteins and are not true organisms in that they lack cellular structure and are incapable of functioning without association with living cells. Bacteria are single-cell organisms, but without a nucleus. Yeasts and molds have a cellular structure, including a nucleus. Foodborne parasites may be protozoa, which are single celled, or helminths, which are multicellular intestinal worms. Insects, or more properly arthropods, contaminating foods are complex multicellular organisms.

In the irradiation of a food in order to inactivate contaminating organisms it is desirable from several points of view to use the minimum dose that is effective. In establishing this dose it is necessary to define what constitutes inactivation of the organism. For viruses, bacteria, yeasts, and molds, loss of ability to produce viable progeny which also can reproduce is a satisfactory definition and is termed "reproductive death." For more complex organisms reproductive death may not be adequate in terms of accomplishing a desired result in treating a food. Loss of reproductive ability for foodborne parasites or insects, for example, may be an insufficient effect.

Due to large differences among the various foodborne organisms, each type will be considered separately.

1. Inactivation Dose

The dose for inactivation of an organism, to a very large degree, is a characteristic of the organism. There is need, therefore, to determine this dose for each kind of organism. There are two principal procedures for this determination: (1) end point method and (2) survival curve method.

2. End-Point Method

In this method each of a number of food units, each containing the same number of organisms, is irradiated in a way so as to provide a series of units, each of which has received a different dose. All units are examined for the presence or absence of survivors. The lowest dose showing no survivors is the inactivation dose. While this method provides information on the radiation sensitivity of the organism, the inactivation dose it provides varies with the initial population of the organism, a circumstance which greatly limits its use for estimating dose requirements for populations different from the one used to establish the inactivation dose.

3. Survival Curve Method

In a given population of organisms, as dose is increased, the fraction of survivors becomes smaller. In the concept of the "target theory," a radiation-induced event such as an ionization in or close to some particular entity of the

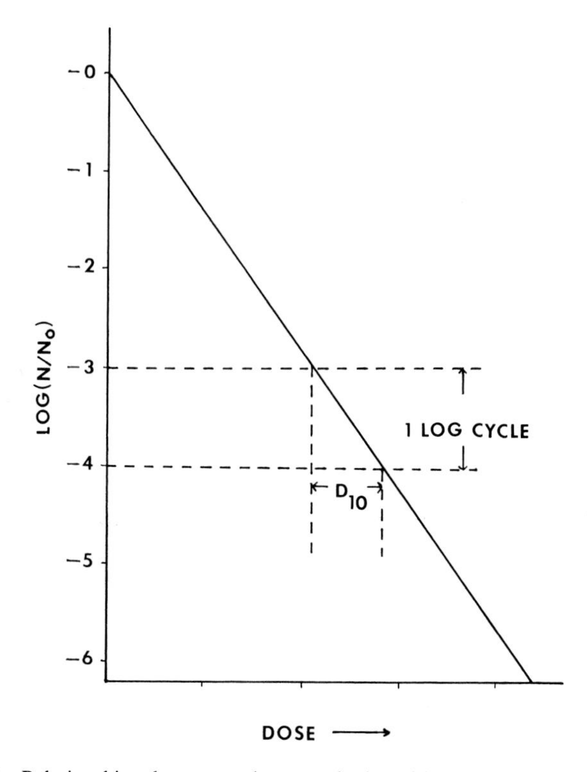

Figure 4.1 Relationship between dose and logarithm of surviving fraction of microorganisms.

organism causes its inactivation. The effectiveness of the radiation in producing lethality is dependent upon it striking the "target." With low doses only some of the organisms present will receive "hits." With increasing doses more will be inactivated in direct proportion to the amount of radiation. With further increases in dose, some targets previously hit will be hit again and the efficiency of the radiation in producing organism inactivations will be reduced. While the actual number of organisms inactivated per dose increment diminishes, the same fraction of organisms present is inactivated with each successive increment. The relationship between dose and surviving fraction is exponential, and a plot of dose versus the logarithm of the surviving fraction is a straight line, as shown in Fig. 4.1.

While the dose to effect a 100% mortality of organisms present in a food could be an objective of irradiation, it will be recognized that this dose is dependent upon the initial number of organisms present. Further, determination of the exact dose to effect 100% mortality of a bacterial population, for example, is very difficult experimentally. Plate-counting procedures for bacteria, for example, are not effective when the counts are below 10 per milliliter. To provide an index of the radiation sensitivity of an organism and to estimate the dose needed for a particular quantitative effect in treating a food, the D_{10} value has been devised. If N_0 is the initial number of the organisms present and N the number after application of a dose D, and D_{10} is the dose needed to reduce the population to 10% of N_0, then

$$\log_{10}\frac{N}{N_0} = -\frac{D}{D_{10}}$$

The D_{10} value will be recognized as the dose needed to secure one log cycle reduction of the population.* The slope of the regression line of Fig. 4.1 is $-1/D_{10}$.

The D_{10} value is an index of the radiation sensitivity of the organism. It is a characteristic of the organism and enables comparison of the sensitivities of different organisms. A dose equal to one D_{10} value will produce a 90% inactivation of a population. If fewer survivors are desired, then additional increments of dose can be applied: A dose equal to two D_{10} values will yield a 1% survival; three, 0.1%; and so on.

The D_{10} value as just described makes use of the target theory. Essentially it assumes that only direct action of the radiation is involved. It assumes that each organism is acted upon independently. While the D_{10} value concept is very useful, its premises do not always hold, mainly in that indirect action of radiation does play an important role in the inactivation of some organisms under some circumstances. For this reason the relevant circumstances applicable to the D_{10} value must be considered in its use.

B. Viruses

Despite the possibility that foods can carry viruses, it is the more unusual situation that foodborne viruses cause health problems for humans. A somewhat different situation exists with animals. Certain foodborne viruses, such as the foot-and-mouth disease virus, are the cause of serious animal diseases, and for this reason foods contaminated with them often are embargoed.

Viruses reproduce only by parasitizing a living cell. As a consequence, when they occur as a food contaminant, generally they do not multiply. They can, however, persist in an infectious state for extended time periods. A given virus generally is capable of infecting only certain particular organisms and this usually limits the hazard the presence of a virus occasions.

* The dose for 37% survival (D_{37}) also has been used. D_{10} equals 2.303 D_{37}.

Viruses may be simple entities made of relatively short single strands of nucleic acid and two or three proteins which exist as a kind of enclosure or capsid. Some viruses are more complex in that they may be as large as a small cell and contain a number of proteins and lipid material. Some viruses have been crystallized. Viruses have diameter dimensions in the range 16–200 nm. The radiation dose for inactivation increases inversely with virus size. Both direct and indirect effects of radiation are involved in virus inactivation.

In accordance with the postulate that simpler (and smaller) organisms have greater resistance to radiation, inactivation doses for viruses are relatively high. Determination of inactivation doses is complicated by a number of factors. First there is the need to define what constitutes inactivation, since irradiation can alter a number of attributes of viruses. Of the possible changes, from the standpoint of food irradiation, an appropriate criterion of inactivation is a change in the ability of the virus to infect an organism. Since the changes causing inactivation can result from indirect as well as direct action of radiation, environmental factors such as medium composition and irradiation temperature need to be specified and considered.

Although there is a variation among the different viruses, the available information suggests that a D_{10} value of the order of 5 kGy is a reasonable working figure and is fairly representative of the sensitivity of many viruses as they exist in foods and at irradiation temperatures above freezing.

A knowledge of the D_{10} value, however, is not all the information needed to determine the dose for an adequate treatment of a food contaminated with a virus. In addition, there is need to know the initial level of the virus contamination. What constitutes an acceptable residual level after irradiation also must be known. Accurate information in these two areas presently is not generally available.

Estimates of dose requirements to ensure safety from viral infection solely through the use of irradiation have ranged from 20 to 100 kGy. These estimates, even if only approximately correct, suggest that the dose requirements for virus inactivation in foods are so high as to make irradiation an unlikely choice as a treatment procedure.

C. Bacteria

1. General

Bacteria are single-cell organisms. They are prokaryotic; that is, they possess no nucleus. Reproduction is by mitosis (cell division).

Bacteria are principal causes of food spoilage and most preservation methods operate so as to control or inactivate bacteria. Some foodborne bacteria are pathogenic to humans. Generally they cause illness or death by one of two mechanisms: infection or toxin formation.

There are a great many different kinds of bacteria. Morphological and physiological characteristics differentiate bacteria and enable their classification and identification. With foodborne bacteria, grouping into two major types is important: vegetative and spores. Spores generally are more resistant to any inactivation process than are vegetative bacteria. For this reason particular measures must be taken to inactivate spores. In food irradiation this generally requires larger doses than for vegetative bacteria.

There are, however, some vegetative bacteria of extremely high radiation resistance comparable with or exceeding that of spores. Among these are a number which are associated with foods, such as *Micrococcus radiodurans* and *Moraxella acinetobacter* isolated from meat, *Micrococcus roseus* No. 248 and *Micrococcus radiophilus* isolated from fish, and *Pseudomonas radiora* isolated from rice. The extraordinarily high radiation resistance of these bacteria is ascribed to an ability to repair radiation damage to DNA. Fortunately, these radiation-resistant bacteria have not been found to have a role in food spoilage nor are they pathogenic. They are easily inactivated by heat or low salt concentrations. In general their presence in foods in relation to irradiation can be ignored.

The standard criterion for death of a bacterium is its inability to form a colony under suitable conditions on a suitable nutrient medium. This can be considered to be reproductive death. Sublethal doses can result in mutations. Both direct and indirect effects of radiation can be involved in inactivation of bacteria.

Each bacterium is characterized by a particular D_{10} value, reflecting its inherent sensitivity to radiation. Since indirect action of radiation enters into the inactivation of a bacterium, different D_{10} values may be obtained with different relevant environmental circumstances. The D_{10} values for a number of bacteria are shown in Table 4.2. These values should not be considered necessarily precise but may be used as a guide.

The target theory concept explains the inactivation of bacteria on the basis of a single "hit" within the cell boundaries. In this situation, D_{10} values are independent of indirect action and associated environmental factors, and the relationship shown in Fig. 4.1 holds. Indirect action, however, does occur and it is necessary, therefore, to specify the particular circumstances which apply to a given D_{10} value.

2. Factors Affecting D_{10} Value

Among the more important environmental factors affecting inactivation of bacteria are the following.

Table 4.2

D_{10} Values of Bacteria

Organism	D_{10} (Gy)	Medium	Temperature (°C)	References[a]
Vegetative bacteria				
Achromobacter spp.	850–1760	Broth	RT[b]	1
Acinetobacter calocetícus	4050–8140	Raw beef	−30	2
Brevibacterium	485–642	Raw beef	−30	2
Campylobacter jejuni	140–160	Raw beef	18–20	3
Escherichia coli	210	Broth	5	4
E. coli	430	Low fat beef	5	4
E. coli	420	High fat beef	5	4
E. coli	350	Oysters	—	5
E. coli	140	Crabmeat	—	5
Lactobacillus brevis NCD110	1200	Buffer	—	6
Lactobacillus planterium NCDO-343	80	Buffer	—	6
Micrococcus radiodurans R_1	2500	Raw beef	5	7
M. radiodurans U_1	2785	Raw beef	5	7
M. radiodurans R_4R	3080	Raw beef	5	7
M. radiodurans R_w	2750	Raw beef	5	7
Moraxella nonliquifaciens	5390–5830	Raw beef	−30	2
Moraxella osloensis	4770–10,000	Raw beef	−30	2
Proteus vulgaris	100	Broth	—	5
P. vulgaris	200	Oysters	—	5
Pseudomonas fluorescens	50	Broth	5	4
Ps. fluorescens	130	Low fat beef	5	4
Pseudomonas geniculata	50	Broth	—	8
Salmonella enteriditis	250	Broth	—	4

(*continued*)

Table 4.2 (*Continued*)

Organism	D_{10} (Gy)	Medium	Temperature (°C)	References[a]
S. enteriditis	700	Low fat beef	—	4
S. enteriditis	490	High fat beef	—	4
Salmonella paratyphoid B	550	Broth	—	5
S. paratyphoid B	850	Oysters	—	5
S. paratyphoid B	1000	Crabmeat	—	5
Salmonella pullorum	250	Broth	—	5
Salmonella typhimurium	280	Broth	5	4
S. typhimurium	640	Low fat beef	5	4
S. typhimurium	530	High fat beef	5	4
S. typhimurium	1740	Fish meal	—	9
Shigella dysenteriae	350	Broth	—	5
Shigella sonnei	200	Broth	—	5
Staphylococcus aureus	240	Broth	5	4
S. aureus	580	Low fat beef	5	4
S. aureus	300	High fat beef	5	4
S. aureus	1500	Oysters	—	5
S. aureus	800	Crabmeat	—	5
S. aureus	1900	Shrimp	—	5
Streptococcus faecium α 21	900	Buffer	5	10
S. faecium α 21	2400	Buffer	−30	10
S. faecium α 21	3300	Buffer	−80	10
S. faecium α 21	3700	Buffer	−140	10
S. faecium α 21	3800	Buffer	−196	10
Streptococcus faecalis	500	Broth	—	5
	1000	Oysters	—	5
Streptococcus pyogenes	450	Broth	—	5
Vibrio parahaemolyticus	~50	Seawater	24	11

Yersinia enterocolitica IP 107	109	Broth	25	12
Y. enterocolitica	195	Beef	25	12
Spore-forming bacteria				
Bacillus coagulans (*thermoacidurans*)	1290	Buffer	2–12	13
Bacillus pumilus E 601	1550–2110	Dried from buffer	RT[b]	9
Bacillus subtilis	2600	Saline	—	14
B. subtilis	350	Pea puree	—	14
B. subtilis	1700	Evaporated milk	—	14
Clostridium botulinum				
Type A				
33	3450–3600	Cooked beef	25	15
33	3730–3850	Cooked beef	0	15
33	4300–4340	Cooked beef	−50	15
33	5770–5950	Cooked beef	−196	15
36	3360	Buffer	10	16
62	2240	Buffer	10	16
12885	2410	Buffer	10	16
Type B				
9	2270	Buffer	10	16
40	3170	Buffer	10	16
41	3180	Buffer	10	16
51	1290	Buffer	10	16
53	3290	Buffer	10	16
Type E				
Alaska	1370	Beef stew	RT[b]	17
VH	1280	Beef stew	RT[b]	17
Beluga	1360	Beef stew	RT[b]	17
8E	1380	Beef stew	RT[b]	17
1304E	1310	Beef stew	RT[b]	17

(continued)

Table 4.2 (*Continued*)

Organism	D_{10} (Gy)	Medium	Temperature (°C)	References[a]
Iwanai	1250	Beef stew	RT[b]	17
Clostridium perfringens				
Type C				
3181	2100	Aqueous	RT[b]	18
Type D				
8503	1800	Aqueous	RT[b]	18
Type E				
8084	1200	Aqueous	RT[b]	18
Clostridium sporogenes				
PA 3679/S_2	2200	Aqueous	RT[b]	19

[a] References: 1, W. Schmidt-Lorenz and J. Farkas, *Kozlemenyei* **4,** 24 (1961); 2, R. B. Maxcy, D. B. Rowley, and A. Anellis, "Radiation Resistance of Asporogenous Bacteria," Tech. Rep. 76-43-FSL. U.S. Army Natick Res. and Dev. Command, Natick, Massachusetts, 1976; 3, J. A. Tarkowski, S. C. C. Stoffer, R. R. Beumer, and E. H. Kampelmacher, *Int. J. Food Microbiol.* **1,** 13 (1984); 4, R. B. Maxcy and N. R. Twari, *In* "Radiation Preservation of Food." Int. Atomic Energy Agency, Vienna, 1973. 5, D. J. Quinn, A. W. Anderson, and J. F. Dyer, *In* "Microbiological Problems in Food Preservation by Irradiation." Int. Atomic Energy Agency, Vienna, 1967; 6, P. DuPuy and O. Tremeau, *Int. J. Appl. Radiat. Isot.* **11,** 145 (1961); 7, D. E. Duggan, A. W. Anderson, and P. R. Elliker, *Appl. Microbiol.* **11,** 398 (1963); 8, Anonymous, "Training Manual on Food Irradiation Technology and Techniques," 2nd Ed., Tech. Rep. Series No. 114. Int. Atomic Energy Agency, Vienna, 1982; 9, F. J. Ley, *In* "Manual on Radiation Sterilization of Medical and Biological Materials," Tech. Rep. Series No. 149. Int. Atomic Energy Agency, Vienna, 1973; 10, A. Anellis, D. Berkowitz, and D. Kemper, *Appl. Microbiol.* **25,** 517 (1973); 11, J. R. Matches and J. Liston, *J. Food Sci.* **36,** 339 (1971); 12, Y. A. El-zawahry and D. B. Rowley, *Appl. Environ. Microbiol.* **37,** 50 (1979); 13, A. Anellis, C. J. Cichon, and M. M. Rayman, *Food Res.* **25,** 285 (1960); 14, B. E. Proctor, S. A. Goldblith, E. M. Oberle, and W. C. Miller, Jr., *Radiat. Res.* **3,** 295 (1955); 15, N. Grecz, A. A. Walker, A. Anellis, and D. Berkowitz, *Can. J. Microbiol.* **17,** 135 (1971); 16, A. Anellis and R. B. Koch, *Appl. Microbiol.* **10,** 326 (1962); 17, C. F. Schmidt, W. K. Nank, and R. V. Lechowich, *J. Food Sci.* **27,** 77 (1962); 18, T. A. Roberts, *J. Appl. Bacteriol.* **31,** 133 (1968); 19, T. A. Roberts and M. Ingram, *J. Food Sci.* **30,** 879 (1965).

[b] RT, Room temperature.

a. Medium Composition

i. Water Content In foods water may or may not be present. If present, indirect action of radiation can occur through the radiolytic products of water, commonly through the action of the OH· radical. Freezing of the water in a food produces an effect similar to water removal. The consequence of either freezing or water removal is to increase the resistance of the bacterium to radiation. Both effects are illustrated in Figs. 4.2 and 4.3.

ii. Food Components The food components other than water also affect indirect action of radiation on bacteria. This is illustrated in Fig. 4.4. The food components may be regarded as competing with the bacteria for interaction with the active radiolytic products of water. Because of the compositional complexity of most foods, this competitive action is highly significant in the D_{10} value, and the application of D_{10} values obtained with simpler systems to foods can lead to error. It is commonly necessary to measure the D_{10} value of a bacterium in the particular food of interest in order to define a process for irradiation.

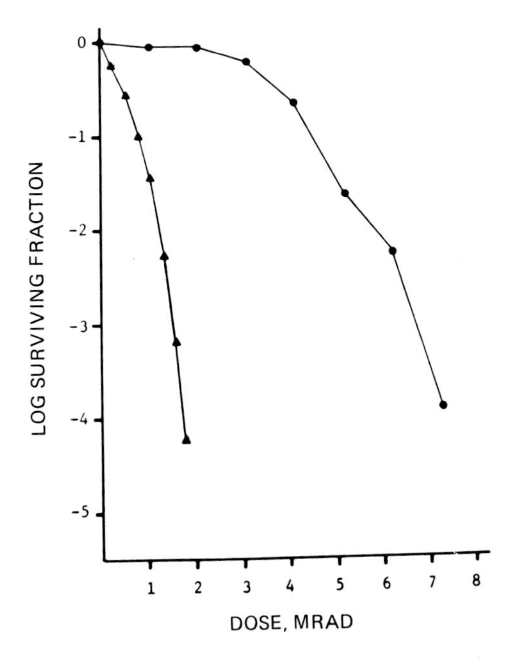

Figure 4.2 Comparative radiation resistance of *Micrococcus radiodurans* in ground beef at −30°C (●) and at ambient temperatures (▲). From M. A. Bruns and R. B. Maxcy, *J. Food Sci.* **44**, 1743 (1979). Copyright © by Institute of Food Technologists.

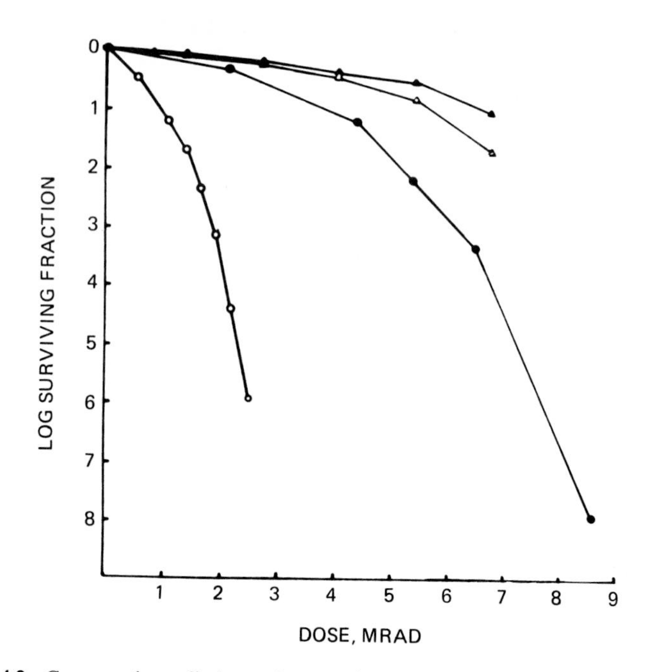

Figure 4.3 Comparative radiation resistance of *Moraxella* spp. in broth at ambient temperature (○) and at −30°C (●). △, Resistance of the same *Moraxella* spp. isolate from beef lyophilized in broth before irradiation at ambient temperatures. ▲, Same, except irradiation was carried out at −30°C. From M. A. Bruns and R. B. Maxcy, *J. Food Sci.* **44,** 1743 (1979). Copyright © by Institute of Food Technologists.

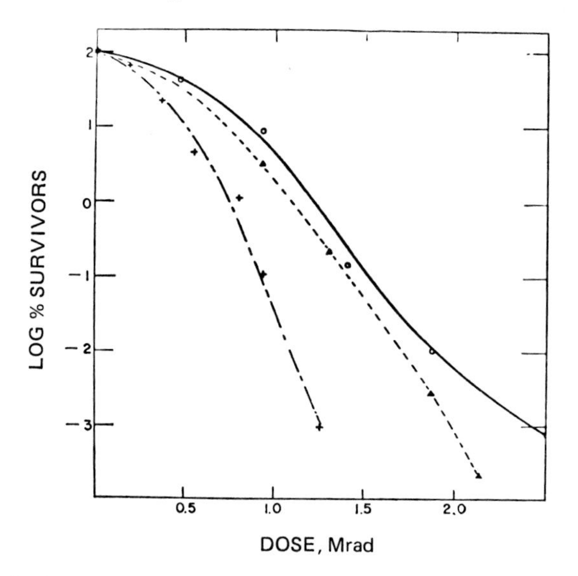

Figure 4.4 Survivor curves for spores of *Clostridium botulinum* in different substrates, irradiated in the frozen state. ○, Pork; ▲, peas; +, phosphate. From C. B. Denny, C. W. Bohrer, W. E. Perkins, and C. T. Townsend, *Food Res.* **24,** 44 (1958). Copyright © by Institute of Food Technologists.

iii. pH The pH of the medium can affect free-radical formation and, in this way, the indirect action. It can also change certain aspects of the bacterium. This is illustrated in Fig. 4.5 for *Clostridium botulinum* 33A spores in borate buffer solution. The explanation for the change in radiation sensitivity with pH is based on (1) the effect of pH on the radiolysis of water and (2) the effect of pH in the bipolymer nature of the spores of *C. botulinum.* As may be anticipated, the role of water in these effects is temperature dependent and is related to the mobility of radicals as determined by the physical state of the water at different temperatures.

iv. Oxygen Due to the nature of the oxygen molecule in that it has unpaired electrons, oxygen adds on to many radicals, thereby preventing radical recombination and dimerization. In this way, under given conditions, the presence of oxygen enhances the indirect action of radiation, increasing its effectiveness and causing a reduction in the D_{10} value. This effect is illustrated in Table 4.3. It should be noted, however, that the usefulness of oxygen in enhancing radiation effectiveness as a bactericide is not always appropriate in irradiating foods. Increased indirect action on the food itself may be undesirable from other aspects, such as oxidative changes in lipid components. Also, oxygen may have an undesired effect on the food, independent of irradiation, and this may require the absence of oxygen.

An initial oxygen level in a food may be reduced prior to irradiation as a result of respiration activities of microorganisms. As discussed below, the use of a high dose rate also can affect the oxygen level in a food.

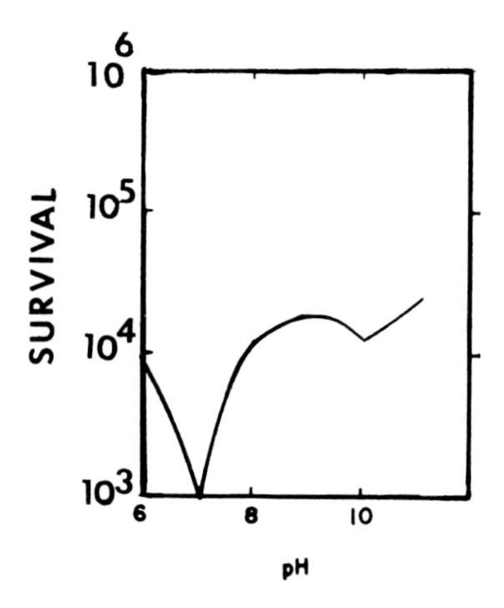

Figure 4.5 Relationship of the survival of spores of *Clostridium botulinum* 33A with pH. Spores suspended in buffer solution and irradiated at −50°C with 9 kGy. From J. Upadhya and N. Grecz, *Can. J. Microbiol.* **15,** 1419 (1969).

Table 4.3

Effect of Oxygen on D_{10} Values of Four Strains of Salmonella[a]

Strain	D_{10} (Gy) Aerated	D_{10} (Gy) Anoxic
Salmonella gallinarium	132	363
S. senftenberg	130	389
S. typhimurium	208	619
S. paratyphi B	190	659

[a] From M. J. Thornley, Microbiological aspects of the use of radiation for the elimination of salmonellae from foods and feed stuffs. *In* "Radiation Control of Salmonellae in Food and Feed Stuffs," Tech. Rep. Series No. 22. Int. Atomic Energy Agency, Vienna, 1963.

Table 4.4

Change of D_{10} Value of *Streptococcus faecium* $A_2 1$ in Phosphate Buffer with Temperature[a]

Irradiation temperature (°C)	D_{10} (kGy)
−196	3.8
−140	3.7
−80	3.3
−30	2.4
5	0.09

[a] From A. Anellis, D. Berkowitz and D. Kemper, *Appl. Microbiol.* **25,** 517 (1973).

b. Conditions during Irradiation

i. Temperature In foods having an appreciable water content, it can be anticipated that temperature will affect the radiation sensitivity of bacteria that are present. The change of D_{10} value of *Streptococcus faecium* $A_2 1$ in phosphate buffer with temperature is shown in Table 4.4. The change of D_{10} value with temperature for the spores of *Clostridium botulinum* 33A in ground beef is shown in Fig. 4.6. The change of D_{10} values with temperature demonstrates the importance of the indirect action in high-moisture foods. It also emphasizes the necessity for specifying the temperature of irradiation in measuring the radiation sensitivity of a bacterium.

Noteworthy also is the precipitous drop in the D_{10} value for the *Clostridium botulinum* 33A spores at temperatures close to 100°C. This is not surprising in that at some temperature near 100°C a thermal inactivation process joins the irradiation process. Although heat and radiation undoubtedly operate by very different mechanisms, the end result, namely spore inactivation, is the same, and the total effect of lethality can be the sum of the two separate effects.

ii. Dose Rate In systems as complex as foods, dose rate is not likely to have a significant effect on the radiation sensitivity of bacteria. This is so largely because radical interaction with food ccomponents will predominate over a wide range of dose rates. Radiation effects on bacteria involving oxygen, however, can be altered by high dose rates, such as those obtained with machine sources. This effect may be seen in Fig. 4.7, in which the dose to secure

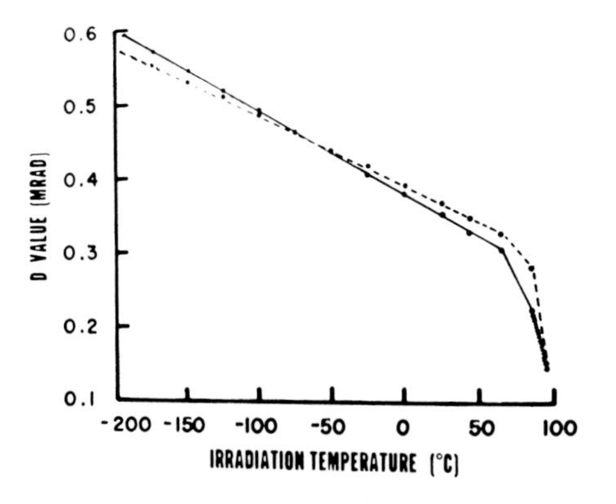

Figure 4.6 Change with irradiation temperature of D_{10} values of spores of *Clostridium botulinum* 33A in ground beef. ----Approximately 10^7 spores/dose; ——approximately 10^9 spores/dose. From N. Grecz, A. A. Walker, A. Anellis, and D. Berkowitz, *Can. J. Microbiol.* **17,** 135 (1971).

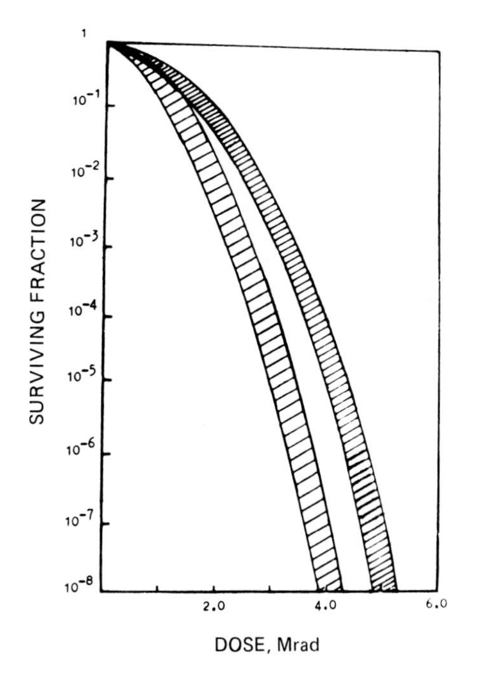

Figure 4.7 The difference in dose requirements for the same surviving fraction of *Streptococcus faecium* A_21 observed with γ (wide shading) and electron (narrow shading) irradiation. The bacteria were dried from broth on glass and equilibrated with air of 10 to 50% R.H. From C. Emborg, *Acta Pathol. Microbiol. Scand., Sect. B.* **80,** 367 (1972).

a given surviving fraction of *Streptococcus faecium* A_21 bacteria is larger with electrons than with γ radiation. At very high dose rates the oxygen of the system can be depleted at a rate greater than it can be replaced by diffusion processes transferring atmospheric O_2 into the food, and the resulting anoxic condition can reduce the lethality effect of the radiation.

c. Preirradiation Conditions

A number of preirradiation treatments of bacteria have been associated with their radiation sensitivity, but generally they have little relevance to food irradiation, since generally there is no opportunity to control the bacteria present in a food prior to irradiation. In some cases, however, it may be possible to limit the use of irradiation to foods that have a relatively low bacterial content, either as a result of a minimized contamination or appropriate handling prior to irradiation that limits growth. Consideration of such measures is supported by the fact that the prior history can affect the radiation sensitivity of bacteria. This is illustrated in Fig. 4.8.

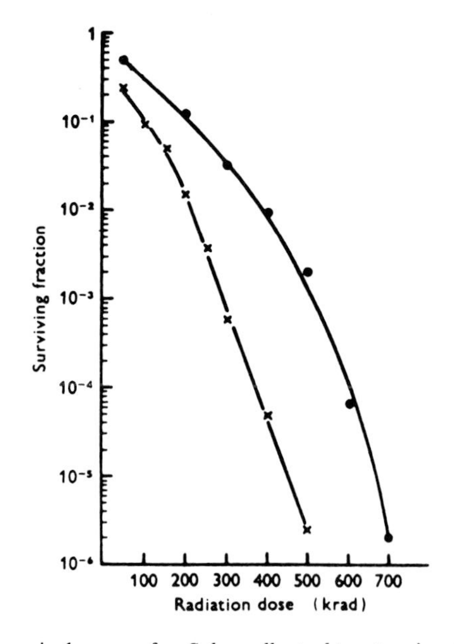

Figure 4.8 Dose–survival curves for *Salmonella typhimurium* irradiated in horse meat at $-15°C$. ●, Following inoculation and growth in the meat at 37°C for 2 days; ×, following inoculation but without preirradiation growth in the meat. From F. J. Ley, T. S. Kennedy, K. Kawashima, D. Roberts, and B. H. Hobbs, *J. Hyg.* **68**, 293 (1970).

3. Ways to Increase Radiation Sensitivity

Reduction of dose requirements for inactivation of bacteria can be advantageous in a number of ways. This is one of the objectives of combination processes (see Chapter 11).

Certain substances added to the medium have been shown to affect the radiation sensitivity of bacteria. Utilization of such materials in food irradiation has not occurred so far, due to difficulties with their compatibility with foods or with the lack of effective means of incorporation in foods.

Other ways of increasing sensitivity of bacteria to radiation have been sought. For example, hydrostatic pressure above 500 atm applied to a food containing spores of *Bacillus pumilus* leads to their germination and, thereby, to an increase in sensitivity to radiation.

4. Determination of the Dose Required for Desired Reduction of Numbers of Bacteria in a Food

The D_{10} value is a useful index of the sensitivity of bacteria to radiation. It provides a convenient quantitative measure for comparing different bacteria. It enables estimation of the dose needed to attain a particular objective. Yet, because of the effects of indirect action of radiation, its use is fraught with problems.

The simple straight-line relationship between dose and the logarithm of the fraction of surviving bacteria given in Fig. 4.1, does not fully portray what actually occurs. As can be seen from actual data on the survival of spores of *Clostridium botulinum* in Fig. 4.9, a straight line fits only part of the data. At the low-dose end there is a "shoulder" and at the high-dose end there is a "tail." The tail is extended in samples containing higher initial numbers of spores. An idealized form of the sigmoidal curve required to fit the data of Fig. 4.9 is shown in Fig. 4.10.

It is clear that a D_{10} value taken from the linear portion of the sigmoidal curve could not predict accurately the dose needed for inactivation of *all* the bacteria present in the samples having the larger populations—that is, to obtain sterility. In situations in which sterility of a product is the desired effect, other procedures for determining the needed dose may be required. This is discussed in connection with radappertized (radiation sterilized) meats (see Chapter 6).

In situations where the desired effect may not be so critical, the D_{10} value is useful in estimating dose, but nonetheless, it should be used with due consideration for the various environmental factors which may alter it.

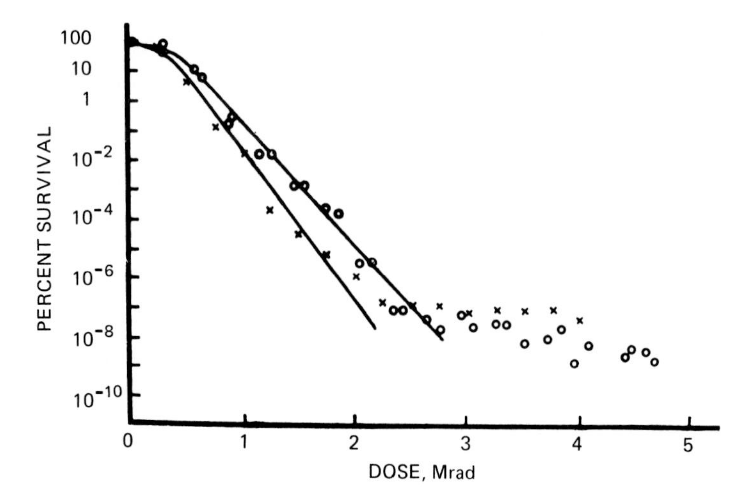

Figure 4.9 Survival of spores of *Clostridium botulinum* subjected to γ irradiation. ○, Pork–pea broth in air; ×, phosphate in air. From E. Wheaton and G. B. Pratt, *J. Food Sci.* **27,** 327 (1962). Copyright © by Institute of Food Technologists.

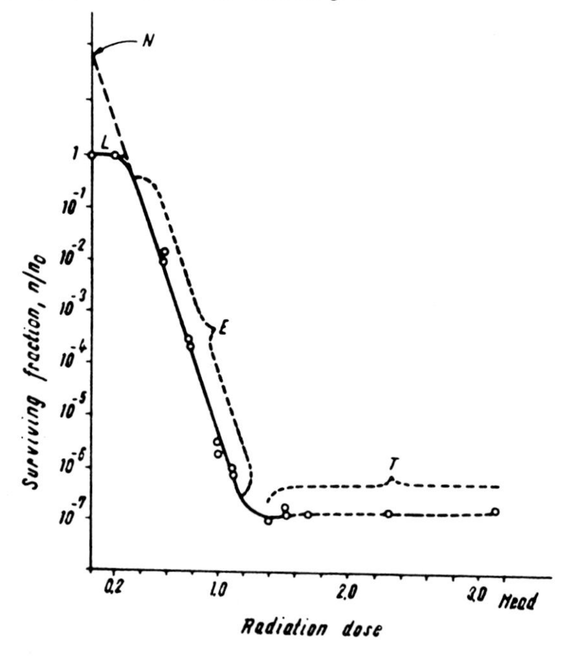

Figure 4.10 Component parts of a sigmoidal radiation survival curve of spores of *Clostridium botulinum* 33A. N, Extrapolation or target number; L, lag or shoulder; E, exponential decline portion; T, tail portion. The exponential decline portion yields the conventional D_{10} value. From N. Grecz, C. Wiatr, E. Durban, and T. Kang, *J. Food Process. Preserv.* **2,** 315 (1978).

5. Radiation-Induced Changes in Surviving Bacterial Flora

In some applications of food irradiation the intended technical effect does not require that all of the bacteria present in a food be inactivated. As a consequence, there are surviving bacteria. From several points of view it is reasonable to have an interest in these survivors.

The survivors may be no different from the other bacteria initially present, and the fact of their survival may reflect simply the use of a dose that was not great enough to inactivate the total initial population. It may be, however, that the survivors are survivors because they have a radiation sensitivity less than the other bacteria present initially; that is, they are more radiation resistant. Such survivors may be more resistant members of a species or a strain present or they may be a different species or strain which has greater radiation resistance. Both possibilities require further consideration.

In the case in which the survivors are more radiation-resistant members of a particular species or strain, irradiation can become a procedure for selectively favoring such naturally resistant bacteria. Alternatively, enhanced radiation resistance may be acquired by radiation-induced mutation of the initially present bacteria. Mutation of bacteria has been observed but only with repeated irradiation through several life cycles. As normally practiced, food irradiation is a "once-through" process. Repeated irradiation of the same food item is not usual. It follows, therefore, that radiation-resistant mutants are not likely to be formed in practice.

Regardless of how enhanced radiation-resistant bacteria may occur—that is, through selection or mutation—there does not appear to be any likely pathway for such organisms to be transferred to other foods and in this way to become their common flora. On this basis, the presence of survivor radiation-resistant bacteria in a food presents no unusual problem beyond possibly one of adjustment of the dose employed.

In the situation in which the survivors are a more resistant species there can be an outgrowth pattern for the food on storage that is different from the normal. This may or may not be important. If the spoilage of the irradiated food due to the altered bacterial outgrowth is sufficiently different from the usual, consumers may have difficulty recognizing the spoilage and this may be unacceptable, since they may be misled as to the state of the food. If the altered postirradiation outgrowth involves a pathogen, the food could become unhealthful in an atypical manner so that the consumer is not aware of the hazard. While unacceptable outgrowth patterns postirradiation are possibilities, there also are situations in which a different outgrowth pattern has little or no meaning, or even can have an advantage.

In view of the various possibilities it is important that the postirradiation bacterial outgrowth pattern be identified and evaluated. In cases where a problem exists, it may be necessary to use additional measures to safeguard the consumer. (For an example of a possible such requirement, see Chapter 7.)

6. Bacterial Toxins

Some foodborne bacteria (e.g., *Staphylococcus* and *Clostridium botulinum*) produce toxins. These toxins are proteins and as they occur in the complex systems of foods, they require doses for inactivation that are very great. The D_{10} values in buffer solutions of the toxins of *C. botulinum* types A and B have been observed to range from 6.2 to 31 kGy. Values 100 times higher were obtained in broth and even greater values in cheese. These values indicate that inactivation of bacterial toxins by radiation is not feasible. For this reason irradiation of foods must be based on treating only foods that do not contain preformed bacterial toxins, unless other measures that are effective in inactivating the toxins are used additionally.

D. Yeasts and Molds

Potentially, yeasts and molds can be contaminants of all foods. Ordinarily, however, present-day food-handling practices limit contamination primarily to fruit and vegetables and to cereal grains, legumes, and baked products. These fungi, especially molds, are aerobic and are distinguished by their ability to tolerate low levels of water content of foods.

Yeasts are single-celled organisms and reproduce through a form of cell division termed budding. Yeast cells generally are much larger than bacterial cells. Molds are multicellular, usually having growth of a filamentous character. The felt-like mass of interlacing filaments is called the mycelium. Reproduction may be sexual or asexual. Spore formation occurs in the reproduction process. Some molds produce a mycotoxin, e.g., *Aspergillus flavus*, which forms aflatoxin. As with toxins of bacterial origin, the dose to inactivate mycotoxins is too large to be usable in food irradiation.

Although sufficient radiation can kill the cells of yeasts and molds, an adequate effect from the standpoint of food irradiation is obtained if the parent cell is prevented from producing progeny which are able to carry out reproduction; that is, "reproductive death" is an adequate result.

Due to the structure of mold growths, determination of cell numbers is difficult, and for this reason, the survival curve method of measuring radiation sensitivity (which yields D_{10} values) has not been used. Instead, the end point method has been used for both yeasts and molds to determine the inactivation or lethal dose. Since values so obtained are related to the size of the initial population of cells, they cannot be used as would D_{10} values for estimating dose requirements to secure a particular effect.

Table 4.5 lists doses effective in attaining either an 80% or a 100% kill of certain yeasts and molds associated with foods, primarily fruits. Only one population level, 10^7 organisms per milliliter, was used with yeasts in this determination. The doses for molds are given either for 10^4 spores and an 80%

Table 4.5

Doses for Inactivation of Yeasts and Molds

Organism	Lethal dose (Gy)		References[a]
Yeasts[b]			
Candida krusei	5,500		1
Candida parapsilosis	5,600		1
Candida solani	7,300		1
Candida tropicalis	10,300		1
Candida zeylanoides	5,600		1
Hanseniaspora valbyensis	7,400		1
Kloekera apiculata	4,650		1
Pullularia pullulans	20,000		1
Rhodotorula glutinis	4,720		1
Saccharomyces florentinus	5,400		1
S. carlsbergensis	5,800		1
Torulopsis spp.	5,900		1
	Cell suspension 80% kill (10^4 cells)	Agar slant 100% kill	
Molds			
Alternaria spp.	—	6,000	1
	4,400	—	2
Aspergillus niger	—	2,500	1
Botrytis cinerea	—	5,000	1
Cladosporium spp.	—	6,000	1
Cladosporium spp.	5,500	—	2
Gliocladium roseum	—	2,500	1
Helminthosporium spp.	—	6,000	1
Penicillium digitatum	1,600	—	2
Penicillium digitatum	—	2,500	1
Penicillium expansum	1,300	—	2
Penicillium italicum	1,200	—	2
Rhizopus nigricans	—	2,500	1
Rhizopus stolonifer	4,500	—	2

[a] References: **1,** G. D. Saravacos, L. P. Hatzipetrou, and E. Georgiadou, *Food Irradiat.* (*Saclay*) **3,** A6 (1962); **2,** N. F. Sommer and R. J. Fortlage, *Adv. Food Res.* **15,** 147 (1966).

[b] Lethal dose given for 10^7 cells/ml.

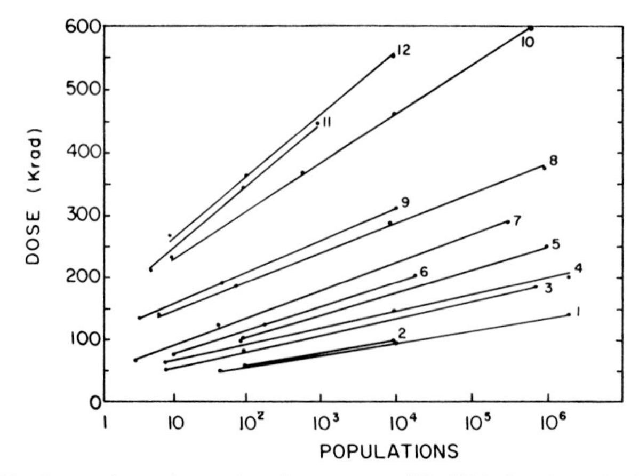

Figure 4.11 Approximate dose to inactivate every cell in 80% of various-sized populations of postharvest disease fungi of fruits and vegetables. Data are for spores in vitro. 1, *Trichoderma viride*; 2, *Phomopsis citri*; 3, *Penicillium italicum*; 4, *Penicillium expansum*; 5, *Penicillium digitatum*; 6, *Geotrichum candidum*; 7, *Monilinia fructicola*; 8, *Botrytis cinerea*; 9, *Diplodia natalensis*; 10, *Rhizopus stolonifer*; 11, *Alternaria citri*; 12, *Cladosporium herbarum*. From N. F. Sommer and R. J. Fortlage, *Adv. Food Res.* **15,** 147 (1966) and N. F. Sommer, E. C. Maxie, and R. J. Fortlage, *Radiat. Bot.* **4,** 309 and 317 (1964). Reprinted with permission. Copyright (1964), Pergamon Press, Ltd.

kill (taken from the curves of Fig. 4.11 or from an agar slant, 3 × 3 mm surface culture). These figures can be regarded as only indicative of doses that are required. They do, however, demonstrate that both yeasts and molds have substantial resistance to radiation, a circumstance that limits the use of irradiation for their inactivation with some foods, especially fruits.

Relatively little information exists on the effects of environmental factors on the radiation resistance of yeasts and molds. Irradiation has been shown to be more effective in the presence of oxygen. It is probable that other environmental factors similar to those that affect the irradiation of bacteria also operate with yeasts and molds.

E. Foodborne Parasites

While a number of parasitic organisms afflict humans, only a relatively few are foodborne. Foods that carry parasites include meat, fish, fruits, and vegetables. Of the foodborne parasites, there are two general groups, protozoa and helminths. The former are single-celled organisms and the latter are intestinal worms.

Of the foodborne parasitic protozoans, only two are considered important. Cysts of *Toxoplasma gondii* occur in the muscles of swine, beef cattle, sheep, and chickens, and can cause the disease toxoplasmosis in humans. *Entamoeba histolytica* is basically waterborne but can be carried by polluted raw fruits and

Table 4.6
Doses for Foodborne Parasites

Organism	Dose (Gy)	Effect
Toxoplasma gondii	90	Not infectious
T. gondii	300	Death
Entamoeba histolytica	251	Death (cyst)
Trichinella spiralis	6300	Death
T. spiralis	150–300	No reproduction
T. spiralis	150–300	Not mature
Taenia saginata	5000	Death
Taenia solium	5000	Death
Fasciola hepatica	185	No excystation (death?)
Anisakis	≥6000	Death

vegetables.

Foodborne helminths include *Trichinella spiralis*, found in pork, *Taenia saginata*, the beef tape worm, *Taenia solium*, the pork tape worm, and *Fasciola hepatica*, the sheep liver fluke, which also occurs on vegetables such as watercress. *Anisakis* and related genera are found in fish.

All of these organisms have a number of forms during their life cycle, but generally occur in foods in only one form, namely as larvae. Vulnerability to radiation depends to a large degree upon the particular form at the time of irradiation. Additionally, the disease potential to humans can be related to the nature or degree of radiation damage sustained by the parasite. This requires consideration of the course of the disease in humans in determining what effect (and, therefore, what dose) is to be employed in treating a food. Generally, a relatively large dose is needed to secure the death of the organism as it occurs in the food. Sublethal doses can control infectiousness and reproductive capability and in some cases may be adequate as a health measure.

Table 4.6 lists doses possibly applicable to the treatment of foods associated with the indicated parasite. Generally, however, the dose to secure death is high, and, for some foods, would not be tolerable from a number of considerations.

F. Insects

Of all the contaminating organisms to which food irradiation is directed, insects are the most complex. Insects are a class of largely land arthropods and are articulate animals having jointed limbs. Their bodies are made up of three parts: head, thorax, and abdomen. Each part has a complex structure related to particular body functions. The number of insect species probably exceeds a million. Of these, only a relatively few are considered to be pests, about 500 species being of major importance, largely in connection with agriculture and its products. Another class of arthropods, the arachnids (mites, ticks, spiders, etc.) include a few species which contaminate foods.

During their life cycle insects undergo metamorphosis. Following egg hatch there is the larval period in which feeding and growth are the principal activities. The pupal stage (or in some species, the prepupal stage) follows; this is a period involving cell differentiation into distinctive adult organs. The last is the adult or imaginal stage, which is concerned with migration, mating, and reproduction. In this manner growth, differentiation, and reproduction are completely separated.

Insects contaminating a food may be at any life stage and their radiation sensitivity varies with each life stage. Generally, radiation sensitivity of cells is in direct proportion to their reproductive activity. With insects very little cell division activity takes place after egg hatch, growth during the later stages occurring primarily through cell enlargement. Differentiated cells are more radiation resistant, and here, also, in insects differentiation at the cellular level occurs in the egg stage. The occurrence of cell division and differentiation in the egg results in greater radiation resistance in the later life forms. In the adult, however, the cells of the gonads carry out cell division, and, as a consequence, they are more sensitive to radiation. This, in part, explains the fact that lower doses prevent reproduction than are needed for adult death. Regardless of the life stage of the insect at the time of irradiation, however, radiation, in appropriate amounts, produces either sublethal or lethal effects on all insects.

Of practical significance also is the fact that radiation of the proper energy level can penetrate a food and reach insects below its surface.

Other factors affect the radiation sensitivity of insects. Some of these are related to their complexity and to the separate life stages. Pre- and postirradiation temperature, as well as temperature during irradiation, for example, can markedly alter requirements for lethal and sterilizing doses. Application of a given dose in a number of fractional exposures causes greater radiation tolerance by providing opportunity for repair of radiation injury between exposures. The same effect can be obtained with very low dose rates. Very high dose rates also increase the tolerance, possibly by reduction of the oxygen content of the tissues.

As a new approach to insect disinfestation of foods, irradiation must be viewed in relation to the established patterns of disinfestation associated with chemical pesticides. Some pesticides act very quickly. To match this speed with irradiation would require very large doses, which generally would be unsuitable from a number of viewpoints. For this reason radiation distinfestation usually is a slower process than that obtained with chemicals.

Therefore, as part of dose specification there is a need to establish the "lethal time." This second irradiation parameter can have important meaning for some applications of radiation disinfestation of foods.

Lethality is not always a necessary effect in radiation disinfestation. In quarantine applications, for example, the important requirement is prevention of survivors arriving at the shipping destination which have the capability of causing infestation. This requirement can be met by using doses that effectively stop reproduction.

Since the dose needed to secure a particular effect is partly dependent upon the size of the population of insects being treated, a food should be given a dose appropriate to its infestation level. Some foods as they move in commerce carry a low level of infestation, and this may provide opportunity for reducing the dose used.

Dose levels for inactivation of selected insects are shown in Tables 4.7 and 4.8. Applied to eggs and larvae, doses of 130 to 250 Gy prevent development to the adult stage. Pupae treated with 1000 Gy do not develop into adults. Sexual sterility is produced in beetles (Coleoptera) with doses of 130 to 250 Gy, but moths (Lepidoptera) require 450 to 1000 Gy. The mite, *Acarus siro* also included in Table 4.8, requires an intermediate dose level of 250 to 450 Gy for sexual sterility.

Table 4.9 provides the time for 99.9% inactivation of a number of grain insects. Immediate lethality requires 3–5 kGy. With doses of about 500 Gy or greater any survivors would be sterile.

Multiple exposures of insects to substerilizing doses do not increase radiation resistance.

III. PLANT FOODS

A. Introduction

Fresh plant foods—in particular fruits, vegetables, cereal grains, and legumes—contain intact biological systems that maintain activity such as respiration after harvest. Unless fruit and vegetables are treated in some manner to limit this activity, changes occur which result in deterioration of the food and which make it unacceptable. Irradiation of such living foods can affect the biological processes involved, sometimes in a beneficial way.

B. Fruits

Fruits can be classified as climacteric or nonclimacteric. Climacteric fruits, picked before the ripe stage, exhibit a declining respiratory rate until ripening

Table 4.7

Minimum Dosage of γ Radiation Completely Arresting Development of Stored-Product Insects and a Mite[a]

Organism	Stage exposed	Stage observed for effect	Dosage (Gy) 132	175	250	400	1000
Lepidoptera							
Plodia interpunctella	Egg	Larva				X···	···X
	Egg	Adult	X···	···X			
	Larva	Pupa					>X
	Larva	Adult	X				
	Pupa	Adult					>X
Sitotroga cerealella	Egg	Larva					>X
	Egg	Adult	X···	···X			
	Larva	Adult	X···	···X			
	Pupa	Adult					>X
Coleoptera							
Tribolium confusum	Egg	Larva					>X
	Larva	Larva[b]	X				
	Larva	Pupa					>X
	Larva	Adult	X[c]				
	Pupa	Adult					>X
Lasioderma serricorne	Egg	Larva					>X
	Larva	Larva[b]				X···	···X
	Larva	Pupa					>X
	Larva	Adult	X[d]				
	Pupa	Adult					>X
Rhyzopertha dominica	Larva	Pupa					>X
	Larva	Adult		X···	···X[e]		
	Pupa	Adult					>X
Attagenus piceus	Egg	Larva					>X
	Larva	Larva[b]			X···	···X	
	Larva	Pupa	X				
	Larva	Adult	X				
	Pupa	Adult				X···	···X
Trogoderma glabrum	Egg	Larva				X···	···X
	Larva	Larva[b]		X···	···X		
	Larva	Pupa	X				
	Larva	Adult	X				
	Pupa	Adult				X···	···X
Acarina							
Acarus siro	Egg	Larva					>X
	Larva	Adult		X···	···X		
	Hypopus	Adult				X···	···X

[a] From G. Golumbic and D. F. Davis, Radiation disinfestation of grain and seeds. *In* "Food Irradiation, Proceedings of Symposium, Karlsruhe." Int. Atomic Energy Agency, Vienna, 1966.

[b] Based on data obtained 21 days after exposure.

[c] One survivor after exposure to 450 Gy.

[d] One survivor after exposure to 1 kGy.

[e] One survivor after exposure to 450 Gy.

Table 4.8
Minimum Dosage of γ Radiation Producing Sterility in 100% Exposed Adult Stored-Product Insects and a Mite[a]

		Dosage (Gy)				
Organism	Sex exposed	132	175	250	450	1000
Lepidoptera						
Plodia interpunctella	M					>X
	F					>X
Sitotroga cerealella	M					>X
	F				X··········	X
Coleoptera						
Tribolium confusum	M and F	X··········	X			
Lasioderma serricorne	M and F		X··········	X		
Attagenus piceus	M	X··········	X			
	F	X··········	X			
Trogoderma glabrum	M		X··········	X		
	F	X				
Acarina						
Acarus siro	M and F			X··········	X	

[a] From G. Golumic and D. F. Davis, Radiation disinfestation of grain and seeds. *In* "Food Irradiation, "Proceedings of Symposium, Karlsruhe." Int. Atomic Energy Agency, Vienna, 1966.

begins. Respiration then rises and peaks at the ripe stage, the peak coinciding with other ripeness indexes such as color and texture. As senescence sets in, respiration declines. Ripening is associated with ethylene production, but whether this causes ripening or is an effect of it is not known. Nonclimacteric fruits exhibit a steady decline in respiration rate after harvest. At harvest they should be fully ripe.

The principal interest in the use of radiation to alter the biological processes of fruits is to delay ripening and/or senescence, and in this manner, to extend their market life. A delay in ripening can be secured only with climacteric fruits, but delayed senescence can be secured with both climacteric and nonclimacteric fruits.

Irradiation produces results that vary with the fruit and the dose applied. Irradiation can increase respiration rate and also cause ethylene production. The exact effect, however, is dependent upon the dose used, as is illustrated in Figs. 4.12 and 4.13. By irradiation in the preclimacteric stage ripening is delayed, but normal ripening occurs only with a suitable dose. In the case of 'Bartlett' pears, for example, doses greater than 2000 Gy (200 Krad) affect texture and flavor adversely.

The manner in which radiation delays ripening and/or senescence of fruits

Table 4.9

Days for 99.9% Inactivation of Grain Insects[a]

Insect	Radiation dose (Gy)											
	100	300	400	500	600	700	800	1000	1500	2000	3000	4000
Calandra granaria L. (grain weevil)	17	16	—	15	15	16	12	6	—	—	—	—
Calandra oryzae L. (rice weevil)	—	—	—	—	16	13	12	4	—	—	—	—
Laemophiloeus testaceus L. (grain beetle)	—	—	7	—	—	5	—	—	—	—	—	—
Tribolium confusum Duv. (confused flour beetle)	24	23	—	20	—	18	—	15	10	—	—	—
T. confusum, larva	26	23	—	20	—	—	—	14	10	—	—	—
Acanthoscelides obsoletus Say. (bean weevil)	—	—	—	—	—	13	—	9	—	—	—	—
Tyroglyphus farinae L. (flour mite)	—	—	—	—	—	—	—	—	—	14	10	3

[a] From L. V. Metlitskii, V. N. Rogachev, and V. G. Krushchev, "Radiation Processing of Food Products," ORNL-IIC-14. Translation of Russian publication. U.S. Atomic Energy Comm., Oak Ridge, Tennessee, 1968.

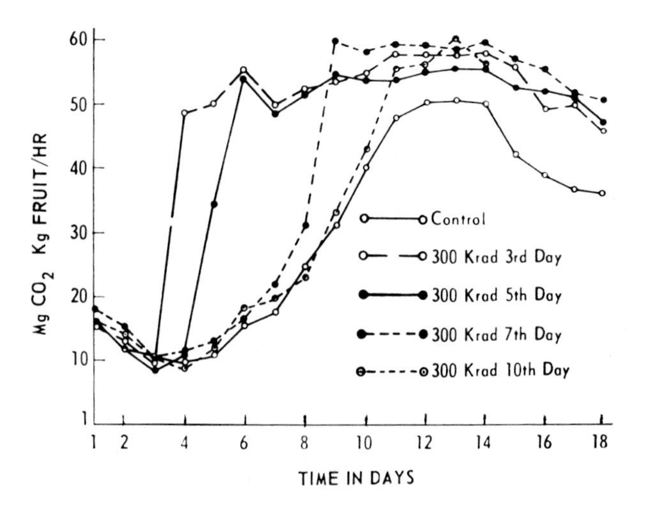

Figure 4.12 Effect of γ irradiation at various stages of the climacteric on the respiratory rate of 'Bartlett' pears. From E. C. Maxie and A. Adel-Kader, *Adv. Food Res.* **15,** 105 (1966) and E. C. Maxie *et al.*, *Plant Physiology* **41.** Copyright (1966) American Society of Plant Physiologists.

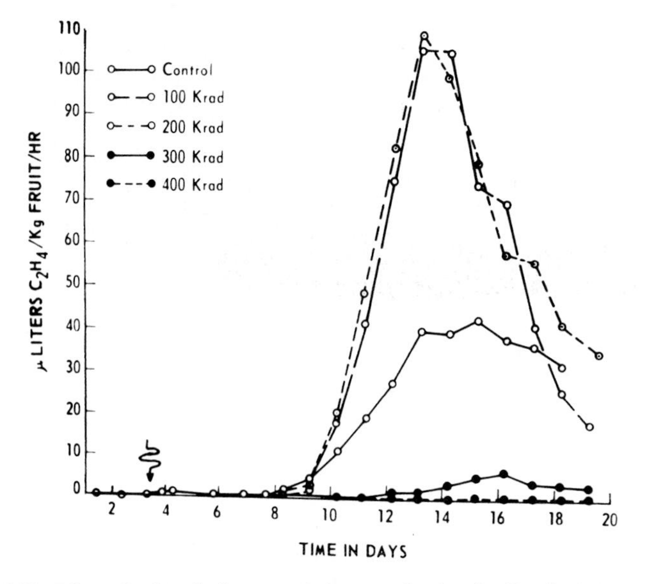

Figure 4.13 Effect of γ irradiation on ethylene production by 'Bartlett' pears. From E. C. Maxie and A. Adel-Kader, *Adv. Food Res.* **15,** 105 (1966) and E. C. Maxie *et al.*, *Plant Physiology* **41.** Copyright (1966) American Society of Plant Physiologists

is not well understood and for this reason, the response of each kind (and variety) of fruit must be examined on an individual basis (see Chapter 8). Knowledge of the postharvest physiology and biochemistry of the fruit is important for the intelligent use of irradiation for this purpose.

Irradiation can also cause compositional changes in fruits. Depolymerization of the structural materials cellulose, hemicellulose, protopectin, and starch can affect texture, often in such a way as to yield unacceptable softening of the fruit. Reduction of acidity may occur. Of the vitamins, only ascorbic acid undergoes sufficient degradation possibly to affect the nutritive value of some fruits. Pigment changes, with some exceptions, generally are negligible. In some cases flavor changes occur.

C. Vegetables

The principal interest in irradiation of raw vegetables is to preserve the optimum quality condition that exists at harvest. This can be lost through the processes of senescence. Of all the vegetables, only a few have been of interest in food irradiation. The largest group are those that are either bulbs, roots, or tubers, such as potatoes or onions. Examples of other vegetables are asparagus and mushrooms.

Sprouting of bulbs, roots, and tubers can be prevented by very low doses in the range of about 20 to 150 Gy. The mechanism of the action of radiation is complex, probably involving factors affecting cell division and growth in "growing points" (meristematic tissues) of bulbs. Suppression of the synthesis of adenosine triphosphoric acid (ATP) and related compounds and of nucleic acids in these tissues by radiation halts all division and growth. The sprouting inhibition is generally irreversible, but can be reversed by treatment with auxins (plant growth hormones) such as indole acetic acid and gibberellic acid.

Irradiation causes variations in the respiration rate and the sugar content of white potatoes. Greening and solanine formation are also inhibited in white potatoes. In order to be effective, irradiation is best done in the dormancy period. Inadvertent effects of irradiation may occur. For example, irradiation also inactivates the wound-healing process of white potatoes. In order to avoid spoilage due to microbial invasion of the potato by entry through wounds, irradiation must be delayed until wounds caused during harvesting have healed.

Irradiation inhibits elongation and curvature of asparagus spears. The opening of the cap and stalk lengthening of mushrooms are inhibited by irradiation. The mechanism of these effects is not understood, but interference with cell division or growth seems likely.

D. Cereal Grains and Legumes

The irradiation of foods which are seeds, such as cereal grains and legumes, affects the viability of the embryo or germ. In various ways this effect can be utilized to advantage (see Chapter 9).

SOURCES OF ADDITIONAL INFORMATION

General

Altman, K. I., Gerber, G. B., and Okada, S., "Radiation Biochemistry." Academic Press, New York, 1970.'

Anonymous, "Radiation Preservation of Food." U.S. Army Quartermaster Corps. U.S. Government Printing Office, Washington, D.C., 1957.

Anonymous, "Elimination of Harmful Organisms from Food and Feed by Irradiation," Proceedings of a Panel. Int. Atomic Energy Agency, Vienna, 1968.

Anonymous, "Training Manual on Food Irradiation Technology and Techniques," 2nd Ed., Tech. Rep. Series No. 114. Int. Atomic Energy Agency, Vienna, 1982.

Bacq, Z. M., and Alexander, P., "Fundamentals of Radiobiology," 2nd Ed. Pergamon, Oxford, 1961.

Casarett, A. P., "Radiation Biology." Prentice-Hall, Englewood Cliffs, New Jersey, 1968.

Ingram, M., and Roberts, T. A., Microbiological principles in food irradiation. *In* "Food Irradiation, Proceedings of Symposium, Karlsruhe." Int. Atomic Energy Agency, Vienna, 1966.

Kuzin, A. M., "Radiation Biochemistry." Translation of Russian publication. Daniel Davey & Co., Inc., New York, 1964.

Lea, D. E., "Actions of Radiation on Living Cells," 2nd Ed. University Press, Cambridge, 1962.

Pollard, E. C., Effect of radiation at the cellular and tissue level. *In* "Preservation of Food by Ionizing Radiation," (E. S. Josephson and M. S. Peterson, eds.), Vol. II. CRC Press, Boca Raton, Florida, 1983.

Viruses

Grecz, N., Rowley, D. B., and Matsuyama, A., The action of radiation on bacteria and viruses. *In* "Preservation of Food by Ionizing Radiation," (E. S. Josephson and M. S. Peterson, eds.), Vol. II. CRC Press, Boca Raton, Florida, 1983.

Johnson, C. D., In-vitro experiments on the radiosensitivity of foot-and-mouth disease virus and other animal viruses to the direct effect of X-irradiation. *In* "Elimination of Harmful Organisms from Food and Feed by Irradiation. "Int. Atomic Energy Agency, Vienna, 1968.

Massa, D., Radiation inactivation of foot-and-mouth disease virus in the blood, lymphatic glands and bone marrow of the carcasses of infected animals. *In* "Food Irradiation, Proceedings of Symposium, Karlsruhe." Int. Atomic Energy Agency, Vienna, 1966.

Massa, D., Radiation inactivation of foot-and-mouth disease virus. *In* "Application of Food Irradiation in Developing Countries," Tech. Rep. Series No. 54. Int. Atomic Energy Agency. Vienna, 1966.

Pollard, E. C., The effect of ionizing radiation on viruses. *In* "Manual on Radiation Sterilization of Medical and Biological Materials." Int. Atomic Energy Agency, Vienna, 1973.

Bacteria

Anonymous, "Radiation control of salmonellae in food and feed products," Tech. Rep. Series No. 22. Int. Atomic Energy Agency, Vienna, 1963.

Anonymous, "Microbiological problems in food preservation by irradiation, proceedings of a Panel." Int. Atomic Energy Agency, Vienna, 1967.

Anonymous, "Microbiological specifications and testing methods for irradiated food," Tech. Rep. Series No. 104. Int. Atomic Energy Agency, Vienna, 1970.

Grecz, N., Rowley, D. B., and Matsuyama, A., The action of radiation on bacteria and viruses. *In* "Preservation of Food by Ionizing Radiation" (E. S. Josephson and M. S. Peterson, eds.), Vol. II. CRC Press, Boca Raton, Florida, 1983.

Ingram, M., "Microbiology of foods pasturized by ionizing radiation," Tech. Rep. Series IFIP-R33. International Project in the Field of Food Irradiation, Karlsruhe, W. Germany, 1975.

Ley, F. J., The effect of ionizing radiation on bacteria. *In* "Manual on Radiation Sterilization of Medical and Biological Materials," Tech. Rep. Series No. 149. Int. Atomic Energy Agency, Vienna, 1973.

Teufel, P., "Microbiological implications of the food irradiation process," Food Irradiation Information No. 11. International Project in the Field of Food Irradiation, Karlsruhe, W. Germany, 1981.

Yeasts and Molds

Saravacos, G. D., Hatzipetrou, L. P., and Georgiadou, E., Lethal doses of gamma radiation of some fruit spoilage microorganisms. *Food Irradiat. (Saclay)* **3**(1-2), A6 (1962).

Sommer, N., The effect of ionizing radiation on fungi. *In* "Manual on Radiation Sterilization of Medical and Biological Materials," Tech. Rep. Series No. 149. Int. Atomic Energy Agency, Vienna, 1973.

Sommer, N. F., and Fortlage, R. J., Ionizing radiation for control of postharvest diseases of fruits and vegetables. *Adv. Food Res.* **15,** 147 (1966).

Parasites

Brake, R. J., Murrel, K. D., Ray, E. E., Thomas, J. D., Muggenberg, B. A., and Sivinski, J. S., Control of trichinosis by low-dose irradiation of pork. *J. Food Saf.* **7,** 127 (1985).

King, F. L., and Josephson, E. S., Action of radiation on protozoa and helminths. *In* "Preservation of Food by Ionizing Radiation" (E. S. Josephson and M. S. Peterson, eds.), Vol. II. CRC Press, Boca Raton, Florida, 1983.

van Kooij, J. G., and Robijns, K. G., Devitalization of cysticerci by gamma radiation. *In* "Elimination of Harmful Organisms from Food and Feed by Irradiation." Int. Atomic Energy Agency, Vienna, 1968.

van Mameren, J., and Houwing, H., Effect of irradiation on Anisakis larvae in salted herring. *In* "Elimination of Harmful Organisms from Food and Feed by Irradiation." Int. Atomic Energy Agency, Vienna, 1968.

Varga, I., The effect of ionizing radiation on animal parasites. *In* "Manual on Radiation Sterilization of Medical and Biological Materials," Tech. Rep. Series No. 149. Int. Atomic Energy Agency, Vienna, 1973.

Insects

Anonymous, "Disinfestation of Fruit by Irradiation." Int. Atomic Energy Agency, Vienna, 1971.

Anonymous, "Irradiation of Grain and Grain Products for Insect Control." Council for Agricultural Science and Technology, Ames, Iowa, 1984.

Anonymous, "Use of Irradiation as a Quarantine Treatment of Agricultural Commodities," IAEA-TECDOC-326. Int. Atomic Energy Agency, Vienna, 1985.

Cornwell, P. B., ed. "The Entomology of Radiation Disinfestation of Grain." Pergamon, Oxford, 1966.

Metlitskii, L. V., Rogachev, V. N., and Krushchev, V. G., "Radiation Processing of Food Products," ORNL-IIC-14. Translation of Russian publication. U.S. Atomic Energy Comm., Oak Ridge, Tennessee, 1968.

Moy, J. H., ed. "Radiation Disinfestation of Food and Agricultural Products." Univ. of Hawaii Press, Honolulu, 1985.

Tilton, E. W., and Brower, J. H., Radiation effects on arthropods. *In* "Preservation of Food by Ionizing Radiation," (E. S. Josephson and M. S. Peterson, eds.), Vol. II. CRC Press, Boca Raton, Florida, 1983.

Raw Plant Foods

Matsuyama, A., and Umeda, K., Sprout inhibition in tubers and bulbs. *In* "Preservation of Food by Ionizing Radiation," (E. S. Josephson and M. S. Peterson, eds.), Vol. III. CRC Press, Boca Raton, Florida, 1983.

Maxie, E. C., and Abdel-Kader, A., Food Irradiation—Physiology of fruits as related to feasibility of the technology. *Adv. Food Res.* **15,** 105 (1966).

Metlitskii, L. V., Rogachev, V. N., and Krushchev, V. G., "Radiation Processing of Food Products," ORNL-IIC-14. Translation of Russian publication. U.S. Atomic Energy Comm., Oak Ridge, Tennessee, 1968.

Romani, R. J., Radiobiological parameters in the irradiation of fruits and vegetables. *Adv. Food Res.* **15,** 57 (1966).

CHAPTER 5

General Effects of Ionizing Radiation on Foods

I. INTRODUCTION

The principal interest in treating foods with ionizing radiation is to secure their preservation. This can involve inactivation of several kinds of microorganisms that may contaminate foods and cause spoilage. A second interest in irradiation concerns inactivation of foodborne pathogenic microorganisms. A third interest, applicable to fresh fruits and vegetables, is to delay ripening or senescence or to inhibit sprouting. A fourth purpose in irradiating foods is to secure decontamination or disinfestation with regard to bacteria, yeasts, molds, and insects. A fifth reason for irradiating foods is to secure a chemical change in the food itself when such a change improves some characteristic of the food, or of its processing.

Because of excessive dose requirements, food irradiation is not used to inactivate foodborne viruses or indigenous enzymes. It must be recognized also that irradiation does not directly affect spoilage of foods which is due to chemical or physical action on the food caused by environmental agents other than those associated with living organisms. For example, irradiation cannot be used to control degradation of a food due to the chemical action of atmospheric O_2.

The general uses of food irradiation, as outlined above, can be designated more specifically as described in the following sections.

II. CONTROL OF MICROBIAL SPOILAGE

Food spoilage caused by bacteria, yeasts, and molds can be controlled by inactivation of either some or all of the spoilage microorganisms present.

A. Inactivation of All Spoilage Microorganisms Present

Such total inactivation is sterilization and, when radiation is used, the process has been termed *radappertization.* Radappertization is

> Treatment of food with a dose of ionizing radiation sufficient to reduce the number and/or activity of viable microorganisms to such a level that very few, if any, are detectable by any recognized bacteriological or mycological testing method applied to the treated food. The treatment must be such that no spoilage or toxicity of microbial origin is detectable no matter how long or under what conditions the food is stored after treatment, provided it is not recontaminated.

Radappertization does not include inactivation of viruses or of bacterial toxins or mycotoxins or of enzymes.

Properly packaged radappertized foods will keep indefinitely without refrigeration. The dose needed to secure sterility is the highest employed in food irradiation, and can be as large as about 50 kGy. To a considerable degree, the exact dose required is governed by the composition of the food. Low-acid foods (pH >4.5) or foods which do not have appropriate levels of salt or nitrite require a dose sufficient to inactivate the spores of *Clostridium botulinum.*

Since the spores of *Clostridium botulinum* are the most radiation-resistant organisms of significance encountered in foods, doses used to secure their inactivation are sufficient to inactivate other spoilage bacteria. Foods which are not low in acid or low in salt or which contain adequate amounts of nitrite can be radappertized without regard to the presence of spores of *C. botulinum* and consequently may be stabilized with smaller doses, provided inactivation of other spoilage microorganisms is secured.

Most of the interest in radappertized foods has been limited to meats and to certain fishery products. These foods tolerate the high radappertization doses applied in the frozen state, without undergoing significant undesirable sensory changes.

B. Inactivation of Some of the Spoilage Microorganisms Present

This is a treatment comparable in effect with heat pasteurization and is termed *radurization*. Radurization is

> Treatment of food with a dose of ionizing radiation sufficient to enhance its keeping quality by causing a substantial reduction in the numbers of viable specific spoilage microorganisms.

The process of radurization is usually added to whatever other handling of a food is normally practiced. Radurized fresh fishery products, for example, are refrigerated. If the product is held long enough, spoilage will occur, since in time the reduced microbial population will build back and ultimately reach numbers characteristic of spoilage. The purpose of radurization, therefore, is simply to gain an extra time period before spoilage occurs and that is useful in the distribution of the food. Doses employed vary with the food and with the kind and number of spoilage microorganisms present. Radurization has been most promising with fresh meats, poultry, and fishery products. Some possibilities exist with fresh fruits and vegetables, but often the sensitivity of these foods to radiation injury at the doses required for radurization limits such use. Doses below 10 kGy, and particularly in the 1- to 5-kGy range, have been found appropriate for radurization.

III. INACTIVATION OF PATHOGENIC NON-SPORE-FORMING BACTERIA

This use of irradiation is termed *radicidation*. Radicidation is

> Treatment of food with a dose of ionizing radiation sufficient to reduce the number of viable specific non-spore-forming pathogenic bacteria to such a level that none is detectable in the treated food when it is examined by any recognized bacteriological testing method.

The quoted material regarding the definitions of radappertization, radurization, and radicidation is taken from "Wholesomeness of Irradiated Food," Series 604, World Health Organization (WHO), Geneva, 1977, page 42. Although these definitions have the sanction of WHO, the term radicidation logically can be extended to include control of toxigenic fungi and to viruses and parasites. In this book, in accord with the usage of other authors, the term radicidation is used to include these organisms in addition to pathogenic non-spore-forming bacteria.

Among the pathogenic bacteria, greatest attention has been directed to the use of irradiation to inactivate salmonellae. Salmonellae contaminate many foods including eggs, poultry, red meats, dairy products, fishery products, frog legs, coconut products, chocolate, and certain spices and vegetables. They also occur in animal feeds which contain products of animal origin such as fish meal and animal by-products.

Doses to control salmonellae contaminations are based upon what is needed to secure a sufficient reduction of the numbers present in order to protect the consumer from salmonellosis. Depending upon the food and the expected level of the salmonellae contamination, reductions of from three- to seven-fold have been estimated as necessary. Doses in the range 2–6.5 kGy have been proposed.

Other pathogenic non-spore-forming foodborne bacteria that may be inactivated by irradiation include *Shigella*, *Neisseria*, *Mycobacterium*, *Escherichia*, *Proteus*, *Streptococcus*, and *Staphylococcus*. It has been estimated that a seven-fold reduction of counts of these organisms would require a dose in the range 5–8 kGy. Additions to the list of pathogenic bacteria that may be considered in connection with radiation include *Vibrio parahaemolyticus*, *Yersinia enterocolitica*, *Campylobacter jejuni*, and *Vibrio cholerae*.

IV. INHIBITION OF SPROUTING AND DELAY OF RIPENING AND SENESCENCE

This use of irradiation is limited to fresh fruits and vegetables. Although it is not concerned with foodborne organisms, it is food preservation, since the market life of these plant foods is extended. Doses fall in the range of about 0.02 to 3 kGy.

V. DECONTAMINATION

This use of irradiation has as its objective the reduction of a microbial population, but not for the purpose of preservation of the food or food material. It may be, for example, the use of irradiation to reduce the microbial population of food product ingredient materials such as spices, vegetable seasonings, starches, and gums. Into this category falls the use of irradiation to obtain essentially sterile diet items for certain hospital patients. In all such cases, preservation through irradiation is not required and normally is obtained by other measures. True sterility likewise is not needed, but the objective is to secure a very low microbial count. In addition to the designation of this use of irradiation as decontamination, the term *hygienization* has been suggested.

Doses in the range 3–30 kGy may be employed.

VI. INSECT DISINFESTATION

The presence of insects in foods is undesirable for at least three reasons: (1) resulting insect damage to the foods, including food losses or reduction of nutritive value, (2) consumer objection to the presence of insects on the basis of aesthetic or other reasons, (3) in the distribution of foods, the spreading of insects from one location to another. The term used to designate the irradiation of foods to control insects is *radiation insect disinfestation.* Doses for insect disinfestation are in the range of about 0.1 to 1 kGy. Such doses do not necessarily cause immediate lethality.

VII. QUALITY IMPROVEMENT

In general the obtaining of an improvement of the quality of a food through irradiation is based upon a chemical change that alters one or more important characteristics of the food. The altered characteristics can be of a *sensory nature*, such as a textural change, or can relate to a *functional property* such as baking performance, or it can improve the *processing* of a food, such as increased drying rate. In most cases one of two kinds of chemical changes is involved: (1) depolymerization of large molecules such as carbohydrates or proteins, or (2) cell injury with consequent easier release of cell contents.

In some cases chemical change in a food can be brought about by using radiation to alter a biological process normal to the food in such a way as to result in changes in the food that are different from the normal or are limited in amount. An example of this use is the effect of irradiation of seed foods to alter the normal germination process. As the seed germinates, the metabolic processes change its composition. Irradiation changes the metabolic processes and in this manner causes compositional changes that are different from the normal.

Doses needed for particular quality improvement effects usually are very specific and are likely to fall in the range 0.1–10 kGy.

VIII. DOSE CATEGORIES

For convenience it has become a practice to group uses of food irradiation according to the size of the dose employed. The commonly accepted groupings for the various types of applications are as follows:

Low dose: up to ~1 kGy
- Inhibition of sprouting
- Delay of ripening or of senescence
- Insect disinfestation
- Quality improvement

Medium dose: 1–10 kGy
  Radurization
  Radicidation
  Delay of ripening or of senescence
  Decontamination
  Quality improvement

High dose: 10–50 kGy
  Radappertization
  Decontamination

SOURCES OF ADDITIONAL INFORMATION

Anonymous, "Radiation Preservation of Food." U.S. Army Quartermaster Corps. U.S. Government Printing Office, Washington, D.C., 1957.

Anonymous, "Training Manual on Food Irradiation Technology and Techniques," 2nd Ed. Int. Atomic Energy Agency, Vienna, 1982.

Anonymous, "Food Irradiation Now—Proceedings of a Symposium," Nijhoff/Junk, The Hague, 1982.

Brownell, L. E., "Radiation Uses in Industry and Science." U.S. Atomic Energy Comm., Washington, D.C., 1961.

Cohen, J. S., and Mason, V. C., "Radappertization (Radiation Sterilization) of Foods," Tech. Rep. TR-77/009. A bibliography. U.S. Army Natick Res. and Dev. Command, Natick, Massachusetts, 1976.

Desrosier, N. W., and Rosenstock, H. M., "Radiation Technology in Food, Agriculture and Biology." AVI, Westport, Connecticut, 1960.

Goresline, H. E., Ingram, M., Macuch, P., Mocquot, G., Mossel, D. A. A., Niven, C. F., Jr., and Thatcher, F. S., Tentative classification of food irradiation processes with microbiological objectives. *Nature* (*London*) **204,** 237 (1964).

Hannan, R. S., "Scientific and Technological Problems Involved in Using Ionizing Radiations for the Preservation of Food." Her Majesty's Stationery Office, London, 1955.

Josephson, E. S., and Peterson, M. S., eds. "Preservation of Food by Ionizing Radiation," Vols. I, II, III. CRC Press, Boca Raton, Florida, 1982–1983.

Kampelmacher, E. H., Prospects of eliminating pathogens by the process of food irradiation. *In* "Combination Processes in Food Irradiation." Int. Atomic Energy Agency, Vienna, 1981.

MacQueen, K. F., "Report on Food Irradiation." Canadian International Development Agency, Quebec, Canada, 1984.

Metlitskii, L. V., Rogachev, V. N., and Krushchev, V. G., "Radiation Processing of Food Products," ORNL-IIC-14. Translation of Russian publication. U.S. Atomic Energy Comm., Oak Ridge, Tennessee, 1968.

Urbain, W. M., Food Irradiation (Review). *Adv. Food Res.* **24,** 155 (1978).

Wierbicki, E., "Radappertization (Radiation Sterilization) of Foods," Tech. Rep. 74-3FL. A bibliography. U.S. Army Natick Labs., Natick, Massachusetts, 1974.

CHAPTER 6

Meats and Poultry

I. GENERAL

The perishability of meats and poultry has led to the development over many millennia of a number of methods for their preservation. Today the principal preservation method is the use of refrigeration. Most meats and poultry reach the consumer as "fresh"; that is, except for cutting and packaging, they are not processed and are preserved simply by being kept cold for the time period between preparation and cooking. Meat curing, a procedure of adding salt and other materials, originally was a preservation process and cured meats did not require refrigeration. Curing often was combined with smoking and drying, which provided additional preservation. Today curing has been modified, mainly by reducing the salt level, and refrigeration is now required for the preservation of cured meats. A few cured and dried sausages, however, still do not need refrigeration. Preservation of meats by thermal canning is practiced today, but for various reasons, is used with only a relatively small fraction of the total volume of meats and poultry. A very small amount of meat and poultry is frozen uncooked.

The spoilage of meats and poultry occurs by a number of pathways. The principal spoilage process is through microbial action, which with today's practices is almost entirely bacterial. Refrigeration, freezing, curing, and canning all are directed primarily at controlling microbial spoilage.

Lipid oxidation can take place when atmospheric O_2 is available, and it can cause rancid flavors. Oxygen can also oxidize the meat pigments, causing discolorations. The red color of fresh meats is due to the pigment oxymyoglobin, which is a loose combination of O_2 and myoglobin, a purple pigment. In order to have a red color, fresh meats must have O_2 available, but true oxidation can occur with time and cause the formation of a brown pigment, ferrimyoglobin. The red color of cured meats is due to the pigment nitric oxide myoglobin (or its denatured derivative), which is formed when sodium nitrite is used in the curing process. It remains red in the absence of O_2. Oxygen also oxidizes the cured meat pigment, especially in the presence of light. In terms of the marketing requirements for meats and poultry, their discoloration is objectionable and is a form of "spoilage," although it is not microbial.

Muscle tissue contains about 70% water. Over a period of time at refrigeration temperatures the cut surfaces of fresh meats and poultry exude a serum-like material, which may be colored with meat pigment. This "weep" or "drip" may be unsightly and, especially with packaged meats, is objectionable. It also can be considered a form of meat "spoilage."

Canning of meats requires processing temperatures to secure stability without refrigeration that are higher than those used for table preparation. This has consequences that put limits on the use of canning in preserving meats. Because of the slow transfer of heat, large cuts of meat are overcooked and display texture degradation. Additionally, release of liquid from the meat into the container can occur, and this is objectionable. Because canning involves wet-cooking, many canned meats do not have the most desirable flavor quality.

Despite the several pathways by which meats "spoil," irradiation can contribute to their preservation *only* by the inactivation of spoilage microorganisms. For "spoilage" due to oxidation and weepage, other control procedures are needed. As an agent for controlling microbial spoilage, irradiation has some important features. Irradiation is a low-energy process and produces only small temperature rises (50 kGy will cause about a 12°C rise in water temperature). As a consequence, raw meat that is irradiated remains raw and, in a gross way, is not noticeably changed. γ rays easily penetrate thicknesses of meat likely to be encountered in food irradiation. This penetration has no time lag, and when the irradiation is done properly, the distribution of energy within a large piece of meat is quite uniform. This is very much different from the uneven heating that occurs in canning. In some canning processes for meats (e.g., bone-in whole chicken), water is added to assist heat transfer. Irradiation does not require this liquid. Dry or irregular-shaped cuts can be sterilized by irradiation, which is not possible with thermal canning.

While it is clear that irradiation offers a number of opportunities for improving the preservation of meats, it must be recognized that it, too, has limitations. Due to excessive dose requirements, it cannot be used to inactivate meatborne viruses and endogenous enzymes. The radiation inactivation of meatborne parasites may or may not be feasible, depending upon circumstances. A somewhat similar situation may exist with the inactivation of pathogenic bacteria carried by meats and poultry.

Table 6.1

Threshold Dose for Various Animal Protein Foods at Which a Slight "Irradiation Flavor" Is Detected[a]

Animal food	Threshold dose (kGy)	Animal food	Threshold dose (kGy)
Turkey	1.50	Turtle	4.50
Pork	1.75	Halibut	5.00
Beef	2.50	Opossum	5.00
Chicken	2.50	Hippopotamus	5.25
Lobster	2.50	Beaver	5.50
Shrimp	2.50	Lamb	6.25
Rabbit	3.50	Venison	6.25
Frog	4.00	Elephant	6.50
Whale	4.00	Horse	6.50
Trout	4.50	Bear	8.75

[a] From S. Sudarmadji and W. M. Urbain, *J. Food Sci.* **37,** 671 (1972). Copyright © by Institute of Food Technologists.

The principal difficulty with the irradiation of meats and poultry has been the development of an undesirable flavor. It resembles scorching, but has been referred to in other more descriptive terms, such as "goaty" or "wet dog." Its formation is dose dependent, the threshold dose for a detectable flavor development varying with the meat species. It is not unique to meats in that a number of other foods also develop it. Table 6.1 lists the "irradiated flavor" threshold doses for a number of protein foods derived from various animals.

Despite many efforts to identify the chemical nature of the "irradiated flavor," this has not yet been done. It has been suggested (Wick *et al.*, 1967) that the following mixture added to a beef slurry approximates the irradiation flavor:

Methional	3.0 ppm
1-Nonanal	0.3 ppm
Phenylacetaldehyde	0.15 ppm

Reference to Table 6.1 indicates that a detectable flavor occurs with relatively low doses in pork, beef, and chicken, when irradiation is done at 5 to 10°C. When the irradiation is done at temperatures below 0°C, however, the flavor development is markedly reduced. This is shown in Table 6.2.

Using gas chromatography and mass spectroscopy, a number of volatile radiolytic products have been identified in radappertized meats. Table 3.8 lists compounds found in beef which was irradiated, thermally sterilized, or frozen. Although it seems probable that the substances causing the irradiated flavor

Table 6.2
Effect of Irradiation Temperature on Flavor of Beef[a]

Irradiation temperature (°C)	Flavor[b]
60	4.1
21	3.3
−40	2.9
−80	2.1
−185	1.5

[a] From G. W. Shults and E. Wierbicki, "Development of Irradiated Beef. 1. Acceptance of Beef Loin Irradiated at Cryogenic Temperatures," Tech. Rep. 74-57-FL. U.S. Army Natick Labs., Natick, Massachusetts, 1974.
[b] Irradiation flavor intensity scale: 1, none; 9, extreme.

are to be found among the radiolytic products listed, it is not known which they are. The data of Fig. 6.1 indicate that the amount of radiolytic products formed and the detectable flavor decrease as the irradiation temperature is lowered. The relationship between flavor intensity and temperature indicates that the flavor formation is due to indirect action of the radiation. Lowering the temperature reduces the interaction of the primary radiolytic products formed by direct action with other meat components or among themselves. The nature and the probable origin of the radiolytic substances listed in Table 3.8, suggests that this effect occurs in both the aqueous and lipid portions of the beef. Table 3.9 lists the volatile radiolytic products formed in protein, lipid, and lipoprotein meat substances.

It is recognized that the development of the irradiated flavor in meats and poultry is a serious limitation on doses that can be employed. In situations where doses above the threshold levels for a detectable flavor are needed, the only solution available is to carry out the irradiation at subfreezing temperatures. The actual irradiation temperature required is related to dose, as can be deduced from the data of Figs. 3.8 and 6.1. What is needed for a given meat is a dose–temperature combination that yields an acceptable flavor and at the same time affects the microbial population in the intended manner.

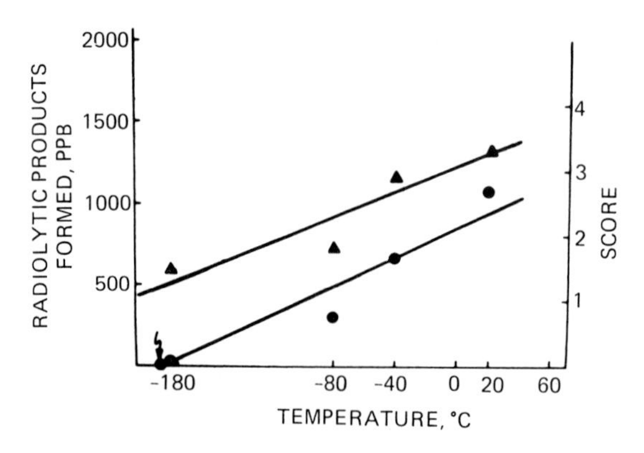

Figure 6.1 Relationship of flavor score (▲) of beef with amounts of volatile radiolytic products formed (●) at various temperatures with a dose of 56 kGy. Arrow denotes value for nonirradiated beef. Flavor intensity score: 1, none, 9, extreme. From C. W. Merritt, Jr., P. Angelini, E. Wierbicki, and G. W. Shults, *J. Agric. Food Chem.* **23,** 1037 (1975). Copyright (1975) American Chemical Society.

II. RADURIZATION

The purpose of radurization of meats and poultry is to obtain an extension of their market life through reduction of the initial microbial population. In this initial microbial population there may be present bacteria, yeasts, and molds, but generally the primary consideration is with bacterial spoilage. Radurization is not a replacement for other preservation methods but is used in addition to the customary procedures, including especially refrigeration. Radurization can be applicable to both fresh and processed meats.

Except for certain dry sausages, processed meats, as presently manufactured, require refrigeration. As part of their processing, usually they are cured and/or cooked. Usually they are packaged into retail units at the packing plant. Since their normal color does not require the availability of O_2 if they are cured, they are generally vacuum packed. This, plus good manufacturing practices and refrigeration provide a market life of as long as 50 days. Although irradiation might be beneficial, the need for a market life longer than 50 days is questionable. For this reason, radurization of refrigerated processed meats does not appear to be useful or needed.

Unlike processed meats, fresh meats and poultry, especially as retail cuts, are highly perishable. A maximum display case life of approximately 3 days is typical of these products. For this reason final cutting, with few exceptions, is done in the retail market. Although this procedure is effective, it involves costs that could be avoided through centralization of final cutting and packaging at some location other than the retail market. With such centralization, retail markets would receive packaged retail cuts.

Of the fresh meats and poultry, beef is perhaps most susceptible to changes that limit its market life. This appears to be related to its relatively heavy pigmentation compared with other meats, since the first evidence of spoilage usually is a discoloration which is more discernible in beef than in other meats. This discoloration is associated with the presence of O_2, which is needed for the initial normal red color. Other meats, including chicken, tend to give first evidence of spoilage by developing odors associated with bacterial activity.

The bacteria that most commonly spoil fresh meats and poultry are *Pseudomonas* and *Achromobacter*. The D_{10} values for *Pseudomonas* are in the range 20–100 Gy, and are among the lowest for bacteria. The *Achromobacter* are somewhat more radiation resistant, with D_{10} values in the 100- to 600-Gy range. It is clear that *low* doses of radiation have the potential for changing the outgrowth flora of fresh meats due to the differences in radiation resistance, virtually eliminating *Pseudomonas* bacteria and favoring outgrowth of *Achromobacter*.

While there is little question that irradiation can reduce the bacterial population of fresh meats and poultry, the value of this by itself in the overall preservation of these foods in terms of marketing requirements is low. Control of microbial spoilage alone is not enough. The other pathways of spoilage also must be controlled. This is an even more important requirement if, in extending the market life beyond the usual few days, opportunity is provided for the nonmicrobial pathways to operate to a greater than normal extent and, hence, to become even more important. For this reason, irradiation of fresh meats must be combined with other measures in order to maintain overall quality. To fulfill this requirement a procedure has been developed which may be described as follows:

1. At a central location, by dipping in a water solution, or by other appropriate means, there is added up to 0.5% condensed phosphate (e.g., sodium tripolyphosphate).
2. The individual cut is wrapped in an oxygen-permeable, moisture-impermeable film.
3. A number of wrapped individual cuts are placed into a bulk container, which is vacuumized.
4. Using doses in the range 1–2 kGy and at temperatures between 0 and 10°C the bulk package of meats is irradiated with γ rays.
5. The bulk package is stored and transported to the retail store at temperatures not exceeding 5°C.

6. About 30 min prior to displaying for sale at the retail store, the bulk package is opened and individual packaged cuts are removed.
7. The retail cuts are displayed at temperatures between 0 and 5°C and sold within 72 hr.

The phosphate is added to control drip. It helps also to maintain color. The double-packaging system provides anaerobic conditions until the retail cut is put in the display case. When removed from the bulk vacuum package, the retail cut is exposed to O_2, which permeates the film in which it is wrapped. The cut has the normal red color at this time. The bulk vacuum package also prevents lipid oxidation during the greater portion of the product market life.

Table 6.3 shows the total plate counts (TPC) obtained with various doses after the indicated days of holding. The dominant outgrowth bacteria are lactobacilli. That this is so is understandable when all aspects of the procedure are considered. *Pseudomonas* and *Achromobacter*, as noted, have low D_{10} values and the 1000-Gy dose employed greatly reduces their numbers. In addition, they are aerobes and the vacuum packaging effectively prevents their growth. The lactobacilli have D_{10} values as high as 1200 Gy and are more resistant as a consequence. This, plus the fact that lactobacilli are both aerobes and anaerobes, favors them to be the outgrowth bacteria.

Table 6.4 shows test panel data on beef handled according to the described procedure and held 24 days after irradiation (21 days in vacuum, 3 in air). These data indicate that the beef was acceptable. Good color was retained and drip was at an acceptable low level below 1%.

Table 6.3

Effect of Various Doses of γ Irradiation on Total Plate Count of Vacuum-Packaged Stored Beefsteaks[a,b]

	Days storage at 4 °C			
Dose (kGy)	0	7	14	21
0	1.6×10^6	1.0×10^7	5.6×10^7	7.0×10^8
0.5	1.8×10^4	6.2×10^4	7.7×10^6	3.0×10^7
1.0	8.0×10^3	3.3×10^5	3.0×10^6	9.0×10^6
2.5	1.0×10^2	2.8×10^2	8.6×10^3	9.8×10^3
5.0	60	1.0×10^2	2.0×10^2	2.2×10^2
10.0	10	10	10	10

[a] From W. M. Urbain, The low-dose radiation preservation of retail cuts of meat. *In* "Radiation Preservation of Food," Proceedings of Symposium, Bombay. Int. Atomic Energy Agency, Vienna, 1973.

[b] Expressed as total plate count per gram.

Table 6.4
Average Scores of 24 Panelists for Radurized Beef[a,b]

Quality parameter	0 days		24 days	
	Not treated	Treated[c]	Not treated[d]	Treated[c,e]
Odor	6.3	6.2	6.3	6.3
Flavor	6.3	6.5	5.5	5.8
Juiciness	6.4	6.0	5.8	6.3
Overall quality	6.2	6.2	5.8	6.2

[a] From W. M. Urbain, The low-dose radiation preservation of retail cuts of meats. *In* "Radiation Preservation of Food," Proceedings of Symposium, Bombay. Int. Atomic Energy Agency, Vienna, 1973.
[b] Hedonic scale: 1, dislike extremely; 9, like extremely.
[c] Phosphated and irradiated (0.1 kGy).
[d] Frozen 24 days.
[e] Stored 21 days in vacuum plus 3 days in air at 5°C.

There are important differences among the several kinds of red meats and poultry. Probably because of its high degree of pigmentation, the drip of beef is more objectionable. Discoloration also is more important with beef. Lamb may present similar difficulties. Pork, veal, and chicken, however, do not, and with these meats some of the procedure outlined for beef may not be needed. Vacuum packaging, whether done for all or part of the holding period, seems desirable.

Retail cuts of meat represent a worst-case situation (other than ground meats), in that the high ratio of surface to mass affords the greatest opportunity for microbial growth. Irradiation of retail cuts, by enabling centralization of final cutting and packaging, provides opportunities for significant cost savings and also for maketing improvements.

Radurization of wholesale cuts, or of sides, or of intact carcasses would avoid some of the difficulties encountered with retail cuts. In view of the success of present practices employing vacuum packaging with these larger product units, irradiation does not appear needed when the time factor for transport and distribution is only a number of days. For circumstances in which the time period involved is a number of weeks, such as may be encountered in international trade, irradiation can be helpful. The value of irradiation in maintaining low bacterial counts of vacuum-packed mutton backstraps stored at 0 to 1°C can be seen from the data of Table 6.5.

The limitation on dose imposed by the sensitivity of meats and poultry to flavor and other changes affecting acceptability prevents the use of sufficient radiation to obtain the maximum benefit in controlling microbial spoilage.

Table 6.5

Bacterial Counts of Vacuum-Packaged Nonirradiated and Irradiated (4.0 kGy) Mutton Backstraps Stored at 0 to 1°C[a]

	Mean bacterial counts ($\log_{10}$ units/cm^2)	
Days	Nonirradiated	Irradiated
7	3.65	1.0
28	6.58	0.8
43	7.26	2.2
57	7.18	1.7

[a] From J. J. Macfarlane, I. J. Eustace, and F. H. Grau. Ionizing energy treatment of meat and meat products. *In* "Ionizing Energy Treatment of Foods," Proceedings of National Symposium. ISBN 085856 0534. Sydney, 1983.

This points to the need to use radurization only with meats that have low initial bacterial counts.

If doses above the threshold value for irradiation flavor are needed, there is always the alternative of irradiating meats and poultry in the frozen state. In doing this the importance of added costs and possible quality changes associated with freezing and thawing need to be considered.

III. RADICIDATION

Fresh meat and poultry regularly are contaminated with salmonellae and are sources of human salmonellosis. The D_{10} value for salmonellae varies with the strain and the medium in which the organism occurs (see Table 4.2). The dose needed to free fresh meats and poultry of salmonellae also is dependent upon the numbers initially present. Early work suggested that a sevenfold reduction of the numbers present was needed, and for this a dose of the order of 5 kGy was required. Subsequent views have held that a reduction of only two- to fivefold and doses as small as 1 kGy will be sufficient. Typical counts of salmonellae in beef and chicken meat are in the range of less than 1 to 20 organisms per gram.

While the principal concern for pathogenic bacteria that contaminate fresh meats and poultry has been with salmonellae, other pathogenic non-spore-forming bacteria all can be present in these foods. These include *Yersinia enterocolitica, Campylobacter jejuni, Staphylococcus aureus, Escherichia coli,* and *Streptococcus faecalis*. The spore-forming *Clostridium perfringens* also can be present. The D_{10} value of these bacteria varies considerably with the species and for some may be greater than those of some salmonellae.

Parasites found in fresh meats are helminths (worms), which can be released into the human gut as larvae upon the ingestion of raw or undercooked meats. The principal meat parasites are *Trichinella spiralis* (in pork), the cause of the disease trichinosis, *Taenia saginata*, the beef tapeworm, and *Taenia solium*, the pork tapeworm. Doses to kill these helminths appear to be too great to be feasible from the standpoint of both costs and radiation-induced flavor changes (see Table 4.6). Flavor changes can be avoided by irradiation at subfreezing temperatures; however, freezing alone inactivates these helminths, making irradiation unnecessary.

In the case of *Trichinella spiralis*, a dose of 150 Gy prevents maturation of the encysted larvae. In the host, death of the immature parasites occurs between excystment and the sixth day. Tests with rats fed large numbers of trichinae treated with a dose of about 200 Gy do not show symptoms of enteric-phase trichinosis. Reproduction and consequently the phases of trichinosis associated with it are prevented by a dose of about 150 Gy. Based on these findings, a dose of 300 Gy has been proposed for the irradiation of fresh pork as a means of making it noninfectious with regard to *T. spiralis*.

IV. RADAPPERTIZATION

Of all foods, meats have been studied most extensively in connection with radappertization. The work on this use of food irradiation was done almost exclusively by the U.S. Army in the period 1953–1981 with the objective of developing shelf-stable combat rations of improved troop acceptance. The army program also included some fish items.

Radappertization of meats requires the largest dose likely to be used in food irradiation. This is the consequence of the fact that, as a group, meats are not acidic (have a pH >4.5) and can have a low salt content. Some also are free of nitrite. Generally they have sufficient moisture to enable microbial growth. In view of these circumstances irradiation used for the preservation of meats must inactivate the spores of *Clostridium botulinum*, the most radiation-resistant bacterium likely to be present. There are asporogenous bacteria such as *Moraxella*, *Acinetobacter*, or *Micrococcus radiodurans* which have greater radiation resistance (see Table 4.2), but they are not involved in the stability and safety of radappertized meats, due to the use of a necessary step of heating the meats to 73 to 77°C to secure inactivation of endogenous autolytic enzymes. This heating also inactivates these radiation-resistant, but heat-sensitive bacteria. Similar considerations apply to viruses and parasites that may be present.

Table 6.6

Consumer Panel Ratings for Beef Loins Irradiated at Various Temperatures[a, b, c]

Irradiation temperature (°C)	U.S. Choice Grade	U.S. Commercial Grade
−40	5.9	6.1
−80	6.1	6.4
−120	6.3	6.5
−185	6.4	6.4
Nonirradiated	6.8	6.4

[a] From G. W. Shults and E. Wierbicki, "Development of Irradiated Beef. 1. Acceptance of Beef Loin Irradiated at Cryogenic Temperatures," Tech. Rep. 74-57-FL. U.S. Army Natick Labs., Natick, Massachusetts, 1974.

[b] Hedonic scale: 1, dislike extremely; 9, like extremely.

[c] Dose, 45–56 kGy.

Early work indicated that at least 20 kGy are needed to provide safety and stability of meats. Later work showed that this dose was too low and doses of the order of 40 to 50 kGy would be required. Such amounts of radiation adversely affect the flavor of most meats and make them unacceptable on a sensory basis. As shown by the data of Table 6.6, undesired flavor development can be reduced greatly by irradiation at subfreezing temperatures. The data of Fig. 6.1 indicate correlation of flavor intensity with amounts of radiolytic products formed at the various irradiation temperatures. Consumer panel ratings of beef irradiated at doses in the range 45–56 kGy at subfreezing temperatures are shown in Table 6.6.

It is clear from the findings on the effect of irradiation temperature on the intensity of the undesirable irradiation flavor that beef of acceptable flavor can be secured by irradiation at subfreezing temperatures. The temperature dependence of the formation of the irradiated flavor clearly points to its origin in the indirect action of radiation.

Figure 4.6 shows the temperature dependence of the D_{10} value for *Clostridium botulinum* 33A. In order to determine what dose is required for the radappertization of meats it is necessary, therefore, that the irradiation temperature be known. In its work, the U.S. Army selected an irradiation temperature of −30°C ± 10°. This temperature is the lowest obtainable with conventional refrigeration and represents a significant cost advantage over cryogenic procedures needed for lower temperatures.

From the standpoint of dose level, the key objective in radappertizing meats is to secure inactivation of the spores of *Clostridium botulinum* with sufficient certainty so as to assure consumer safety in the consumption of such meats. The experience with the thermal canning process has been utilized to serve as a guide in the attainment of this objective. In thermal canning of low-salt, low-acid (pH >4.5) foods, sufficient heat is applied to obtain a $12D_{10}$ reduction in count of the most heat-resistant strain of *C. botulinum.* The excellent record of safety secured over many decades of thermal canning with this standard supports the wisdom of this approach and warrants application of the same concept to the radappertization of meats.

The D_{10} values for *Clostridium botulinum* listed in Table 4.2 indicate that types A and B are the most radiation resistant. Based on the listed D_{10} values, a $12D_{10}$ dose is about 50 kGy (irradiation at about $-50°C$).

While application of the $12D_{10}$ concept has been made to radappertization, irradiation and thermal sterilization are not entirely parallel processes. The spores of *Clostridium botulinum* are not the most heat-resistant organisms found in foods, and while their inactivation is needed in order to assure safety, thermal processes applied to low-acid, low-salt foods usually are determined by the amount of heating required to inactivate other more heat-resistant bacteria, which if not inactivated would cause product spoilage. As a consequence, from the standpoint of botulism hazard, the thermal process has a built-in safety factor which adds to the assurance of inactivation of *C. botulinum* spores.

With radappertization such a safety factor does not exist. The spores of *Clostridium botulinum* are the most radiation-resistant bacteria encountered. Therefore, unless an amount greater than the minimum dose for inactivation of this organism is intentionally employed, there is no safety factor such as exists with thermal processing. This circumstance requires that the minimum dose be known accurately and that it be applied with precision.

As noted earlier (Chapter 4), the use of the D_{10} value to compute doses to secure a desired reduction of a bacterial population is based upon relationships which do not always hold. In particular, the curve representing the relationship of the logarithm of the surviving fraction of a bacterial population to dose is not always a simple straight line, showing exponential relationship, but is sigmoidal, having both a "shoulder" and a "tail" in addition to a straight-line portion (see Fig. 4.10). If the part of the curve that is a straight line is used to compute a dose to yield a value of 10^{-12} for $\log_{10} N/N_0$, the value so obtained would be erroneous.

An improved method for establishing the minimum radiation dose (MRD) has been devised. An "inoculated pack" of the product is prepared according to the procedure of the example for beef as shown in Table 6.7. The MRD is

Table 6.7

Inoculated-Pack Procedure for Radappertized Beef[a,b]

Procedural aspect	Description
Clostridium botulinum strains	Mixture of 33A, 36A, 62A, 77A, 12885A, 9B, 40B, 41B, 53B, 67B
Spore inoculum amount	10^6 per strain; 10^7 per can
Containers	211 × 101.5 (epoxy enamel) metal cans
Weight of food contained per can	40 ± 5 g
Number of cans per dose	100 each replicate
Vacuum seal pressure	16 kPa
Radiation source	^{60}Co γ rays
Radiation doses (kGy)	14, 18, 22, 26, 30, 34, 38, 42, 46, 50
Radiation temperature	−30°C ± 10°
Incubation	6 months at 30 ± 2°C
Analysis	Swelling: daily, month 1; weekly, months 2–6 Botulinal toxin: month 7 Recoverable *C. botulinum*: month 7

[a] From A. Anellis, D. B. Rowley, and E. W. Ross, Jr., *J. Food Prot.* **42,** 927 (1979).
[b] Beef formulated with 0.75% NaCl and 0.38% sodium tripolyphosphate.

based on the following:

1. Any can containing one or more viable spores after irradiation and incubation is not sterile.
2. Those data obtained with the single most resistant strain placed in the specific product and processed in the specified manner.
3. A shifted straight-line portion of the log N/N_0 dose curve. The shift provides a shoulder at the low-dose end. The slope and the shoulder are determined by using the binomial confidence limit method. The intercept at the abscissa of this shifted straight line at a value of $\log_{10} = 10^{-12}$ provides the $12D_{10}$ dose in the regular manner (see Fig. 4.1).

This method makes allowance for the shoulder and provides a better procedure for fitting a straight line to the data derived from the inoculated-pack study. It may be regarded as providing D_{10} values which likely are higher than the true values and, therefore, provides a margin of safety. This, plus the safety aspects of the $12D_{10}$ concept, which are based on providing a treatment to handle spore populations that are unrealistically high for the foods in question, add to the safety of the radappertization process from the standpoint of a botulism hazard.

Table 6.8 lists the MRDs for a number of radappertized foods. The dose given is specific for each product and the indicated irradiation temperature. In

Table 6.8
Minimum Radiation Doses for Radappertization of Certain Meats and Codfish Cakes[a]

Food	Irradiation temperature (°C)	MRD (kGy)
Bacon	5–25	25
Beef[b]	-30 ± 10	37
Ham[c]	5–25	31
Ham[d]	-30 ± 10	33
Pork	5–25	43
Codfish cake	-30 ± 10	32
Corned beef	-30 ± 10	24
Pork sausage	-30 ± 10	27

[a] From E. Wierbicki, A. Brynjolfsson, H. C. Johnson, and D. B. Rowley, "Preservation of Meats by Ionizing Radiation. An Update." European Meeting of Meat Research Workers, Bern, 1975.
[b] Contained 0.75% NaCl and 0.375% sodium tripolyphosphate.
[c] $NaNO_2$ and $NaNO_3$ (156 and 700 mg/kg, respectively) added.
[d] $NaNO_2$ and $NaNO_3$ (25 and 100 mg/kg, respectively) added.

this way variations of the D_{10} value related to product composition and irradiation temperature are provided for. The use of 10 strains of *Clostridium botulinum* provides also for variations of the resistance of botulinum strains in different foods. Subfreezing irradiation temperatures are employed only when needed to secure a satisfactory flavor.

Meats contain enzymes which cause proteolysis and the formation of free amino acids, with resultant texture and flavor changes. Doses in the range of the MRDs for inactivation of botulinum spores are inadequate to prevent this proteolysis, as may be seen from the data of Table 6.9. The only practical effective way to secure enzyme inactivation is to use heat. Table 6.10 provides data showing that a temperature of 77°C is effective, whereas 66°C is not. It is clear, therefore, that stable radappertized meats cannot be raw but must be moderately heated. The heating done for enzyme inactivation is adequate also to inactivate radiation-resistant vegetative bacteria, viruses, and parasites.

Appropriate packaging of radappertized foods is needed for the purposes of (1) preventing recontamination by spoilage microorganisms, (2) preventing

Table 6.9

Free Amino Acids in Raw Beef Irradiated at 50 kGy and Stored at 38°C[a,b]

Amino acid	Days of storage		
	0	60	200
Leucine	0.7	8.9	91.7
Phenylalanine	Trace	6.8	14.1
Valine	0.7	7.7	45.3
Methionine	0	8.2	14.6
Alanine	0.6	17.6	65.6
Threonine	0	14.7	17.9
Glycine	Trace	9.7	16.4
Serine	0	9.7	19.3
Glutamic acid	1.4	11.6	46.9
Aspartic acid	0	0.2	0
Lysine	0	2.4	13.0
Tyrosine	0.7	25.1	142.0
Histidine	0	41.0	124.0
Arginine	3.2	22.1	109.0
Tryptophan	0	3.4	Trace
Proline	1.8	13.1	52.8

[a] From R. F. Cain and A. F. Anglemier, Enzymatic stabilization of radiated meats. *In* "Enzymological Aspects of Food Irradiation." Int. Atomic Energy Agency, Vienna, 1969.

[b] Expressed as mg% dry-weight basis.

Table 6.10

Effect of Temperature of Heating Prior to Irradiation on Total Soluble Nitrogen of Irradiated (30 kGy) Beef Stored at 22°C[a]

Temperature (°C)	Total soluble (% dry weight basis) nitrogen			
	0 days	60 days	120 days	180 days
54	2.19	2.79	3.03	3.33
66	1.66	1.82	1.96	1.92
77	1.17	1.23	1.30	1.32

[a] From O. G. Atar, J. C. R. Li, and R. F. Cain, *Food Technol.* (*Chicago*) **15,** 488 (1961). Copyright © by Institute of Food Technologists.

chemical deterioration, mainly by atmospheric O_2, (3) preventing moisture loss, and (4) excluding O_2 during irradiation in order to prevent radiation-catalyzed oxidation by O_2 or O_3 (the latter formed by action of radiation on O_2).

Tinplate rigid containers with inside epoxy-phenolic enamels and end-sealing compounds made of blends of either cured and uncured butyl elastomers or of neoprene and butadiene styrene elastomers or of neoprene and uncured butyl rubber elastomers are satisfactory.

A laminated flexible package, consisting of chemically bonded Mylar and medium-density polyethylene as the food contactant layer, aluminum foil as the middle layer, and nylon-6 as the outside layer, is available and satisfactory for radappertized meats.

The containers are filled and sealed prior to irradiation. The enzyme heat inactivation step is done prior to placing the product in the container. Before sealing, the container is vacuumized.

Irradiation produces gases which are found in the container head space and which cause some loss of the initial vacuum. They constitute about one-half of the initial head-space gas. The radiation-formed gases are (1) mainly H_2, (2) lesser amounts of CO_2, and (3) small amounts of CH_4 and CO. (In one beef sample irradiated at $-40°C$ and 47–71 kGy, analysis showed 35% H_2, 2.6% CO_2, 0.7% CH_4, and traces of CO.) The radiation-formed gases occur mixed with the residual air in the head space of the container. An initial 63 cm of vacuum in a rigid metal container is reduced to about 38 cm due to radiation-formed gases.

Unlike thermal sterilization, irradiation does not change the water-holding capacity of meat significantly. There is no release of liquid, as can occur with the thermal process. Dry products such as roasted or fried meats may be radappertized. The addition of liquid, such as is done with some thermally canned products (e.g., whole chicken) in order to aid in heat transfer, is not needed. The energy distribution in products irradiated with γ or X rays is very much more uniform than is obtainable with heat sterilization. Because of this uniformity, the unit size of product has no effect on quality, a difference from what occurs with heat sterilization that greatly favors radappertization.

As noted (see Fig. 3.10), irradiation increases the water solubility of collagen, the principal constituent of connective tissue and whose presence is related to the tenderness of meat. The increase in solubility of collagen leads to increased meat tenderness. Table 6.11 shows the tendering action of irradiation on beef at 45 to 56 kGy. Lowering the irradiation temperature reduces the tendering action, indicating it is a result of indirect action. Storage of radappertized beef leads to additional tendering, as may be seen from the data of Table 6.12. These tenderness changes have significance in consumer acceptance of radappertized meats but are substantially less than those

Table 6.11

Effect of Irradiation Temperature on Textural Characteristics of U.S. Commercial Grade Beef Loin[a, b]

Irradiation temperature (°C)	Panel ratings[c]	
	Mushiness	Friability
60	5.3	5.0
21	3.4	3.0
−40	2.5	1.9
−80	2.0	1.8
−185	2.0	1.9

[a] From G. W. Shults and E. Wierbicki. "Development of Irradiated Beef. 1. Acceptance of Beef Loin Irradiated at Cryogenic Temperatures," Tech. Rep. 74-57-FL. U.S. Army Natick Labs., Natick, Massachusetts, 1974.
[b] Dose, 45–56 kGy.
[c] Panel rating scale: 1, none; 9, extreme.

experienced with thermal canning, and this difference favors radappertization significantly.

Table 6.13 gives expert panel ratings of sensory characteristics of U.S. Choice and Commercial grades of beef loins stored at 21°C for up to 1 year. These data indicate changes of sensory characteristics with storage time. Lowering the irradiation temperature below −80°C, although reducing the

Table 6.12

Effect of Storage on Panel Ratings for Tenderness of Beef Irradiated at 45 kGy at Several Temperatures[a,b]

Irradiation temperature (°C)	U.S. Choice Grade		U.S. Utility Grade	
	0 Months	24 Months	0 Months	24 Months
20	7.2	8.6	6.0	6.6
−20	7.1	8.4	5.7	6.3
−80	7.0	8.3	5.5	6.1
−196	6.9	8.3	5.1	5.7

[a] From F. L. Kauffman and J. W. Harlan. "Effect of Low Temperature Irradiation on Chemical and Sensory Characteristics of Beef Steaks," Tech. Rep. 69-64-FL. U.S. Army Natick Labs., Natick, Massachuseets, 1969.
[b] Scale: 1, extremely tough; 10, extremely tender.

Table 6.13

Effects of Storage on Sensory Characteristics and Preference Ratings of U.S. Choice Beef Loin Irradiated at Various Temperatures[a, b, c]

Irradiation temperature (°C)	Storage time (months)	Discoloration[d]	Off-odor[d]	Mushiness[d]	Irradiated flavor[d]	Off-flavor[d]	Preference[e]
−40	0	—	2.06	1.71	1.85	3.00	6.42
	1	—	2.25	2.25	2.62	2.12	6.00
	3	1.87	2.00	3.00	2.50	2.25	6.00
	12	2.12	1.62	2.62	2.00	1.50	5.50
−80	0	—	1.71	3.00	1.57	2.42	6.57
	1	—	1.87	1.75	2.12	2.00	6.25
	3	1.62	1.87	2.50	2.25	2.37	6.12
	12	3.42	2.12	2.00	2.75	1.12	6.00
−120	0	—	1.57	3.28	2.28	2.57	6.00
	1	—	1.62	1.87	2.00	1.75	6.25
	3	2.00	1.73	2.87	2.00	2.75	6.12
	12	1.87	1.75	2.00	1.75	1.50	6.00
−185	0	—	1.57	2.28	1.28	2.57	7.00
	1	—	1.37	1.75	1.37	1.25	7.37
	3	1.62	1.62	3.12	1.87	2.75	6.50
	12	2.12	1.87	2.38	2.00	2.38	6.00
Frozen control (non-irradiated)	0	—	1.14	1.42	1.00	1.14	8.14
	1	—	1.00	1.37	1.00	1.50	7.75
	3	1.12	1.12	1.75	1.00	1.62	7.75
	12	1.12	1.25	1.12	1.00	1.62	7.00

[a] From G. W. Shults and E. Wierbicki, "Development of Irradiated Beef. 1. Acceptance of Beef Loin Irradiated at Cryogenic Temperatures," Tech. Rep. 74-57-FL. U.S. Army Natick Labs., Natick, Massachusetts, 1974.

[b] 21°C Storage (irradiated samples); −29°C storage (nonirradiated).

[c] Dose 45–56 kGy.

[d] Intensity scale: 1, none; 9, extreme.

[e] Hedonic scale: 1, dislike extremely; 9, like extremely.

Table 6.14

Troop Acceptance of Radappertized Meats, Poultry, and Seafoods[a]

Item	Irradiated		Nonirradiated	
	Number of evaluators	Rating[b]	Number of evaluators	Rating[b]
Ham	1,657	5.87	1,437	6.66
Chicken	583	6.07	548	6.36
Pork	391	5.71	458	6.85
Beef	589	5.99	644	6.61
Bacon	25,656	6.16	—	—
Shrimp	539	6.09	849	6.43
Codfish cakes	531	5.40	578	6.30

[a] From W. M. Urbain, Radiation update. *In* "Feeding the Military Man." U.S. Army Natick Labs., Natick, Massachusetts, 1970.

[b] Hedonic scale: 1, dislike extremely; 9, like extremely.

amounts of radiolytic products formed, does not result in statistically significant different scores for irradiated flavor intensity.

Table 6.14 shows U.S. Army and Air Force personnel ratings of radappertized meats and poultry. All ratings above 5 on the nine-point hedonic scale are considered satisfactory ratings for combat rations.

Uncured cooked meats, radappertized in sealed evacuated containers, display a pink color immediately upon opening the container. In the presence of air the brown or gray color normal for such meats quickly develops. The same phenomenon may be observed with thermally canned uncured meats which are heat processed in sealed evacuated containers from the raw state. In the case of radappertized meats, the sequence of color changes is associated with the reducing action of radiolytic products formed in the meat and produces a reduced denatured myoglobin pigment that is red and is easily oxidized by O_2 to the usual brown pigment. Because this pink color changes rapidly to the normal brown or gray after the container is opened, it is regarded as not significant in consumer acceptance of radappertized uncured meats.

Each of the radappertized products listed in Table 6.8 has associated with it particular technology so as to meet commercial production requirements and product characteristics. Such technology in general, however, is not necessarily specific to radappertization. There can be substantial latitude in the preparation and formulation of radappertized meats and poultry.

While the use of subfreezing irradiation temperatures unquestionably yields improvement in the sensory properties of radappertized foods, trained expert evaluators can note an "irradiation flavor" in some of the products listed in Table 6.8. As a consequence, concerns about acceptability for civilian markets have been expressed. Army experience indicates that products scoring 5 or higher on the nine-point hedonic rating scale are acceptable as rations. Because the ratings for radappertized foods developed by the army have exceeded the value of 5, their acceptability for combat rations has been indicated. It appears that the determination of the acceptability of these products for the civilian market must await their availability in the marketplace. In this connection it should be noted that radappertized meats have obviously superior texture characteristics and do not undergo moisture release, as is the case with thermal sterilization. It is also significant that many common foods undergo flavor changes as a result of processing and yet obtain a high degree of consumer acceptance.

Irradiation can be used to replace nitrite in cured meats in relation to its function to control *Clostridium botulinum* bacteria. Nitrite in cured meats also functions to provide the typical cured (red) color and flavor of these meats. The amount of nitrite needed for this latter function, however, is substantially less than that needed for *C. botulinum* control. When a reduced amount of nitrite is used the potential for nitrosamine formation is lessened, and this is the principal purpose of this use of irradiation.

Table 6.15 lists radiation–nitrite combinations which enable reduction of nitrite over the use of nitrite alone. Normal cured colors are obtained. The dose requirements are similar to those for radappertization, although sterilization of the products is not necessarily a requirement.

Table 6.15
Radiation–$NaNO_2$ Combinations for Cured Meats[a]

Product	$NaNO_2$ added (mg/kg)		Dose (kGy)
	Nonirradiated[b]	Irradiated	
Bacon	120	20	30
Ham	156	25[c]	32
Corned beef	156	50	26
Frankfurter	156	50	32

[a] From E. Wierbicki, Technological feasibility of preserving meat, poultry and fish products using a combination of conventional additives, mild heat treatment and irradiation. *In* "Combination Processes in Food Irradiation." Int. Atomic Energy Agency, Vienna, 1981.
[b] U.S.D.A. maximum allowed.
[c] Plus 25 mg/kg $NaNO_3$.

SOURCES OF ADDITIONAL INFORMATION

Anellis, A., Rowley, D. B., and Ross, E. W., Microbiological Safety of Radappertized Beef. *J. Food Prot.* **42**(12), 927 (1979).

Anonymous, "Trichina-safe Pork by Gamma Irradiation Processing. A Feasibility Study," Contract No. DE-AC 04-83AL19411. U.S. Dept. of Energy, Albuquerque, New Mexico, 1983.

Brake, R. J., Murrell, K. D., Ray, E. E., Thomas, I. D., Muggenberg, B. A., and Sivinski, J. S., Control of trichinosis by low-dose irradiation of pork. *J. Food Saf.*, **7**, 127 (1985).

Josephson, E. S., Radappertization of meat, poultry, finfish, shellfish and special diets. *In* "Preservation of Food by Ionizing Radiation" (E. S. Josephson and M. S. Peterson, eds.), Vol. III. CRC Press, Boca Raton, Florida, 1983.

Ouwerkerk, T., Salmonella control in poultry through the use of gamma irradiation. *In* "Combination Processes in Food Irradiation." Int. Atomic Energy Agency, Vienna, 1981.

Urbain, W. M., "Meats and Poultry," Food Irradiation Information, No. 8. Int. Project in the Field of Food Irradiation, Karlsruhe, West Germany, 1978.

Urbain, W. M., Radurization and radicidation of meat and poultry. *In* "Preservation of Food by Ionizing Radiation," (E. S. Josephson and M. S. Peterson, eds.), Vol. III. CRC Press, Boca Raton, Florida, 1983.

Urbain, W. M., and Campbell, J. F., Meat preservation. *In* "The Science of Meat and Meat Products." Food and Nutrition Press, Westport, Connecticut, in press.

Urbain, W. M., Belo, P. S., and Giddings, G. G., "Centralized Processing of Fresh Meats and Poultry Including Radiation Pasteurization. A Bibliography," ORNL-IIC-20. U.S. Atomic Energy Comm., Oak Ridge, Tennessee, 1969.

Urbain, W. M., Belo, P. S., and Giddings, G. G., "Bibliography on the Centralized Processing of Fresh Meat and Poultry Including Radiation Pasteurization (Updated)," TID-25998. U.S. Atomic Energy Comm., Washington, D.C., 1971.

E. L. Wick, E. Murray, J. Mizutani, and M. Koshika, Irradiation Flavor and the Volatile Components of Beef. *In* "Radiation Preservation of Foods," Advances in Chemistry Series 65. Am. Chem. Soc., Washington, D.C., 1967.

Wierbicki, E., Technological feasibility of preserving meat, poultry and fish products by using a combination of conventional additives, mild heat treatment and irradiation. *In* "Combination Processes in Food Irradiation." Int. Atomic Energy Agency, Vienna, 1981.

CHAPTER 7

Marine and Freshwater Animal Foods

I. GENERAL

Because they are highly perishable, marine and freshwater animal foods present problems in their distribution. Although they are not as subject to seasonality of production as are some foods which as a consequence, require preservation over periods of time, generally fresh marine and freshwater animal foods, due to their perishability, are available only in localized areas, as, for example, coastal waters. This circumstance normally either limits their geographic distribution or requires preservation measures in order to reach interior or other distant markets.

Marine and freshwater animal foods undergo spoilage in ways that parallel, to a degree, those of fresh meats. Spoilage pathways for fishery products associated with postmortem holding include microbial action and chemical and physical changes resulting in flavor and color alteration and in "drip" formation. In the course of regular commercial handling, bacteria invade muscle tissue after death and in time cause spoilage. Fishery products, however, are more susceptible to proteolysis due to endogenous enzymes* than are meats and undergo this form of spoilage in addition to that caused by bacterial growth. To a significant degree, spoilage by both enzyme and bacterial action can be delayed by the use of low temperatures during storage and distribution. Because of limitations imposed by spoilage, commercial

* According to some investigators, the enzymes causing proteolysis are not endogenous but instead are of bacterial origin.

practices generally restrict the shipment of *fresh* products preserved by refrigeration to markets only a few hundred miles from the landing location, unless air transport is employed.

In addition to refrigeration, other methods of preservation are used. Drying and salting and/or smoking have been used extensively in the past and still are practiced in various parts of the world. While of limited use, thermal canning is employed for certain products, notably salmon and tuna fish. Freezing is commonly used for a variety of products, although loss of quality may occur with some.

To some degree the microflora of marine and freshwater animal foods reflect the location of their origin. Bacteria are present on the surfaces, in the gills, and in the intestine as fish are taken from the water. Usually they are stored aboard the fishing vessel in ice until landed, a procedure which causes additional microbial contamination, the sources of which may be both the ice used and the boat holding chests. Processing such as gutting, filleting, and steaking causes further contamination. Generally the organisms causing spoilage are psychrotrophic and psychrophilic bacteria of freshwater origin.

The principal bacteria found in fresh fish in northern waters are *Pseudomonas* and *Achromobacter*. Others include *Acinetobacter*, *Corynebacterium*, *Flavobacterium*, and *Micrococcus*. Spoilage of fresh fish stored in air usually is due to the outgrowth of *Pseudomonas* bacteria. The spoilage of oysters is associated with the growth of Lactobacteriaceae. Cooked products such as crab meat spoil due to the growth of *Achromobacter*. Pathogenic bacteria may be present in marine and freshwater animal foods and may include *Staphylococcus aureus*, *Vibrio parahaemolyticus*, enterococci, and *Clostridium botulinum*. Yeasts with proteolytic capability can occur. Pathogenic viruses, usually due to sewage contamination of the waters in which the foods occur, may be present.

Exemplary microbial flora of three nonirradiated U.S. Pacific Coast seafood products are shown in Table 7.1. As mentioned, the microflora vary considerably with the location in which the fish are caught, and the data given in Table 7.1 may not be representative for products of other originating locations.

Parasites occur in certain fish. They include *Diphyllobothrium latum*, the broad fish tapeworm, *Clonorchis sinensis*, a trematode and important pathogen of Asia, and *Anisakis marina* found in North Sea herring and elsewhere. Other parasites also occur.

As noted, some fish are preserved by drying. For such foods, insect infestation can be a major problem.

From what is indicated above, it is clear that opportunities exist for radurization, radicidation, and radappertization as well as insect disinfesta-

Table 7.1
Microflora of Nonirradiated U.S. Pacific Coast Seafoods[a]

Microbial group	Percentage of total microflora		
	Dungeness crab meat	Oysters	Dover sole
Coliforms	0	1.1	1.1
Pseudomonas spp.	5.3	34.8	58.2
Achromobacter spp.	77.0	3.3	7.4
Flavobacterium spp.	1.6	2.8	15.8
Bacillus spp.	2.7	55.0	9.5
Gram-positive pigmented rods	8.2	1.1	2.6
Micrococcus spp.	0	0	1.0
Yeasts	0	0	1.6
Unclassified	4.9	1.8	2.6

[a] From R. O. Sinnhuber and J. S. Lee, "Effect of Irradiation on the Microbial Flora Surviving Irradiation Pasteurization of Seafoods," RLO-1950-1. U.S. Atomic Energy Comm., Washington, D.C., 1966.

tion. While convenience dictates the grouping of all marine and freshwater animal foods in discussing their irradiation it is recognized that important differences exist among the various kinds of these foods and that these differences can have significant effects on the utility of irradiation. Without seeking to catalog such differences according to kind of food and species of animal, major differences will be noted where possible.

II. RADURIZATION

As is to be expected, the exact dose used in the radurization of marine and freshwater animal foods relates to the initial microbial population, the D_{10} values of the organisms present, the needed or desired reduction of population, the postirradiation handling conditions, including atmosphere, and the product life extension sought. Important in the maximum dose that can be employed is the sensitivity of the food to the development of irradiated flavor; Table 6.1 indicates the flavor sensitivity of certain seafood products to radiation. The doses needed to obtain a significant delay of spoilage are not likely to exceed 5 kGy. The exact dose employed, however, depends upon the particular product and related conditions as indicated above.

How soon after catch or harvest irradiation is carried out, as well as the nature of the general sanitation employed, enter into the effectiveness of irradiation. These variables involve practical aspects of commercial fishing, such as how long the catching vessel remains at sea. To cope with such

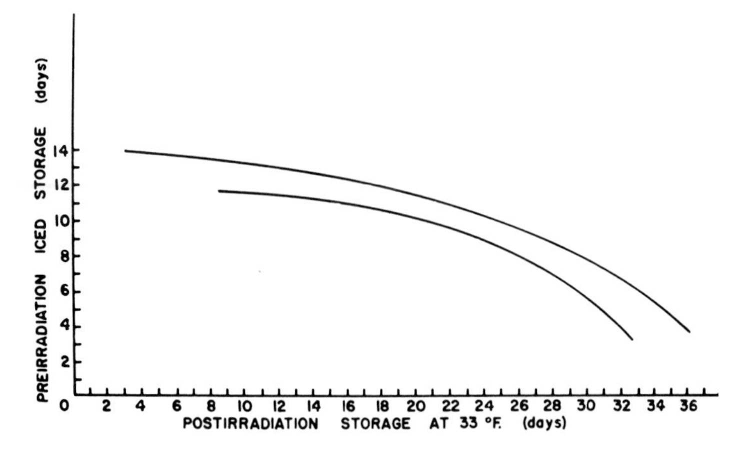

Figure 7.1 Effect of the preirradiated quality of iced, eviscerated fish on postirradiation product life of fillets stored at 1°C (33°F). Upper curve, cod (one experiment); lower curve, haddock (average of three experiments). From J. W. Slavin, L. J. Ronsivalli, and J. D. Kaylor, *Act. Rep. Res. Dev. Assoc. Mil. Food Packag. Syst.* **17** (2) (1965).

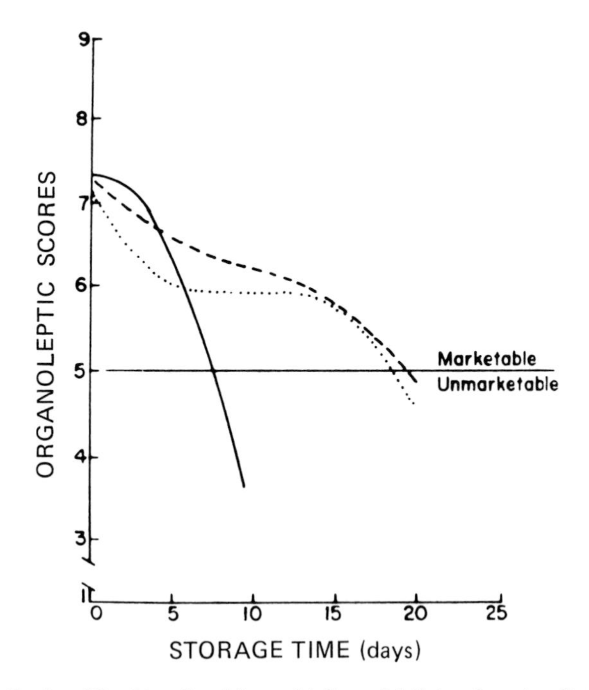

Figure 7.2 Product life of irradiated (––––1 kGy;···2 kGy) and nonirradiated (——) haddock fillets held at 1 to 2°C. Score: 1, inedible; 5, borderline; 9, excellent. From L. J. Ronsivalli, J. D. Kaylor, E. J. Murphy, L. J. Learson, and M. S. Schwartz, Studies in petition-oriented aspects of radiation pasteurized fishery products. *In* "Preservation of Fish by Irradiation." Int. Atomic Energy Agency, Vienna, 1970.

situations consideration has been given to carrying out the irradiation of freshly caught fish at sea, with the possibility of a second irradiation after landing. Figure 7.1 shows the relationship between product life and the preirradiation storage period, and the increasing loss of effectiveness as fish are held prior to irradiation.

The payoff for radurization lies in the preservation of the product to enable its distribution as fresh rather than frozen and without resorting to excessively expensive shipment procedures. Figure 7.2 shows the extension of market life as measured by panel sensory scores, of fillets of Atlantic haddock irradiated with doses of 1 and 2 kGy.

The dominant bacteria in fish not irradiated and stored in air under

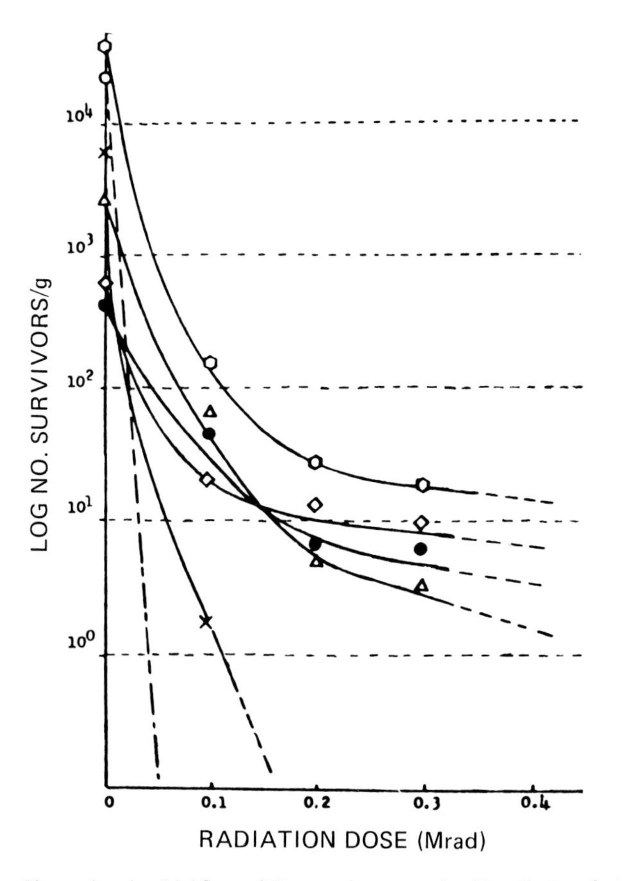

Figure 7.3 Change in microbial flora of Dover sole as a result of irradiation. ⬡, Total count; ◇, yeasts; ●, *Micrococcus*; △, *Achromobacter*; ×, *Flavobacterium*; —-—- *Pseudomonas*. From R. O. Sinnhuber and J. S. Lee, "Effect of Irradiation on the Microbial Flora Surviving Irradiation Pasteurization of Seafoods," TID-22289. U.S. Atomic Energy Comm., Oak Ridge, Tennessee, 1965.

refrigeration usually are pseudomonads. Fish irradiated and stored similarly may have *Achromobacter* as the principal but not sole outgrowth bacteria. This shift in bacterial flora correlates with the greater sensitivity of pseudomonads to radiation (see Table 4.2). Because pseudomonads are aerobes, a similar shift occurs when the product is vacuum packaged. Bacteria other than *Achromobacter* occur as outgrowth organisms with radurized fishery products.

Figure 7.3 shows the change in microbial flora in Pacific Dover sole as a result of irradiation at several doses. Figure 7.4 shows the percentage distribution of the bacterial flora of Atlantic haddock filleted 2 days after

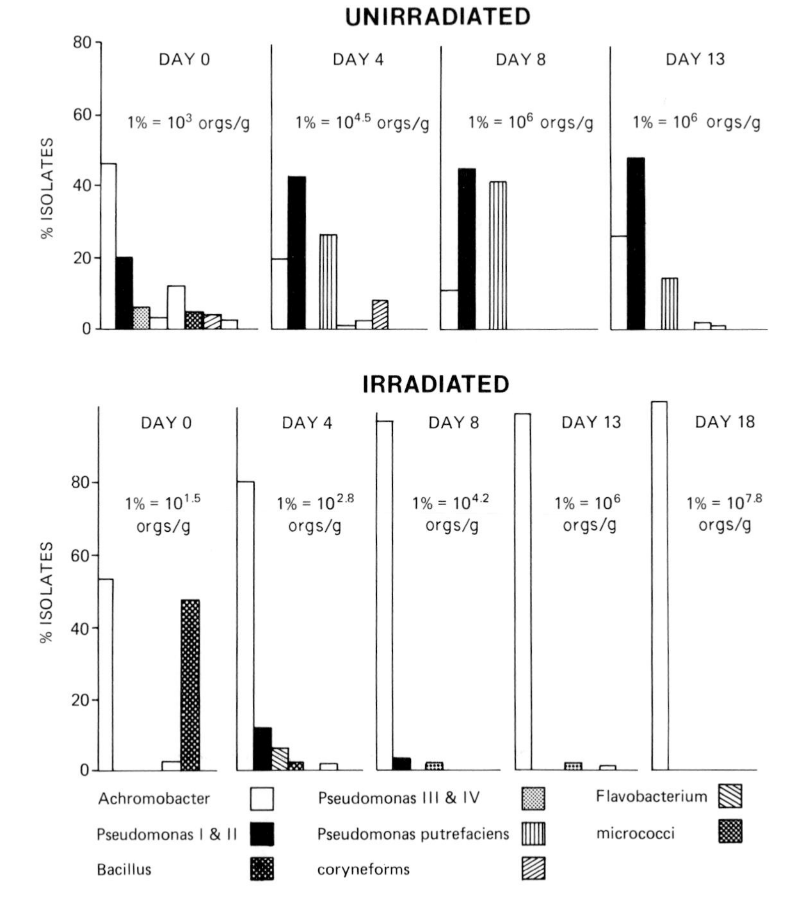

Figure 7.4 Percentage composition of bacterial flora of 2-day fish during storage at 3°C. From A. Laycock and I. W. Regier, The effect of gamma irradiation on the microbial spoilage patterns on fish in relation to initial quality. *In* "Preservation of Fish by Irradiation." Int. Atomic Energy Agency, Vienna, 1970.

Table 7.2

Total Bacterial Counts of Irradiated Petrale Sole Fillets Stored at 1°C in Air[a, b]

		Irradiation dose (kGy)	
Days	Nonirradiated	2	3
0	7.9×10^5	4.0×10^4	1.8×10^4
7	1.4×10^6	1.0×10^4	4.0×10^3
11	5.5×10^6	—	—
14	—	7.3×10^4	4.0×10^3
21	—	3.7×10^7	2.2×10^4
28	—	4.4×10^7	3.5×10^5
35	—	1.3×10^8	3.3×10^7
42	—	2.6×10^8	1.9×10^7
49	—	—	8.8×10^8

[a] From J. Spinelli, M. Eklund, N. Stoll, and D. Miyauchi, *Food Technol. (Chicago)* **19,** 1016 (1965). Copyright © by Institute of Food Technologists.

[b] Expressed as total number of bacteria per gram.

catch with and without irradiation and stored at 3°C. The data of these two figures demonstrate the changes in the microflora of fish on storage as a result of irradiation.

Table 7.2 lists total plate counts for Pacific petrale sole irradiated at 0, 2, and 3 kGy and stored at 1°C in air. The relationship of bacterial counts to sensory scores for these same petrale sole fillets is shown in Fig. 7.5. These data point out that, at the time of observable sensory spoilage, the total bacterial counts of irradiated fish generally are higher by a factor of 10^1 or 10^2 than those for nonirradiated fish. This difference has been ascribed to the difference in kind of microflora and is based on observations that pseudomonads, particularly *Pseudomonas putrefaciens*, are more effective in producing substances such as trimethylamine, which yield undesirable odors and flavors associated with spoiled seafoods.

Figure 7.6 shows the growth of aerobic–facultative bacteria in Pacific Dover sole irradiated at 0, 1, 3, and 5 kGy and stored in the presence of air at 6°C. After irradiation, the population of bacteria either remains constant or decreases for a period, the length of which is greater with increasing dose. Following this period, a rapid increase in numbers occurs. One can consider the extension of the product sensory life as largely determined by the length of the period in which the bacterial population does not increase.

While the reduction of the microbial population aids in the preservation of fishery products, other changes which affect quality and are unrelated to the action of microorganisms can take place. These may include formation of drip

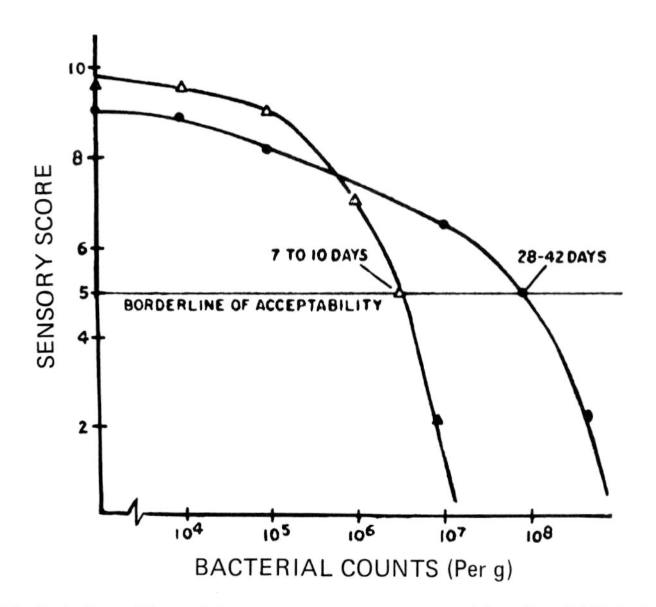

Figure 7.5 Relation of bacterial counts to sensory scores of irradiated (●, 2–3 kGy) and nonirradiated (△) fillets of petrale sole stored at 1°C. Ten-point difference-rating sensory scale using fresh-frozen reference sample. From J. Spinelli, M. Eklund, N. Stoll, and C. Miyauchi, *Food Technol.* (*Chicago*) **19,** 1016 (1965). Copyright © by Institute of Food Technologists.

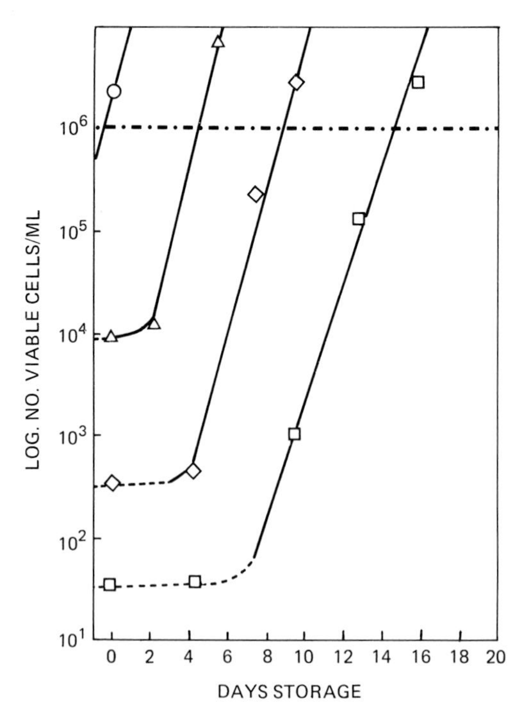

Figure 7.6 Aerobic–facultative counts of irradiated Dover sole during storage at 6°C. △, 1 kGy; ◇, 3 kGy; □, 5 kGy; ○, nonirradiated. From R. O. Sinnhuber and J. S. Lee, "Effect of Irradiation on the Microbial Flora Surviving Irradiation Pasteurization of Seafoods," SAN-100-1. U.S. Atomic Energy Comm., Oak Ridge, Tennessee, 1964.

and changes in color, texture, and flavor. Because of the longer product life obtained with irradiation, these changes may be more important than otherwise.

Drip loss may be controlled by treating fish with solutions of sodium tripolyphosphate (Na-TPP). Figure 7.7 shows the relationship between drip loss for Pacific ocean perch fillets and concentration of the dipping solution. The phosphate concentration must be at least 5% to be effective. Figure 7.8 shows the drip loss for several kinds of fin fish irradiated with a dose of 2 kGy and stored in vacuum at 1°C. The addition of 2% salt to the phosphate solution increases its effectiveness. Phosphate also improves other sensory characteristics, including color and texture. This may be seen from the data of Table 7.3. Fillets not treated with phosphate and stored for more than 2 weeks at 1 to 2°C suffer loss of the normal "bright hue" and tend to darken when cooked. Phosphate effectively counteracts these changes.

Treatment of Bombay duck with Na-TPP inhibits the activity of the muscle lysosomal hydrolases and reduces the amount of endogenous proteolytic end products formed during storage at 0 to 4°C, as may be seen from the data of Table 7.4. The effect of this with irradiated Bombay duck is a twofold increase in product life.

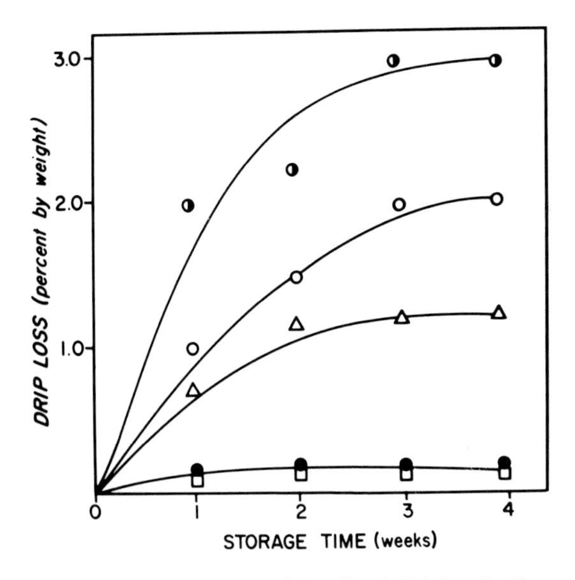

Figure 7.7 Loss of drip during storage of irradiated (2 kGy) Pacific ocean perch fillets pretreated with various concentrations (○, 2.5%; △, 5.0%; ●, 7.5%; □, 10%; ◑, control) of Na-TPP containing 2% NaCl. From J. Spinelli, G. Pelroy, and D. Miyauchi, *Fish. Ind. Res.* **4,** 37 (1967).

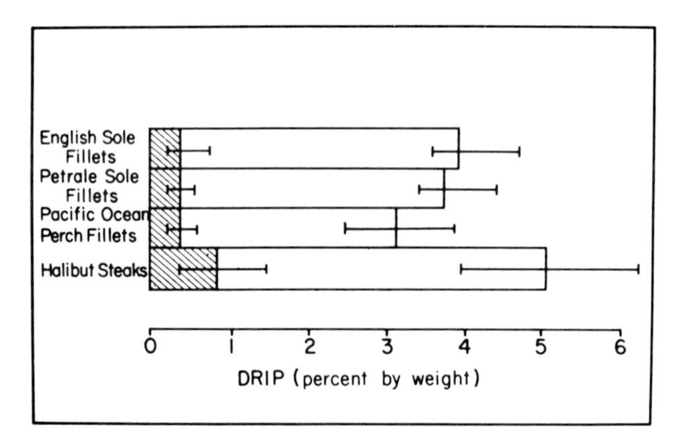

Figure 7.8 Loss of drip from several kinds of fin fish irradiated with 2 kGy and stored 3 weeks at 1°C. "Treated" (hatched bars) signifies treated with a solution of Na-TPP. Empty bars, untreated. From J. Spinelli, G. Pelroy, and D. Miyauchi, *Fish. Ind. Res.* **4,** 37 (1967).

Some irradiated fishery products such as haddock, shrimp, and king and Dungeness crab can be stored in air. Fatty fish, such as petrale sole and flounder, become rancid when irradiated and stored in air. Ozone formed from atmospheric O_2 during irradiation may be a significant cause of such rancidity development. Exclusion of O_2, both during irradiation and also in the subsequent handling, is indicated for fatty fishery products. The use of N_2 or CO_2 atmospheres during irradiation to replace air is ineffective, apparently due to the presence of active radiolytic products resulting from these gases during irradiation. The use of helium, however, is effective. The data of Table 7.5, showing sensory scores and TBA numbers for English sole fillets support these conclusions.

Color changes in some irradiated marine and freshwater animal foods occur and can affect the utility of irradiation. The flesh of salmon and lake trout undergoes color loss with doses as low as 1 kGy and may be completely bleached at high doses. Lobsters and shrimp undergo melanosis, which often occurs as "black spots." This phenomenon is complex and involves the action of the enzyme phenolase on tyrosine. With very fresh shrimp, irradiation reduces melanosis during subsequent refrigerated storage. It has been postulated that radiation inactivates phenolase through indirect action resulting from radiolytic products of water and O_2, which have diffused through skin joints between segments of the shrimp. On the other hand, the irradiation of shrimp of sufficient postmortem age, such as is often obtained in common commercial handling, has an effect opposite to what is obtained with very fresh shrimp. In the less-fresh shrimp some melanosis has already occurred. It has been theorized that quinone-like products of the reaction

Table 7.3

Overall Sensory Quality and Texture Panel Scores for Fish Fillets[a, b]

Species	Storage time (weeks)	Overall sensory score[c]		Texture score[c]	
		Untreated fillets	Phosphate-treated fillets	Untreated fillets	Phosphate-treated fillets
Pacific ocean perch	0	—	—	9.0	9.0
	1	7.2	7.8	8.3	9.0
	2	7.0	7.8	7.4	8.5
	3	7.0	7.5	7.1	8.7
Petrale sole	0	—	—	9.0	9.0
	1	8.5	8.0	9.0	8.1
	2	8.0	8.0	8.2	8.2
	3	7.2	6.8	7.2	7.8
English sole	0	—	—	9.0	9.0
	1	8.7	7.3	8.8	8.3
	2	8.3	7.2	8.2	8.2
	3	6.7	7.3	7.4	8.0

[a] From J. Spinelli, G. Pelroy, and D. Miyauchi, *Fish. Ind. Res.* **4,** 37 (1967).

[b] Fillets were irradiated with 2 kGy and stored at 1°C after treatment with 10% Na-TPP solution containing 2% NaCl.

[c] Ten-point sensory quality score: 9, best; 0, poorest.

sequence act as competitive inhibitors for the melanosis reaction and also function to protect the phenolase from radiation inactivation. Acceleration of melanosis is the result of radiation destruction of the quinone-like products. For crustaceans that are not strictly fresh, blanching with hot water so as to inactivate phenolase has been found to be effective in controlling melanosis, although it may alter the product adversely by lightening shell color and cooking the product somewhat.

The bacterium *Clostridium botulinum* occurs in certain marine and fresh waters. It is believed that rivers wash the organism from soils into lakes and marine waters and deposit it in coastal muds, from which it is transferred to the water creatures through their feeding. Types A, nonproteolytic B, C, E, and nonproteolytic F have been found in such muds. Types A and C require temperatures above 10°C for growth, and consequently cause no hazard for products that are refrigerated. Types B, E, and F grow and produce toxin at refrigeration temperatures as low as 3.3°C. Types B and F are only infrequently associated with fishery products. Type E, on the other hand, is often found in these products.

The incidence of *Clostridium botulinum* in fishery products is extremely variable. Some geographic areas have a high incidence, whereas other areas

Table 7.4

Effect of Na-TPP-Induced Inhibition of Proteolysis during Storage of Bombay Duck Fillets at 0 to 4°C[a]

Treatment		Activity[b]				
		0 Days	3 Days	5 Days	10 Days	20 Days
None	Cathepsin D	167.58	158.20	176.07	208.40	—
	Products	225.0	—	289.0	526.0	—
10% Na-TPP	Cathepsin D	55.89	47.93	40.95	41.35	—
	Products	168.0	—	197.0	214.0	—
Combination of 10% Na-TPP and irradiation (1 kGy)	Cathepsin D	26.96	20.12	27.93	22.70	33.80
	Products	116.0	—	131.0	148.0	220.0

[a] From V. Ninjoor, S. N. Doke, and G. B. Nadkarni, Storage stability and improved quality of fish products by enzyme suppression and gamma irradiation. *In* "Combination Processes in Food Irradiation." Int. Atomic Energy Agency, Vienna, 1981.

[b] Cathepsin D activity is expressed in terms of specific activity. The products of proteolysis (amino acids and peptides) are measured in terms of tyrosine equivalents (nmoles of tyrosine per milligram of protein).

Table 7.5

Sensory Score and TBA Numbers for English Sole Fillets Packed under Various Atmospheres, Irradiated at 2.5 kGy, and Stored at 1°C[a,b]

Storage time (days)	Air		O_2		H_2		CO_2		He		Vacuum	
	Sensory score	TBA no.	Sensory score	TBA no.	Sensory score	TBA no.	Sensory score	TBA no.	Sensory score	TBA no.	Sensory score	TBA no.
0	9	2.3	9	2.3	9	2.3	9	2.3	9	2.3	9	2.3
3	7	3.4	2	4.6	9	2.4	7	2.1	9	2.3	9	2.0
7	2	4.2	2	9.0	7	3.0	5	4.6	8	2.3	6	2.6
10	0	9.5	0	12.0	5	4.5	4	4.8	8	2.7	9	1.5
15	0	7.0	0	12.8	4	3.7	4	3.6	6	2.3	7	2.6
28	0	5.6	0	10.8	2	3.3	2	3.9	5	2.1	7	2.3
35	0	—	0	—	0	—	0	—	4	—	6	—

[a] From D. Miyauchi, J. Spinelli, G. Pelroy, and N. Stoll, "Application of Radiation Pasteurization Processes to Pacific Crab and Flounder," TID 25515. U.S. Atomic Energy Comm., Washington, D.C., 1965.

[b] Ten-point sensory quality scale: 9, best; 0, poorest. TBA number, mg malonaldehyde/kg.

have none or a very low incidence. Most positive findings have been in the northern hemisphere; this, however, may reflect fewer observations in the southern hemisphere, rather than restriction of the distribution of *C. botulinum* to northern waters.

When found, the incidence in marine and freshwater foods can vary from as little as 1 to over 50% and occasionally to 100% of the samples examined. The organism is found in the intestine of the fish, presumably carried there by the ingested food. The organism does not establish itself in the intestine. It is believed that contamination of the flesh occurs during postmortem product handling and processing. Cross-contamination among individual products is likely to occur at this time. The level of contamination has not been well determined but has been estimated to be low, not more than one spore per gram of fish.

Outbreaks of botulism associated with marine and freshwater animal foods have occurred over the past several decades. The preponderance have been due to type E. There is, however, no record of fresh (refrigerated) or frozen product being implicated. This can be ascribed partly to the fact that most raw products are cooked before consumption, which would cause inactivation of any toxin present. Additionally, endogenous spoilage bacteria would cause noticeable sensory spoilage before toxin production would occur. They, by their presence, may also inhibit growth of *Clostridium botulinum*, or through enzyme production, inactivate any toxin that is formed. Processing of these foods short of thermal sterilization or radappertization, however, may interfere with the just-noted circumstances that afford protection with unprocessed products. Radurization of marine and freshwater animal foods, by altering the normal content of spoilage bacteria, either simply numerically or by kind, has the potential of increasing the hazard from botulinum toxin.

Radurization doses are insufficient to accomplish inactivation of botulinum spores. A most critical factor, also, is the fact that types B, E, and F can grow and produce toxin at refrigeration temperatures, that is, as low as 3.3°C. Ordinary commercial handling of fishery products employs temperatures in the range 5–10°C. Because of the possibility of increasing the hazard of botulism through radurization of fishery products, a great deal of research has been carried out, but without results that provide a complete resolution of the problem.

The important question, the answer to which can affect the safety of radurized fishery products, is whether recognizable sensory spoilage occurs prior to toxin formation. An affirmative answer would make the hazard from radurized fishery products no different from that obtained without irradiation. If, on the other hand, recognizable sensory spoilage does not occur prior to toxin formation, it is clear that a potential hazard exists.

Figure 7.9 shows the maximum product life of cod fillets, both irradiated and not irradiated and inoculated with spores of *Clostridium botulinum* type E.

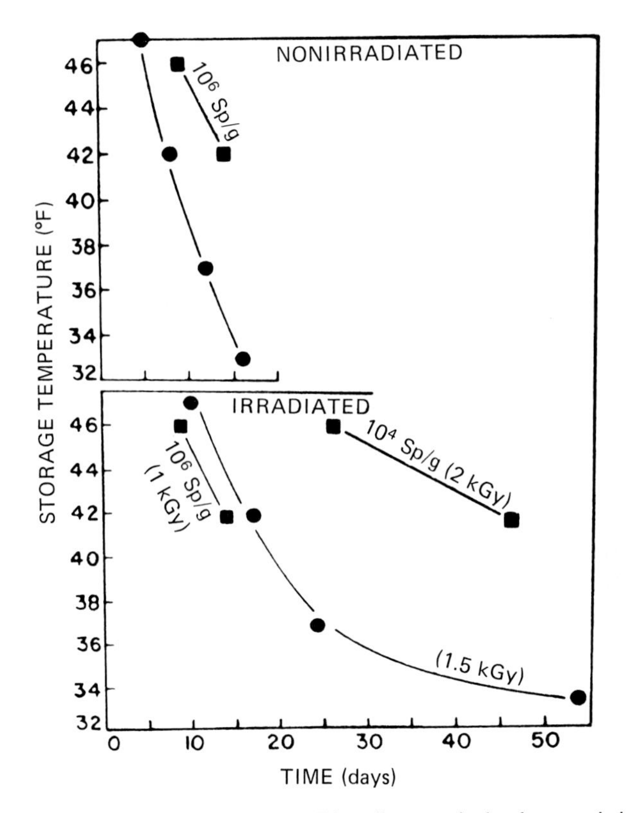

Figure 7.9 Maximum product life and type E botulinum toxin development in irradiated and nonirradiated cod fillets. ●, Shelf-life based on odor of raw fillet; ■, toxin development for indicated inoculum level in spores per gram. From J. W. Slavin, L. J. Ronsivalli, and T. J. Conners, Status of the research and development studies on radiation pasteurization of fish and shellfish in the United States. *In* "Food Irradiation," Proceedings of Symposium, Karlsruhe. Int. Atomic Energy Agency, Vienna, 1966.

Maximum product life was determined by subjective judgment of odor of raw product. Also shown is the storage time at which toxin was detected. In this experiment the spore inoculum level of 10^4 or 10^6 spores per gram presumably was substantially greater than the naturally occurring levels of contamination. The data indicate that, without irradiation, even with 10^6 spores per gram, spoilage preceded toxin formation. With irradiation (1 kGy), however, the same level of inoculation produced toxin *prior to spoilage*. The 10^4 level did not follow the same pattern, toxin formation lagging behind spoilage considerably. Noteworthy also is that toxin formation was not observed at temperatures below 5.6°C (42°F), regardless of inoculum level or the use of irradiation.

The data of Table 7.6 show the effect of dose on the time relationship between product life and toxin formation in haddock fillets. Important in this relationship is the storage temperature. At 5.6°C toxin formation lags behind spoilage by a substantial margin. At 7.8°C a similar margin of safety exists for irradiation with 1 kGy, but for 2 kGy the margin is significantly reduced, and with the higher inoculum level of 10^4 spores per gram, toxin formation precedes spoilage. Whether an O_2-impermeable film for packaging the food is used or not appears to be unimportant in these relationships.

Table 7.7 shows similar data for petrale sole fillets. Fillets stored at 5.6°C and irradiated with a dose of 2 kGy become toxic about the same time as spoilage occurs. A moderate safety margin was found for fillets irradiated with 1 kGy. Lowering the storage temperature to 3.3°C removed the toxin hazard. Regardless of other factors, petrale sole fillets have been found to provide a better substrate than haddock for growth of *Clostridium botulinum.*

Factors which affect the relationship between product life and toxin formation include dose, temperature of storage, level of spore contamination, species of animal food, and possibly packaging as it affects the O_2 content of the surrounding atmosphere. The use of a dose sufficient to secure a somewhat large extension of product life can lead to toxin formation within the period of the product life, provided other relevant factors, such as storage temperature, permit toxin formation. Because of this role of dose, it has been proposed that only low doses of the order of 1 kGy be employed. Hopefully, this would allow storage temperatures above 3.3°C.

Attention has been directed to the importance of the level of spore contamination. Criticism has been made of conclusions based on experiments employing inocula of the order of from 100 to 1 million times the natural level. The evidence is clear that the time for toxin formation increases significantly as the spore population is smaller, and if it is correct that the natural contamination level is less than one spore per gram, toxin formation in commercial fishery products may, in fact, not precede spoilage, even with doses greater than 1 kGy. It is clear that additional information is needed in this area.

Exclusion of O_2, as may be obtained by vacuum packaging, causes several effects. For high-fat fishery products it lengthens the product life by controlling rancidity development. It prevents the growth of aerobic organisms, and conversely, aids the growth of anaerobes. (*Clostridium botulinum* is an anaerobe.) Vacuum packaging appears necessary for at least some fishery products and, consequently, when used must be regarded as a factor that can increase the potential of a botulinum hazard.

The difference among species that affect growth of *Clostridium botulinum* appears to be related to composition. Fatty fishes contain the sugars ribose and glucose, which are essential for the low-temperature growth of *C. botulinum* type E. It is believed that this compositional difference accounts for the species variation observed.

Table 7.6

Relationship between Maximum Product Life and Time for *Clostridium botulinum* Type E Toxin Development in Inoculated Haddock Fillets[a]

Dose (kGy)	Inoculum[b]	Packaging[c]	Storage temperature (°C)	Days for maximum product life	Days for earliest toxin development
0	10^2	P	7.8	11	48
0	10^4	P	7.8	11	22
0	10^4	I	5.6	18	>55
1	10^2	P	7.8	20	>50
1	10^4	P	7.8	20	23
1	10^4	I	5.6	38	>55
2	10^2	P	7.8	30	40
2	10^4	P	7.8	30	<12
2	10^4	I	5.6	60	>60

[a] From M. W. Ecklund, *Food Technol.* (*Chicago*) **36,** 107 (1982). Copyright © by Institute of Food Technologists.
[b] Expressed as number of spores per gram.
[c] P, Permeable to O_2; I, impermeable to O_2.

Table 7.7

Relationship between Maximum Product Life and Time for *Clostridium botulinum* Type E Toxin Development in Inoculated Petrale Sole Fillets Stored in O_2-Impermeable Film[a, b]

Dose (kGy)	Storage temperature (°C)	Days for maximum product life	Days for earliest toxin development
0	3.3	18	>40
1	3.3	32	>70
2	3.3	43	>70
0	5.6	12	>42
1	5.6	20	32
2	5.6	30	28

[a] From M. W. Ecklund, *Food Technol.* (*Chicago*) **36,** 107 (1982). Copyright © by Institute of Food Technologists.
[b] Inoculum amount: 10^2 spores per gram.

While, as noted, research has not provided a universally accepted resolution of the concerns for the safety of radurized fishery products, as related to botulinum toxin, each of the following two conditions appear to be generally regarded as providing safe products: (1) restriction of radurization to products secured in locations that have been demonstrated to be free of contamination by *Clostridium botulinum* and (2) handling of product post-irradiation at temperatures below 3.3°C. Other possibilities may include (1) limitation of the time before consumption to less than that needed to produce toxin and (2) restriction to species demonstrated to be poor media for toxin development.

In some locations refrigeration may not be available. To meet the need for preservation of fishery products, short of thermal sterilization and radappertization, irradiation combined with another procedure has been proposed. Heat and radiation along with suitable packaging, or heat, radiation, and partial drying provide shelf-stable shrimp (see Chapter 11). Chub mackerel, boiled in saturated brine for 5 to 10 min and subsequently irradiated, keep significantly longer without refrigeration (0 kGy, 2 days; 1 kGy, 9 days; 2 kGy, 12 days; and 3 kGy, 14 days). The salt content of the boiled chub mackerel ranges from 3.5 to 4.6%, somewhat below the 4.87% found to be needed to prevent the outgrowth of *Clostridium botulinum* type E spores. It seems unlikely, however, that the spores of types B, E, and F, which are relatively less heat resistant than other types, would survive the brine-boiling process. Also the irradiation process would reduce the spore population by as much as one log cycle. To provide safety from other types of *C. botulinum*, especially A and B, storage below 10°C has been recommended, a requirement that conflicts with the objective of obviating the use of refrigeration.

Radurization of fresh fishery products combined with refrigeration appears to provide beneficial effects in extending product life. The optimal doses proposed for various fishery products are given in Table 7.8. The optimal dose in considered to be that which provides the greatest product life without changing the normal characteristics of the product.

From a microbiological viewpoint it is clear that the irradiation process should be applied as soon after catch as possible. Since catching vessels do not always return promptly to port, this suggests that there is value in performing the irradiation at sea. Depending upon the length of stay at sea, a second irradiation on arrival at the port might also prove of value.

It is clear, also, that fishery products likely to contain spores of *Clostridium botulinum* types B, E, or F must be handled after irradiation in a manner so as to avoid any added hazard of toxin formation.

The extension of product life of dried or of dried and salted fishery products through irradiation has been shown to be possible. Such products contain both molds and bacteria. At water activities (A_w) that exist in dried and cured fishery products, mold growth can occur. A dose of about 5 kGy is needed to prevent such growth. The use of fungistatic agents such as sorbic acid enables reduction of the dose requirement.

Table 7.8

Optimal Radiation Dose Levels and Shelf-Life at 1°C for Fish and Shellfish Aerobically Packed in Hermetically Sealed Cans[a]

Seafood	Optimal dose (kGy)	Product life (weeks)
Oysters (shucked)	2.0	3–4
Shrimp	1.5	4
Smoked chub	1.0	6
Yellow perch fillets	3.0	4
Petrale sole fillets	2.0	2–3
Pacific halibut steaks	2.0	2
King crab meat (cooked)	2.0	4–6
Dungeness crab meat (cooked)	2.0	3–6
English sole fillets	2–3	4–5
Soft-shell clam meats	4.5	4
Haddock fillets	1.5–2.5	3–4
Pollock fillets	1.5	4
Cod fillets	1.5	4–5
Ocean perch fillets	1.5–2.5	4
Mackerel fillets	2.5	4–5
Lobster meat (cooked)	1.5	4

[a] From J. W. Slavin, J. T. R. Nickerson, L. J. Ronsivalli, S. A. Goldblith, and J. J. Licciardello, *Isot. Radiat. Technol.* **3**(4), 366 (1966).

Bacterial spoilage can occur in products of intermediate moisture levels (20–40%). Doses of 1 to 4 kGy are effective in controlling bacterial growth at storage temperatures of about 30°C. Product life is about doubled.

III. RADICIDATION

Pathogenic non-spore-forming bacteria occur in fishery products and are the cause of a substantial amount of food poisoning. Of the total number of reported cases of food poisoning, the following percentages have been identified as originating with fishery products: in the United Kingdom, 4%; United States, 14%; Japan, 60%. The high incidence in Japan has been associated with the common consumption of raw fish.

Doses of 1.5 to 2.5 kGy applied to crab meat are effective in inactivating coagulase-positive staphylococci, coliform, fecal coliform, and *Escherichia coli* bacteria. *Vibrio parahaemolyticus*, a principal cause of food poisoning in countries such as Japan, has D_{10} values in the range 30–160 Gy and, therefore, also is amenable to inactivation by low doses.

The use of low doses, such as are employed in radurization of marine and freshwater animal foods, probably does not completely clear these products of the pathogenic non-spore-forming bacteria. The foods, being complex in chemical nature, apparently exert a protective action which causes a "tailing effect" at higher doses, as illustrated in Fig. 7.10 for *Streptococcus pyogenes* in several shellfish. There is evidence, however, that a further reduction of the numbers of a pathogen may occur during postirradiation refrigerated storage, as may be seen from the data of Fig. 7.11 for *Salmonella typhimurium* in oysters.

The effect of doses in the radurization range on non-spore-forming pathogens in fresh fishery products stored at refrigeration temperatures requires further study.

Almost no information has been secured on the use of irradiation in connection with parasites carried by marine and freshwater animal foods. *Anisakis* larvae in salted herring are not inactivated by a dose of 6 kGy. This dose and larger ones needed to secure inactivation are too great to provide herring of acceptable flavor.

The inactivation of pathogenic viruses in fishery products also requires doses that are too high to be usable.

Radicidation of fish meal is of interest due to its use in animal feeds. Fish meal is a principal source of salmonellae bacteria which cause infection of domestic meat animals and subsequently cause human infection through the meats derived from these animals. This subject is covered in Chapter 10 (Section VII, F).

In connection with some fishery products, particularly shellfish, is the

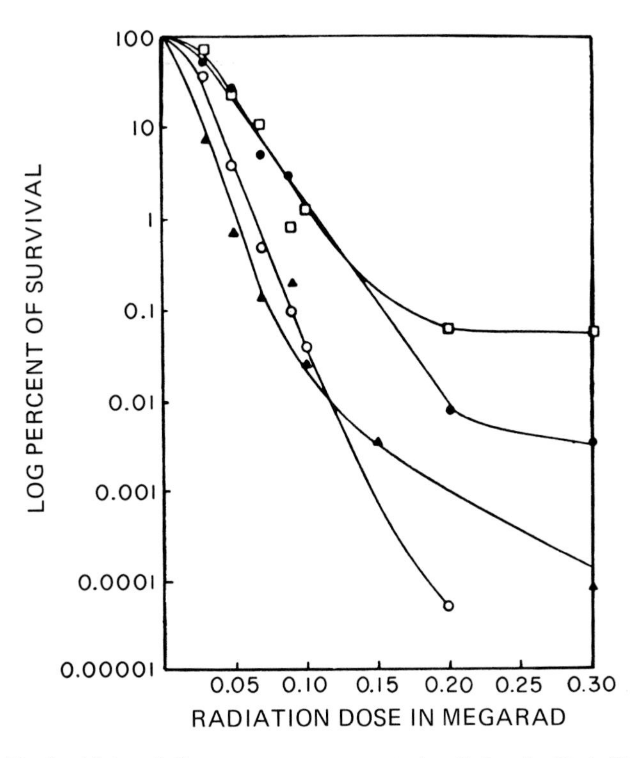

Figure 7.10 Sensitivity of *Streptococcus pyogenes* to irradiation in Hartsell's broth (○), oysters (□), crab (▲), and shrimp (●). From D. J. Quinn, A. W. Anderson, and J. F. Dyer, The inactivation of infection and intoxication microorganisms by irradiation in seafood. *In* "Microbiological Problems in Food Preservation by Irradiation." Int. Atomic Energy Agency, Vienna, 1967.

controversial use of irradiation to secure a reduction in numbers of microorganisms for the purpose of enabling compliance with government or private industry regulatory microbial standards. While irradiation can reduce the number of bacteria in these foods, doses likely to be employed will have little effect on viruses, which also may be present.

IV. RADAPPERTIZATION

The same general technology developed for the radappertization of meats has been utilized in efforts to develop radappertized fishery products. As with meats, the key aspects of the procedures used are (1) basing the dose for sterilization on the inactivation of spores of *Clostridium botulinum* and on the $12D_{10}$ concept, (2) inactivation of endogenous enzymes by heat, and

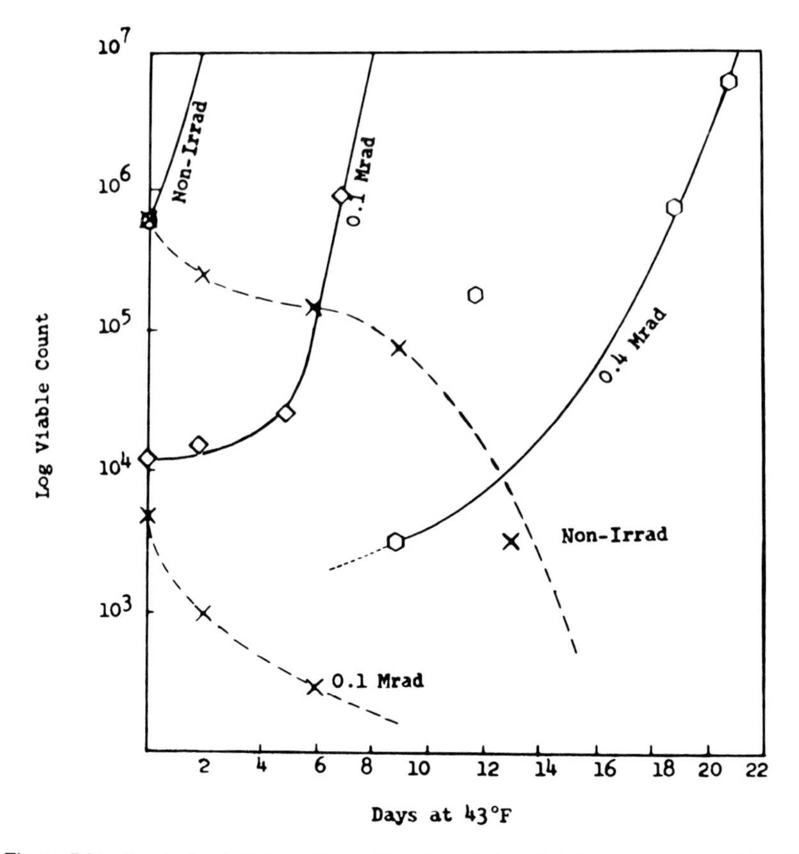

Figure 7.11 Survival of *Salmonella typhimurium* in irradiated oysters. —— Microbial count; –––– *Salmonella* count. From R. O. Sinnhuber and J. S. Lee, "Effect of Irradiation on the Microbial Flora Surviving Irradiation Pasteurization of Seafoods," RLO-1950-1. U.S. Atomic Energy Comm., Oak Ridge, Tennessee, 1966.

(3) irradiation at subfreezing temperatures. Other aspects, such as packaging, also have been borrowed from the meat procedures.

The results obtained have not been fully satisfactory. While product of an initial acceptable quality can be made, deterioration occurs on storage at ambient temperatures. Deterioration is evidenced mainly by marked browning. The browning is due to nonenzymatic Maillard-type reactions involving endogenous sugars such as ribose and glucose. Various measures to prevent this browning, such as leaching of the fish to remove the sugars and the use of sulfur dioxide or calcium chloride to block the reaction, have not proved to be adequately effective. Figure 7.12 shows data on the results on color and panel acceptance obtained with haddock fillets irradiated at $-80°C$ with a dose of

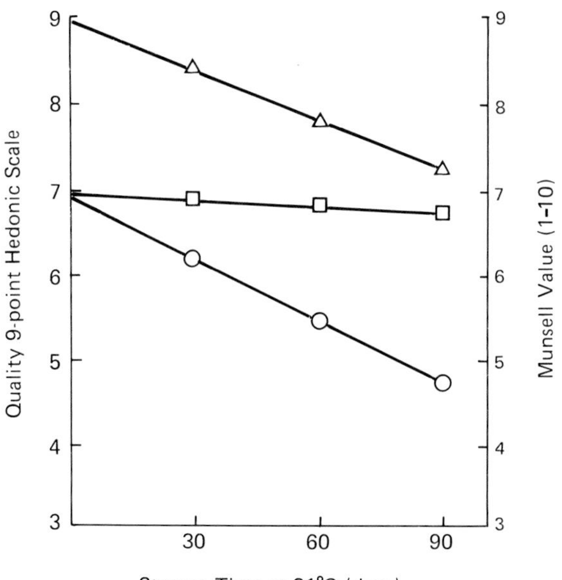

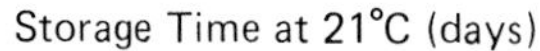

Figure 7.12 Effect of storage at 21°C on panel acceptance and the color of prefried (bisulfited) radappertized haddock fillets. Irradiation at −80°C and with a dose of 45 to 59 kGy. □, Acceptance scores of frozen control; ○, acceptance scores of irradiated fillets; △, color values of irradiated fillets. From L. J. Ronsivalli, "Development of Radiation Sterilized Fish Items," Tech. Rept. 72-42-FL. U.S. Army Natick Labs., Natick, Massachusetts, 1972.

45 kGy and stored at 21°C. Reduction of panel acceptance with storage time was due principally to color change, although flavor deterioration also occurred.

Problems have been encountered in the preparation of certain fishery products, especially steaks and fillets. These problems have involved the obtaining of satisfactory structure and texture. Tuna, salmon, and cod fish cakes, as well as tuna and salmon loaves have been prepared.

It is clear that additional development is needed in order to secure satisfactory radappertized fishery products.

V. DISINFESTATION

Although preservation of fishery products by drying is a very ancient method, it is very little used presently in developed countries. In less developed countries, however, especially those with warm climates, it remains an important food preservation method. Drying is frequently combined with salting and/or smoking.

Dried and smoked fish, and to a lesser extent, salted fish are subject to insect infestation. In Thailand, for example, 60–70% of commercially processed dried and salted fish were found to be infested. In the regions of the Middle Niger and the Chad Basin in Chad, losses of dried and smoked fish due to insects may be as much as 25 to 30%. Infestation may occur at various locations as the product is produced and distributed.

The kind of insect that infests dried fishery products varies with the location. The following have been reported:

Dermestes (genus)
Necrobia (genus)
Piophila casei (L.)
Phaenicia (*Lucilia*) *cuprina* (Wied.)
Lucilia illustris (Meigen)
Chrysomya megacephala (Fab.)
Sarcophagidae (family)
Lasioderma spp.

The dose for lethality varies with the species, with the stage of the insect at the time of irradiation (whether egg, larva, or pupa), and with the moisture content of the product. In dry fish (<20% moisture), 0.15 kGy prevents insect development. The larvae remain alive, but with reduced feeding rates. If pupation occurs, there is no further development. With a dose of 0.25 kGy, larvae and adult insects are inactive and do not feed. A dose of 0.3 kGy has been proposed for insect disinfestation of fish of less than 20% moisture. A higher dose of about 0.5 kGy is needed for the disinfestation of products of intermediate moisture (20–40%). In general a dose of 0.5 kGy has been recommended.

The success of radiation disinfestation of dried fishery products is necessarily tied to concomitant use of packaging that is effective in preventing reinfestation.

SOURCES OF ADDITIONAL INFORMATION

Anonymous, "Preservation of Fish by Irradiation," Proceedings of a Panel. Int. Atomic Energy Agency, Vienna, 1970.

Anonymous, "Radurization of Scampi, Shrimp and Cod," Tech. Rep. Series No. 124. Int. Atomic Energy Agency, Vienna, 1971.

Anonymous, FAO/IAEA Advisory Group Meeting on Radiation Treatment of Fish and Fishery Products. *Food Irradiat. Newsl.* **2**(2), 27, Int. Atomic Energy Agency, Vienna, 1978.

Anonymous, "Recommended Code of Technological Practice for Insect Disinfestation of Dried and Cured Fish by Irradiation." Int. Atomic Energy Agency, Vienna, in press.

Eklund, M. W., Significance of *Clostridium botulinum* in fishery products preserved short of sterilization. *Food Technol.* (*Chicago*) **36,** 61 (1984).

Giddings, G. G., Radiation Processing of Fishery Products. *Food Technol.* (*Chicago*) **38,** 61 (1984).

Hannesson, G., Objectives and Present Status of Irradiation of Fish and Seafoods. Food Irradiation Information No. 1. Int. Project in the Field of Food Irradiation, Karlsruhe, West Germany, 1972.

Hobbs, G. *Clostridium botulinum* in irradiated fish. Food Irradiation Information No. 7. Int. Project in the Field of Food Irradiation, Karlsruhe, West Germany, 1977.

Josephson, E. S., Radappertization of meat, poultry, finfish, shellfish and special diets. *In* "Preservation of Food by Ionizing Radiation," (E. S. Josephson and M. S. Peterson, eds.), Vol. III. CRC Press, Boca Raton, Florida, 1983.

Licciardello, J. J., "Code of Practice for Irradiation of Seafoods." Int. Atomic Energy Agency, Vienna, in press.

Nickerson, J. T. R., Licciardello, J. J., and Ronsivalli, L. J., Radurization and radicidation of fish and shellfish. *In* "Preservation of Food by Ionizing Radiation," (E. S. Josephson and M. S. Peterson, eds.), Vol. III. CRC Press, Boca Raton, Florida, 1983.

Suhadi, F., and Thayib, S. S., Toxin formation by *Clostridium botulinum* type B in radurized fish. *In* "Wholesomeness of the Process of Food Irradiation," IAEA-TECDOC-256. Int. Atomic Energy Agency, Vienna, 1981.

CHAPTER 8

Fruits, Vegetables, and Nuts

I. FRUITS

A. Introduction

The irradiation of fruits may be undertaken for various purposes, namely (1) radurization, (2) delay of ripening, (3) inhibition of senescence, (4) radicidation, (5) radappertization, and (6) insect disinfestation. Of these possible purposes, all except radicidation have been studied.

Except for radappertization, the general purpose of irradiation is to maintain these foods in the fresh state. Irradiation employed to delay ripening or to inhibit senescence operates not on microbial contaminants but on the foods themselves and accomplishes the desired result by acting upon one or more biological processes of the still "live" fruits or vegetables. Except for radicidation and some situations involving insect disinfestation, the general objective of irradiation of fruits and vegetables is to secure preservation.

In the attainment of this objective, the intent is the retention of the normal characteristics of the fruits and vegetables. If there is a single principal problem in securing this result, it is the effect of irradiation to cause softening and loss of the normally firm texture. This softening has been ascribed to

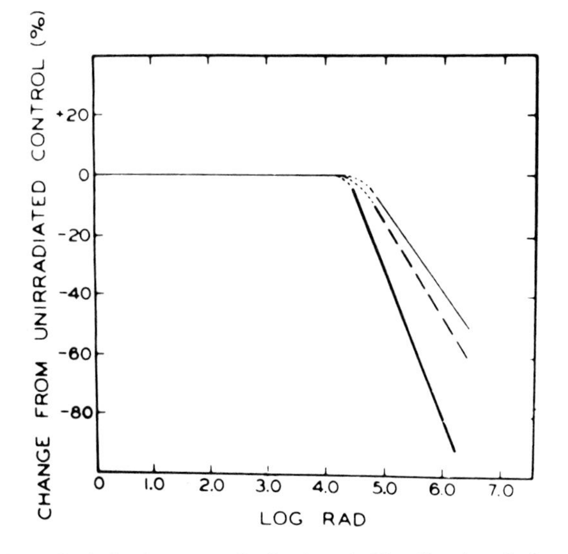

Figure 8.1 Overall relation between softening (——) of irradiated apple tissue as measured by "punch" test (crushing load) and degradation of tissue pectins (----) and cellulose (——) as measured by changes in viscosity. From L. M. Massey, Jr., Tissue texture and intermediary metabolism of irradiated fresh fruits and vegetables. *In* "Preservation of Fruits and Vegetables by Radiation." Int. Atomic Energy Agency, Vienna, 1968.

degradation of pectin and cellulose components of fruits and vegetables, as is illustrated by the data of Fig. 8.1 on apple tissues. Radiation-induced release of bound calcium from the plant tissues occurs and has been related to softening through disturbance of the calcium–pectin association normally present. Soaking blueberries, sweet cherries, or strawberries in a 0.5% solution of $CaCl_2$ for 30 min prior to irradiation causes a rapid refirming of the fruit after irradiation. With some fruits and vegetables the softening effect occurs at doses below those needed to obtain the desired technical effect, such as reduction of microbial population. Primarily because of this problem, which prevents retention of a very important normal characteristic, the irradiation of a number of fruits and vegetables to secure some technical effects is not possible.

The unusual aspect of fresh fruits and vegetables, namely that they are "live" foods, plus the fact that the desired technical effect of irradiation in certain uses is related to securing changes in the life processes of these foods, leads to an indirect effect of irradiation that is different from the usual, not only because of the nature of the change, but also in that the change may be delayed timewise to a great extent.

In discussing the effects of irradiation on fruits and vegetables, individual treatment often is needed, since the response to irradiation varies considerably and grouping of foods is not possible.

In the succeeding pages it can be observed that there are difficulties in using irradiation in the treatment of fresh fruits and vegetables, which considerably limit this use. Unquestionably, these are real difficulties which to a large degree are inherent in the nature of fresh fruits and vegetables, and, therefore, largely insurmountable. Further research, however, can be expected to yield improvements which will overcome at least some of the difficulties. The following comments are offered as suggestions as to what may be helpful in this regard:

1. It is essential to have knowledge of the postharvest physiology of fruits and vegetables in order to obtain the best advantages from irradiation. As part of this, an understanding of how irradiation affects the food also is important. Such understandings are especially needed in connection with the following:

   a. Delay of ripening
   b. Delay of senescence
   c. Changes in the food caused by irradiation which make it more susceptible to microbial decay

2. Often a new procedure is expected to require no change in previous practices. Yet, adjustments may be needed to accommodate to the new procedure. With irradiation some modification of harvesting, handling, storage, and marketing practices usually employed with fresh fruits and vegetables may be needed in order to fit in with the needs and opportunities associated with irradiation. For example, in some cases how the food is packaged may greatly affect the benefits that can be derived from its irradiation. Packaging used previously may not be suited for the irradiated food and an appropriate change may be needed.

3. It is not uncommon that certain varieties of fruits are better suited than others to certain types of processing. That this is so with irradiation is evident in some cases. Recognition of this aspect is important in selecting fruits and vegetables for irradiation.

4. Since irradiation loses its effectiveness in delaying ripening when fresh fruits are irradiated after the onset of the climacteric, it is necessary to be certain that irradiation is in fact being done prior to the beginning of the climacteric. It appears to be insufficient to base this determination on subjective or sensory criteria. Instead, objective criteria, such as respiration rate, may be more useful and accurate.

5. In experimental investigations of particular fruits in connection with microbial decay, it is important to know the kind and amount of infection that is being encountered. Without such information meaningful results may not be obtained.

6. In the commercial use of irradiation, where possible, adjustment of the dose to the nature of the microbial infection encountered in the particular circumstances of the usage can be made. Some more radiation-resistant microorganisms are not universally distributed, and if absent in a particular situation, a smaller dose may be sufficient.

B. Radurization

1. Introduction

The term "postharvest disease" has been used customarily to refer to spoilage of fresh fruits and vegetables caused by microorganisms. Fruit spoilage results principally from infections of filamentous fungi and also of yeast; bacterial infections are of minor importance in fruit spoilage. Radurization of fresh fruit is aimed at control of these spoilage microorganisms.

Contamination of fresh fruits and vegetables with fungi, yeasts, and bacteria can occur in several ways. Field infection can occur during blossoming, and postharvest growth of the organism may occur within the food tissue. The epidermis protects against invasion, but wounds and cuts incurred during harvesting and handling may serve as entry ports. Contact infections can occur in storage and handling and spread infection from infected to noninfected product.

The lethality doses for the various species of yeasts associated with fresh-fruit spoilage lie between 4 and 20 kGy. Fungi are inactivated with doses between 1.5 and 6 kGy (see Table 4.5). In general these doses are too great for many fruits in that they cause unacceptable texture changes.

A somewhat unexpected result can occur in the radurization of fruits. As dose is increased with some fruits, instead of being reduced, spoilage increases. This effect has been explained generally on the basis of tissue damage by the radiation, which makes it more vulnerable to attack by the spoilage microorganisms. While cell rupture or increased permeability of cell membranes, tissue softening, weakening of the skin, and other similar radiation-induced changes can be involved in the increase of spoilage, a different specific mechanism has been proposed. Naturally occurring antibiotic compounds (phytoalexins) have been observed to control fungi that cause spoilage of fruits. It has been postulated that irradiation reduces the capability of a fruit to form these antifungal substances and in this way it is less able to resist attack by fungi. This concept of the action of radiation in larger doses to promote rather than inhibit spoilage is supported by experimental evidence not only for fruits, but also for vegetables.

2. Berries

Generally berries are very perishable foods, and irradiation is directed to the control of spoilage fungi. Unfortunately these foods can tolerate doses no greater than 2 to 3 kGy if excessive softening is to be avoided. Findings obtained in the 1960s indicated that a dose of 2 kGy combined with refrigerated postirradiation handling provided some life extension of strawberries without undue softening.

The relationship between dose and life extension of strawberries is shown in Fig. 8.2. The effect of packaging (0.03-mm perforated polyethylene film) to reduce postirradiation infection also is shown. Such packaging can extend the fruit life by a factor of 3.

Somewhat similar effects occur with other berry fruits such as red and black raspberries.

Part of the problem in evaluating the utility of irradiation of berry fruits is that its effectiveness is greatly influenced by the initial level of mold contamination, which, in turn, is affected by the growing conditions, e.g., moisture level whether wet or dry. In some locations, at least, improved cultivation procedures for strawberries have reduced this level, and in view of this, it appears probable that this use of irradiation should be restudied.

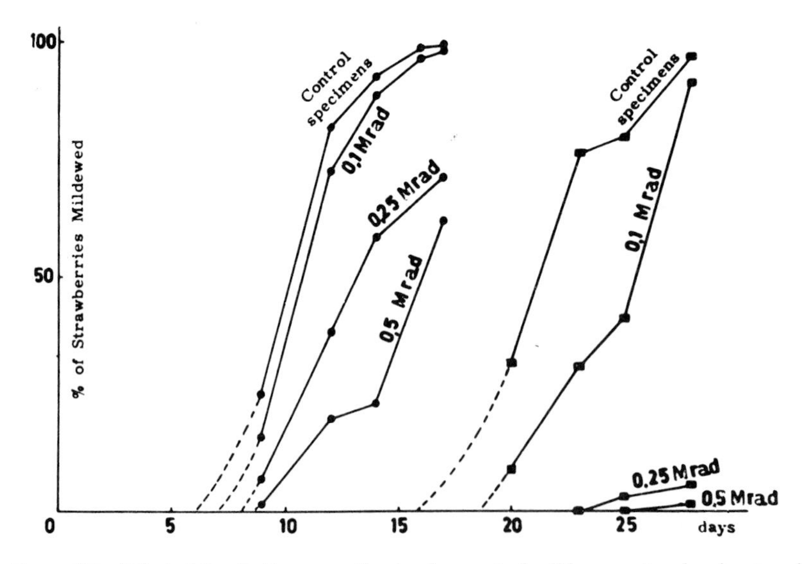

Figure 8.2 Effect of irradiation upon the development of mildew on strawberries stored at 15°C. Secondary infection occurred with berries not protected by 0.03-mm perforated polyethylene film. ●, Secondary infection; ■, no secondary infection. From M. Herregods and M. de Proost, *Food Irradiat.* (*Saclay*) **4**(1–2), A35 (1963).

In addition to softening, irradiation, especially above doses of 3 kGy, can cause color and flavor changes. Increased sweetness of strawberries results from a reduction of the normal acidity.

3. Citrus

The objective of radurization of citrus fruits is the control of stem-end rot due to green and blue molds. The dose required to control this kind of infection is related to the state of the infection at the time of irradiation. Established infections require higher doses than those that are at an incipient stage. The specific organism present also affects the dose requirement. *Alternaria citri*, for example, on tangerines is little affected by doses as high as 7.5 kGy, whereas *Penicillium italicum* and *P. digitatum* are well controlled by doses of 1 to 2 kGy.

Pitting of the peel can be one consequence of radurization. Figure 8.3 shows for 'Shamouti' oranges the percentage of maximum possible damage as related to dose. Pitting damage develops after irradiation and becomes evident in 4 to 5 days with storage at ambient temperatures. This damage may be due to depolymerization of pectins, which causes albedo tissue to collapse. The damage tends to be concentrated in the stem area. Terpene compounds in

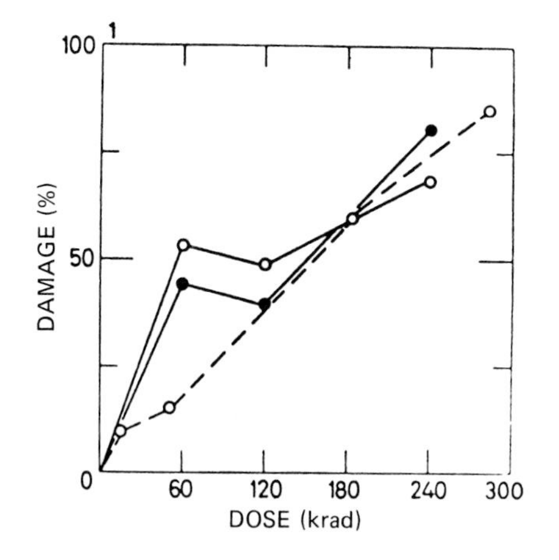

Figure 8.3 Damage to peel of 'Shamouti' oranges indicated as percentage of maximum damage possible, caused by irradiation with various doses 5 days after picking. S and N indicate different groves. ●——●, S—1966; ○——○, N—1966; ○----○, N—1965. From S. P. Monelise and R. S. Kahan, Effect of gamma radiation on appearance, composition and enzymatic activities of citrus fruits. *In* "Preservation of Fruit and Vegetables by Radiation." Int. Atomic Energy Agency, Vienna, 1968.

Table 8.1
Flavor Acceptability Ratings of 'Temple' Orange Juice as Related to Dose and Storage Time at 2°C[a, b]

	Dose (kGy)		
Days	0	1	2
0	2.9	2.8	2.1
7	2.8	2.8	2.2
14	2.6	2.6	2.3
21	2.7	2.6	2.3

[a] From R. A. Dennison and E. M. Ahmed, "Effects of Low Level Irradiation upon the Preservation of Food Products," ORO-680, U.S. Atomic Energy Comm., Washington, D.C., 1971.
[b] Ratings: 3, good; 2, fair; 1, poor.

irradiated 'Ovale' oranges increase with dose and postirradiation holding time. It has been hypothesized that the skin damage associated with irradiation is due to the diffusion of terpene compounds in the exocarp cells. The severity of the peel damage increases with dose, storage time, and storage temperature. Storage at 2°C, either alone or in combination with a 5-min dip in 53°C water, reduces peel injury. The hot-water dip also reduces the incidence of mold infection on storage.

Various other effects on citrus fruits have been noted. Irradiation causes softening, which is somewhat reduced with storage. Doses up to 2 kGy generally do not seriously affect flavor, as may be concluded from the data of Table 8.1, on the acceptability of 'Temple' orange juice. Doses as low as 0.6 kGy, however, produce detectable flavor and odor changes in grapefruit. Reduction of ascorbic acid content seldom exceeds 10%, as demonstrated by the data of Table 8.2, but increases with storage. An increase in weight loss on storage has been related to irradiation at doses above 2 to 3 kGy. Irradiation may increase the respiration rate, but this effect disappears as the fruit is held. Lemons develop internal cavities along the segment walls. This has been ascribed to formation of gas in the fruit. Irradiation of lemons hastens the death of the calyces (buttons) and can lead to *Alternaria* rot. Color changes of various kinds occur. Doses of 2 kGy cause browning of lemons. Lower values accelerate the degreening of lemons, but 0.3 kGy delay the degreening of limes. California navel oranges irradiated with doses of 2 to 3 kGy become more

Table 8.2

Changes in Reduced Ascorbic Acid of Tangerines as Related to Irradiation, Storage Temperature, and Storage Duration[a,b]

Dose (Gy)	0 Days		10 Days		20 Days	
	2°C	28°C	2°C	28°C	2°C	28°C
0	31.19	31.19	28.13	27.65	27.71	24.71
400	32.48	32.48	27.68	26.93	25.51	24.78
800	29.19	29.19	26.26	25.88	23.85	23.10
1600	29.42	29.42	25.23	24.07	23.08	20.20

[a] From R. A. Dennison and E. M. Ahmed, "Effects of Low Level Irradiation upon the Preservation of Food Products," ORO-680. U.S. Atomic Energy Comm., Washington, D.C., 1971.

[b] Expressed as milligrams per 100 ml.

orange in color, but Florida navel and 'Temple' oranges undergo bleaching and become yellow-orange in color. The flesh color of some products becomes brown.

While there is some variation in response with kind and varieties of citrus fruits, the radurization dose to control stem-end rot appears to be no less than 1.5 kGy. At this dose level unacceptable fruit damage occurs, especially to the peel, and for this reason, this use of irradiation is not likely to be feasible.

4. Grapes

Decay of grapes is caused primarily by *Botrytis cinerea.* Other fungi, however, may be present. Doses as low as 1.25 kGy control *B. cinerea.* Doses as high as 3 kGy, however, fail to inactivate *Cladosporium* and *Pullalaria* spp. Doses greater than 3 or 4 kGy accelerate decay. Dipping grapes in water at 50°C for 5 min plus irradiation with a dose of 1 kGy provides life extension both at ambient and refrigeration temperatures.

There is substantial variation of response to irradiation with variety of grape. 'Razaki,' 'Hoanes,' 'Sideritis,' white 'Muscat,' and 'Chasselas' varieties of table grapes do not undergo noticeable changes in color, texture, and other sensory characteristics with doses up to about 2 kGy. 'Thompson' and 'Emperor' varieties, however, become objectionably soft with a dose of 1 kGy. 'Fredonia' grapes exhibit undesirable color and flavor changes with a dose of 1.5 kGy. Doses of 0.2 to 0.3 kGy applied to white 'Muscat' and 'Chasselas' grapes enable their storage at 25°C for 15 to 30 days. These varieties not irradiated have a normal life of only 5–7 days at 25°C.

5. Olives

Fresh green olives may undergo decay due to growth of molds such as *Cladosporium* and *Aspergillus* spp. Doses in the range 1–4 kGy soften fresh green olives and make them more susceptible to decay. Irradiation also causes internal and external discolorations. In view of these undesirable effects, radurization of fresh green olives appears infeasible.

6. Pome Fruits

a. *Apples*

Spoilage of apples results from infections of *Penicillium* spp., *Gloesporium* spp., *Aspergillus niger*, *Botrytis* spp., and other molds. A dose of 2 kGy may be adequate to control these organisms in apples, at least in some situations.

Various adverse effects of irradiation on apples have been observed, including softening and internal breakdown (core flush). The interaction with environmental factors such as storage temperature has a pronounced effect on such changes. Apple variety also is a highly significant factor. Unless other information becomes available radurization of apples appears to be impractical.

b. *Pears*

Botrytis cinerea is the most important decay-causing organism of pears. A dose of 2 to 3 kGy significantly delays the appearance of decay spots on pears stored at ambient temperatures.

A number of adverse side effects of irradiation occur, including softening, blemishes of the skin, and grittiness of the pulp. There is a large variation in effects with different pear species. Based on present knowledge, radurization of pears seems impractical.

7. Stone Fruits

a. *Apricots*

Fungal rot is the cause of decay of apricots. Doses in the range 1–3 kGy afford reasonable control of such decay, as can be seen from the data of Table 8.3. Based on these data, the optimum dose is 2 kGy. Sensory characteristics of the fruit are not changed significantly.

b. *Cherries*

The principal cause of decay of cherries is brown rot, due to organisms such as *Cladosporium herbarum* and *Monilinia* spp. Doses in the range 1–4 kGy

Table 8.3
Some Effects of Irradiation on Apricots[a]

Stage of maturity	Storage time (days)		Dose (kGy)				
	2°C	Ambient	0	0.5	1.0	2.0	3.0
			Decayed (%)				
Firm ripe	0	2	42	40	22	21	16
	21	4	82	44	17	7	6
Immature green	0	5	21	19	17	19	21
	21	3	92	56	42	23	12
			Firmness rating[b]				
	0	2	3.4	3.8	3.8	3.4	3.5
	21	3	1.8	—	3.9	3.0	1.4
			Color rating[c]				
	0	2	3.0	3.4	3.8	4.3	4.1
	21	3	3.6	3.8	3.9	3.6	3.5
			Flavor rating[c]				
	0	4	3.3	3.3	4.0	3.4	2.0
	21	3	2.7	2.5	2.9	3.0	2.1

[a] From R. G. Mercier and K. F. MacQueen, "Gamma-Irradiation to Extend Postharvest Life of Fruits and Vegetables," 1965 Report of the Horticultural Experiment Station and Products Laboratory, Ontario Dept. of Agriculture, Toronto, 1965.
[b] Scale: 0, very soft; 5, firm.
[c] Scale: 0, unacceptable; 3, acceptable; 5, typical of variety.

extend product life. Above 4 kGy radiation decay is increased. At relatively low doses, namely about 2 kGy, softening is detectable by panel testing. At about 4 kGy skin wrinkling can occur. At about 10 kGy a tendency toward change of the normal red color to brown occurs.

The effect of radiation to cause softening is a principal deterrent to radurization of cherries.

c. *Nectarines*

Limited work has shown that irradiation with doses of 2 to 4 kGy controls brown rot. Firmness and color are unaffected. An adverse effect on flavor, however, occurs.

Table 8.4

Percentage Decay of Stored 'Loring' Peaches Irradiated at the Firm Ripe Stage with and without Hot-Water Dip[a,b]

Dose (kGy)	Decay (%)	
	No dip	Dip[c]
0	53.3	4.4
0.75	60.3	2.2
1.5	2.2	0.0
2.25	8.9	0.0

[a] From R. A. Dennison and E. M. Ahmed, "Effects of Low Level Irradiation upon the Preservation of Food Products," ORO-680. U.S. Atomic Energy Comm., Washington, D.C., 1971.

[b] Stored 10 days at 3°C plus 4 days at 20°C.

[c] Hot-water dip for 7 min at 49°C.

d. Peaches

Brown rot, caused by fungi such as *Monilinia* spp. and *Rhizopus stolonifer*, is the principal decay process for fresh peaches. Table 8.4 shows data on the effect of radiation on decay of peaches with doses up to 2.25 kGy, with and without a hot-water dip and with storage for 10 days at 3°C plus 4 days at 20°C. These data indicate an effective control of decay can be secured with the combination of heat and radiation with a dose of only 0.75 kGy, or with irradiation alone with a dose of 1.5 kGy.

Peaches irradiated at the mature green stage soften, but on storage have about the same firmness as nonirradiated fruit. The data of Table 8.5 demonstrate this effect. Fruits irradiated at the firm ripe stage undergo only a small amount of softening. On storage the firmness of the treated fruit is about the same as that of the untreated. The data of Table 8.6 portray these effects.

Although the cause is unclear, sloughing of the peach skin may occur. The skin is more fragile as a result of irradiation, which makes peeling more difficult. It has been suggested that this effect is related to the degree of ripeness at the time of irradiation or to breakdown of some constituents of the skin.

Table 8.5

Firmness of Stored 'Loring' Peaches Irradiated at the Mature Green Stage[a, b]

	Dose (kGy)			
Days	0	1.1	2.2	4.3
0	16.8	12.7	9.9	7.7
6	2.7	1.6	1.6	2.2
12	1.9	1.2	1.3	0.3

[a] From R. A. Dennison and E. M. Ahmed, "Effects of Low Level Irradiation upon the Preservation of Food Products," ORO-680. U.S. Atomic Energy Comm., Washington, D.C., 1971.

[b] Expressed as pounds force, as measured with Magness–Taylor pressure tester.

Flavor differences occur with doses as small as 1 kGy or less. They have been described as "off" or bland. Some of the differences in flavor have been ascribed to odor differences. Other flavor differences have been related to acid and/or other peach components.

Table 8.6

Firmness of 'Loring' Peaches Irradiated at the Firm Ripe Stage with and without Hot-Water Dip and Stored at 2°C[a, b]

	Firmness[c]					
	0 Days		7 Days		14 Days	
Dose (kGy)	No dip	Dip	No dip	Dip	No dip	Dip
0	6.2	3.3	1.8	2.4	1.5	2.0
1.1	5.9	2.6	1.5	2.2	2.4	3.1
2.2	5.7	2.1	2.1	1.6	1.8	2.7
4.3	4.2	2.1	2.6	2.1	2.4	2.1

[a] From R. A. Dennison and E. M. Ahmed, "Effects of Low Level Irradiation upon the Preservation of Food Products," ORO-680. U.S. Atomic Energy Comm., Washington, D.C., 1971.

[b] Hot-water dip for 7 min at 49°C.

[c] Expressed as pounds force, as measured with Magness–Taylor pressure tester.

Table 8.7

Effect of Irradiation on the Percentage of Decay of Precooled and Stored Plums[a,b]

Postirradiation holding time (days)		% Decay at dose (kGy)					
2°C	Ambient	0	1.0	1.5	2.0	2.5	3.0
0	9	14	11	10	9	11	17
28	3	18	0	0	1	3	3

[a] From R. G. Mercier and K. F. MacQueen, "Gamma-Irradiation to Extend Postharvest Life of Fruits and Vegetables," 1965 Report of the Horticultural Experiment Station and Products Laboratory. Ontario Dept. of Agriculture, Toronto, 1965.

[b] Ontario, Canada, varieties: 'Early Golden,' 'Shiro,' 'Czar,' 'Burbank Starking,' 'Delicious,' 'Queenston,' 'Stanley,' 'Italian,' and others.

e. *Plums*

Doses of about 1.5 kGy effect decay control. Precooling of the fruit is important. Plums undergo quality losses at higher doses, the kind and magnitude of which depend considerably upon variety. These changes include inhibition of external color development, softening, and other internal flesh changes such as wateriness and gelatinization, loss of flavor, and skin weakening. By selection of variety and limitation of holding period, a moderate reduction of decay may be secured as is demonstrated by the data of Table 8.7.

8. Tamarillos

The tamarillo, of the same family as the tomato, is subject to field infection with mold such as *Botrytis.* Doses of 0.75 to 1.5 kGy delay stem-end rot of fruit stored 2 weeks or more at 2°C. Radiation injury, which includes softening and flavor changes, however, limit the maximum dose to 1 kGy. At this dose the life extension is only of marginal value.

9. Tomatoes

Irradiation of tomatoes in the pink or riper stage with doses of 2.5 or 3 kGy extends the life of the fruit stored at 12 to 15°C by 4 to 12 days. Doses greater than 3 kGy increase the decay rate. Irradiation at the mature green or breaker stage does not increase fruit life. A dose of 3 kGy inactivates most bacteria present but not fungi, *Alternaria fusarium* and *Cladosporium* spp. being little affected by the radiation at this dose.

10. Tropical and Subtropical Fruits

a. Bananas

For commercial or export trade bananas are picked at the mature green stage. While contamination with spoilage bacteria and molds usually exists, growth of such organisms, with the exception of *Thielaviopsis paradoxa*, does not occur at the green stage unless there is bruising injury or other wounds such as cuts and scratches. The role of bacterial growth in banana spoilage is not clear. The usual spoilage symptoms are associated with the growth of various fungi, including *Gloesporium musarum* (*Colletotrichum musae*), *Botryodiplodia theobromae*, *Nigrospora musae*, *N. sphaerica*, *Rhizopus* spp., *Thielaviopsis paradoxa*, *Ceratocystis paradoxa*, and *Fusarium roseum*. Growth of these fungi occurs when the fruit ripens. The fungi growth patterns cause various types of spoilage, which include anthracnose (dark spots that coalesce to form larger spots), crown rot, black end, squirter, false squirter, and *Thielaviopsis* rot.

The doses needed to inactivate the fungi found on bananas fall in the range 2.5–5 kGy. Irradiation at these doses causes darkening of the peel, softening, and acceleration of spoilage. The response of bananas to radiation with regard to these effects varies substantially with variety. For 'Lactan' bananas irradiated at the mature green stage, the optimum dose has been reported to be 0.25–0.30 kGy. Doses below this range accelerate ripening, while doses above it accelerate decay. Doses as low as 0.05 to 0.10 kGy have been suggested to maximize product life extension.

b. Dates

Fresh dates may undergo microbial spoilage due to fermentation and molding. Fortunately, dates tolerate doses as large as about 5 kGy without impairment of eating qualities. Microbial spoilage is reduced by doses greater than 0.9 kGy. A dose of 5.4 kGy about doubles the time at ambient temperatures for fresh dates to reach a condition in which 5% of the fruit are spoiled.

c. Figs

Fresh figs are subject to spoilage due to the growth of surface molds. Doses in the range 2–4 kGy reduce this spoilage. The response to radiation varies with variety. 'Kadota' figs become brown at these doses, whereas 'Calmyra' and 'Mission' varieties do not. Flavor is unchanged at doses of 2 to 4 kGy or higher. Doses in the range 8–12 kGy cause softening. Dipping fresh figs in water at 50°C for 5 min plus irradiation with a dose of 1.5 kGy extends the life 3–4 days at 20 to 32°C and 8–10 days at 15°C by delaying the growth of *Rhizopus* and *Aspergillus* spp.

d. Guava

Fresh guavas (commercial maturity) irradiated with a dose of 0.3 kGy and stored at ambient temperatures keep a week or more, whereas 50% of similarly stored nonirradiated fruit spoil. Sensory properties are unaffected by this dose.

e. Longan

This fruit is similar to the lychee. A dose of 2 kGy extends its life 4–5 days at 17°C.

f. Lychees

Fungal spoilage control for lychees requires doses in excess of 4 kGy. Fortunately such high doses are tolerated by this fruit. Combination of irradiation with a hot-water treatment enables reduction of dose.

g. Mangoes

The more common types of spoilage of mangoes and the causative fungi are anthracnose, caused by *Colletotrichum gloeosporioides*; stem rot, caused by *Diplodia natalensis*; and soft brown rot, caused by *Hendersonia creberimma* Syd. and Butl.

Without some preservative measure mangoes do not keep well enough for transport to distant markets. Because of the wide distribution and popularity of this fruit, a great deal of work has been done on the irradiation of mangoes. On the whole, it has been shown that radiation used alone at doses that can be tolerated is not sufficiently effective to control fungal spoilage. In some varieties of mangoes doses as low as 0.5 kGy cause fruit injury. As with other fruits, excessively high doses promote rather than inhibit spoilage, possibly through damage to the fruit which make it more vulnerable to attack by fungi. The most effective irradiation treatment devised so far appears to be a combination of a radiation dose of 0.75 kGy with a 5-min dip in water at 50 to 55°C. This type of treatment along with maintenance of a temperature of about 11°C during storage and transit provides a product life of about 30 days, of which 1 week can be at ambient temperature.

When applied to mature green mangoes, this combination treatment in addition to controlling fungal decay provides two other benefits: (1) delay of ripening and (2) inactivation of the mango weevil.

h. Papaya

The papaya is subject to postharvest decay caused by fungi. The following fungi have been identified as causative microorganisms:

Asochyta caricae	*Colletotrichum gloeosporioides*
Rhizopus stolonifer	*Botryodiplodia theobromae*
Fusarium solani	*Phytophothora parasitica*
Phomopsis spp.	

The maximum radiation dose tolerated by papayas is about 1 kGy. Doses above this cause surface scalding and bitter off-flavors and aroma. This dose is too small to effect sufficient inactivation of fungi likely to be present. For this reason irradiation alone cannot be employed. Instead a procedure similar to that devised for mangoes has been developed. The essential steps of this procedure are (1) harvest the fruit at "color break" stage (color turning to one-quarter ripe), (2) immerse in water at 48°C for 20 min (an alternate dip is 20 sec at 60°C), and (3) irradiate with a dose of 0.75 kGy. Irradiation at 0.75 kGy has only a slight effect in decay control. Raising the dose to 1.0 kGy improves decay control, but when used in combination with the hot-water treatment, a slight surface scalding results. If the fungal infection is due to the highly radiation-resistant *Rhizopus*, decay control may not be achieved by this combination procedure.

By delaying ripening, irradiation maintains product firmness longer and provides more desirable product at the time of final sale.

i. Rambutan

This tropical fruit spoils after a few days of storage at ambient temperature. Irradiation with doses in the range 0.8–1.0 kGy extends the product life about 4 days. A dose of 1.2 kGy produces skin darkening.

C. Delay of Ripening and Senescence

1. Introduction

In other preservation uses of irradiation the actual targets of the radiation are the various organisms that contaminate foods. In using irradiation to delay ripening and senescence of fruits, the food itself is the target. Although they have been harvested, fruits treated for this purpose are respiring and carrying out metabolic processes. It is the alteration of these activities which is the objective of irradiation in securing delay of ripening and senescence.

When mature but unripe fruits are harvested, they may undergo one of two processes: (1) a slowly declining respiration rate or (2) first a decline of respiration rate and then an increase which coincides timewise with the beginning of ripening. The respiration rate reaches a peak and subsequently declines. Fruits which follow the first process are termed *nonclimacteric*, and since they do not undergo postharvest ripening, they must be picked in the ripe stage. Fruits which follow the second process are *climacteric* fruits. They can be picked when they are mature but not yet ripe, and with suitable environmental conditions, such as appropriate temperatures, they will ripen in time. Ripe fruits of both classes undergo decline in respiration rate and ultimately exhibit degradation changes commonly identified with senescence, which is a form of spoilage.

Only fruits which are of the climacteric class can undergo delay of ripening by irradiation. Since this delay extends the postharvest life of the fruit, the procedure is properly identified as preservation. In addition to delaying ripening, irradiation also often delays senescence.

Not all climacteric fruits respond to irradiation in the same way. For this reason each kind of fruit and, in some cases a particular variety, needs to be considered on an individual basis. Important to the nature of the response to irradiation also is the physiological stage of the fruit, particularly with reference to the climacteric phase. Application of radiation after the start of the climacteric respiration increase generally is ineffective in influencing the ripening process. Postirradiation conditions, especially temperature, also have significant effects upon the end result.

The production of ethylene is associated with the ripening of climacteric fruits. The role of ethylene in ripening is not adequately understood, even as to whether it is the cause of ripening or simply a product of the ripening process. The effect of irradiation upon ethylene production varies with the kind of fruit. Irradiation of preclimacteric pears, for example, with doses of 1 or 2 kGy increases the rate of ethylene production. Larger doses reduce it. Pears irradiated at the natural peak of ethylene production also exhibit a decrease in rate. Exposing such pears to exogenous ethylene leads to an increase in ethylene production and also to ripening, but the ripened fruit has abnormal characteristics.

Irradiation of other fruits yields different patterns. Peaches and nectarines, for example, are stimulated to ripen, possibly due to increased ethylene production. Bananas, plums, papayas, and mangoes, on the other hand, exhibit delayed ripening. In some cases, the postirradiation ripening results in fruits with abnormal characteristics.

When irradiated, some nonclimacteric fruits, such as citrus, exhibit some of the usual responses of climacteric fruit. The irradiation of mature and ripe oranges and grapefruit results in increased rates of respiration and ethylene production. Irradiated green lemons turn yellow faster than normal. Irradiation of immature green oranges stimulates ethylene production and causes partial de-greening.

The action of irradiation in delaying ripening is complex. The available information indicates that success with this use of irradiation requires an understanding of the postharvest physiological processes of fruits and treatment that is applied only at particular stages of the development of the fruits. In the past, difficulties with obtaining proper timing of irradiation of some fruits with respect to the climacteric phase may have led to incorrect observations and to the reporting of faulty conclusions. Much additional information is needed for effective use of irradiation to delay the ripening of a number of particular fruits.

Delay of senescence often involves retention of fruit firmness longer than is obtained without irradiation. This effect of irradiation appears to be associated with interference with the normal process of conversion of carbohydrate polymers, which are the basis of fruit firmness, to smaller molecules.

The classification of fruits as climacteric or otherwise has some uncertainties. With one or two possible exceptions, the fruits considered in this section ordinarily are considered to be climacteric.

In the determination of the effect of irradiation to delay ripening, a number of criteria which characterize ripening may be used. Among such criteria are

Color change, which often is from green to another color such as yellow or red. It may involve loss of chlorophyll and formation of anthocyanins or carotenoid pigments
Increased softness (or loss of firmness), frequently associated with changes in pectic substances to form soluble pectins
Increased sugar content, as starch is converted to sugar
Increased ethylene production, which is associated with the ripening process of climacteric fruits
Increased respiration rate, which occurs as ripening begins
Flavor and aroma development, a key aspect of ripening that is associated with the formation of volatile substances
Total water-soluble solids, increasing with ripening

2. Berries

Generally, berries have not been examined for the delay of ripening by irradiation.

'Shasta' strawberries, picked at the green-tip stage and irradiated with doses of 0, 1, 2, and 3 kGy, have an unacceptable lack of flavor development. Ripening is uneven. Immaturity and underripeness increase with dose.

3. Pome Fruits

a. *Apples*

Only limited information is available on the irradiation of apples. Doses in the range of about 0.2 to 0.4 kGy decrease ethylene production in 'Rhode Island' greening apples. Irradiation in the preclimacteric stage lengthens storage life. Irradiation causes immediate softening. The threshold dose for this effect varies with the apple variety and lies in the range of about 0.2 to 0.4 kGy. With long-term storage, however, irradiated apples exhibit greater firmness than nonirradiated ones. Irradiation strongly inhibits the formation of the volatiles responsible for flavor and aroma.

Somewhat different effects which indicate that radiation stimulates the ripening process have been observed. Such differences may be due to varietal differences or possibly to carrying out the irradiation after the initiation of the climacteric phase.

b. Pears

'Bartlett' pears irradiated with a dose of 3 kGy in the preclimacteric phase exhibit a marked increase in respiration rate. The fruit, however, does not ripen normally. Irradiation after the onset of the climacteric produces only a small change in the respiration rate and ripening is unaffected. The effects of irradiation on 'Bartlett' pears at various stages of the climacteric may be seen from the data of Fig. 4.12.

Doses as low as 1 kGy inhibit yellow color development. This effect increases with dose. There is, however, a green mottling of the ripened fruit. An initial softening occurs with doses of 2 and 3 kGy. The normal softening associated with ripening, however, is inhibited, and stored irradiated pears are firmer than nonirradiated ones. In some instances grittiness or mealiness occurs with doses of 3 or 4 kGy. Likewise, the characteristic normal flavor of the pear fails to develop.

4. Stone Fruits

a. Apricots

Irradiation with doses up to 3 kGy increases the respiration of apricots and accelerates ripening. Softening, skin brown spots, internal bruising, and mealiness of the pulp occur.

b. Cherries

One study on sweet cherries suggests that irradiation with doses of 2 to 4 kGy can delay ripening when the fruit is stored at 5°C.

c. Peaches

Irradiation with doses in the range of about 0.5 to 3 kGy increases the respiration rate and accelerates ripening. Due to increased anthocyanin formation, enhanced red color results with doses in the range 1.25–3 kGy. Irradiation of mature green fruit causes an immediate softening of the fruit. After completion of ripening, irradiated and nonirradiated fruit exhibit about the same firmness. Irradiation at the firm ripe stage causes only a slight softening. Irradiated peaches after ripening have less flavor and aroma than nonirradiated peaches.

d. *Plums*

Different varieties of plums respond differently to irradiation. Respiration rate is increased. With doses up to 5 kGy, irradiation generally inhibits the development of normal skin color and weakening of the skin may occur. Texture changes extending beyond simple softening result from irradiation.

The importance of variety in affecting the results of irradiation seems unusual. It is clear that only certain varieties may be irradiated with beneficial effects.

5. Tropical and Subtropical Fruits

a. *Avocados*

The reported responses of avocados to irradiation differ greatly with variety—from marked acceleration of ripening, as measured by softening, to marked delay. The response varies also with dose. The 'Nabal' variety, for example, undergoes accelerated ripening with doses up to 0.35 kGy, whereas doses in the range 0.5–1 kGy have no effect. Irradiation causes an immediate increase in the respiration rate but reduces ethylene production.

Doses of as little as 0.1 kGy can cause injury, as evidenced by skin discoloration and blackening of the flesh. Dose rate also affects the amount of injury. A rate of 0.85 kGy/hr is required to minimize damage. Other factors, such as the length of time between picking and irradiation, or whether the fruit is picked early or late in the season, affect the response to irradiation. The patterns of response that have been observed suggest that very close attention to the maturity of the fruit and its state with regard to the climacteric must be given in order to secure a delay of ripening.

A combination process using a water treatment of 46°C for 10 min and a dose of only 30 Gy provides a several-day life extension at ambient temperature for avocados first held for 3 weeks at 6°C.

b. *Bananas*

Irradiation of preclimacteric bananas causes delay of ripening, whereas irradiation after the onset of the climacteric does not. The variables that affect the response of bananas to irradiation include variety, maturity, dose, and storage temperature.

Irradiation does not cause an increase in the respiration rate. The climacteric peak, however, is delayed timewise. This may be seen from the data of

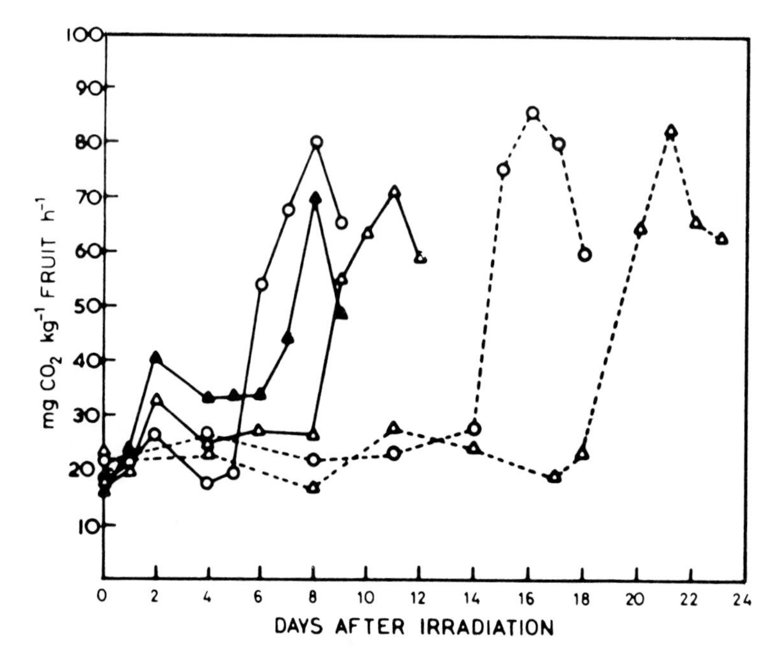

Figure 8.4 Respiratory patterns of nonirradiated and irradiated bananas ('Fill Basket' variety). ——Fully mature; ---- 80% mature. ○, Nonirradiated; △, 0.25 kGy; ▲, 0.50 kGy. From A. Sreenivasan, P. Thomas, and S. D. Dharkar, *In* "Disinfestation of Fruit by Irradiation." Int. Atomic Energy Agency, Vienna, 1971.

Table 8.8

Extension of Product Life at 24 to 29°C by Irradiation in Preclimacteric Condition of Five Varieties of Bananas Harvested at 75% Maturity[a]

Variety	Dose (Gy)		Product life extension over nonirradiated (days)
	Optimum	Maximum tolerated	
Dwarf Cavendish	300	400	8
Giant Cavendish	350	400	7–8
Fill Basket	250	350	8–9
Red	400	500	7–8
French Plantain	200	300	14

[a] From A. Sreenivasan, P. Thomas, and S. D. Dharkar, Physiological effects of gamma irradiation on some tropical fruits. *In* "Disinfestation of Fruit by Irradiation." Int. Atomic Energy Agency, Vienna, 1971.

Fig. 8.4 for the 'Fill Basket' variety. With fully mature bananas a dose of 0.25 kGy delays the peak occurrence by several days. Increasing the dose to 0.50 kGy does not cause such delay. Fruit picked at the 80% mature stage undergo the climacteric later than do fully mature fruit, as also can be seen from Fig. 8.4. Irradiation further delays the climacteric for such fruit. As evidenced by retention of the green skin color, increasing doses up to about 0.3 kGy cause progressive delay of the ripening of the 'Fill Basket' variety. Doses greater than 0.3 kGy cause skin discoloration.

Data on the radiation-induced extension of the life of five varieties of bananas harvested at 75% maturity are given in Table 8.8. Postirradiation storage at 21°C yields increased extension over that obtained with storage at 29 to 32°C, as can be seen from the data of Table 8.9. The data of this table also show that the life extension is greater with less mature fruit.

Doses beyond critical maximum values can cause fruit injury. The maximum tolerance dose for each of the five banana varieties listed is given in Table 8.8. Doses greater than the maximum tolerance dose can cause skin darkening, splitting of the skin and pulp, and softening and mealiness of the pulp. Damage to the skin can be reduced by irradiation in nitrogen.

Although softening of the pulp occurs immediately upon irradiation, there is no concomitant increase in reducing-sugar content. This suggests that starch degradation is not the cause of this immediate softening. The normal softening associated with ripening is inhibited by irradiation and as a

Table 8.9

Effect of Fruit Maturity and Storage Temperature on Delayed Ripening of Bananas Irradiated in Preclimacteric Condition[a]

Variety	Maturity at harvest (%)	Dose (Gy)	Product life extension over nonirradiated (days)	
			29–32°C	21°C
Giant Cavendish	75	350	4–5	9–10
	85	350	3	7–8
Red	70	400	7–8	9–10
	85	400	4–5	7–8
Fill Basket	75	250	—	11–12
	90	250	—	7–8

[a] From A. Sreenivasan, P. Thomas, and S. D. Dharkar, Physiological effects of gamma radiation on some tropical fruits. *In* "Disinfestation of Fruit by Irradiation." Int. Atomic Energy Agency, Vienna, 1971.

Table 8.10
Starch Content Percentage of Irradiated Taiwan Bananas Stored at 25 to 30°C for 2 weeks[a]

	% Starch at dose (Gy)			
Banana origin[b]	0	200	300	500
CHL	2.0	11.6	11.6	6.3
CP	1.4	11.1	10.91	5.3
SP	1.1	9.1	9.8	5.2

[a] From H. Kao, Extension of storage life of bananas by gamma irradiation. *In* "Disinfestation of Fruit by Irradiation." Int. Atomic Energy Agency, Vienna, 1971.
[b] CHL, Central highland (of Taiwan); CP, central plain; SP, southern plain.

consequence, irradiated bananas at the ripe stage are firmer than nonirradiated bananas. This greater firmness can be ascribed to retention of the starch in irradiated bananas, as may be seen from the data of Table 8.10.

Skin browning caused by doses above the maximum tolerable dose has been ascribed to increase in phenoloxidase activity, resulting from cell membrane damage which permits contact between enzymes and substrates.

The action of ripening stimulants such as ethylene is slower on irradiated than nonirradiated bananas. The decreased response of irradiated bananas to ethylene has been suggested to be the basis for the radiation-induced delay of ripening. Also suggested as the mechanism for ripening delay is a shift in the metabolic pathways from the glycolytic pathway to the pentose-phosphate pathway. The finding of increased amounts of glucose 6-phosphate dehydrogenase in irradiated bananas is cited as evidence for this mechanism.

The use of irradiation to delay the ripening of bananas appears to be technically feasible. Certain aspects of this use, however, are highly critical. Only certain varieties may be suited. Irradiation must be applied in the preclimacteric stage. Maturity of the fruit at the time of harvest must be controlled. Less than fully mature fruit seem to provide a greater delay in ripening than do fully mature fruit. Postirradiation storage temperatures must be within limits. A dose appropriate to the variety must be used.

The cooking banana, the plantain, also undergoes delay of ripening with irradiation. Doses in the range 0.20–0.30 kGy with mature plantains stored at 20°C provide a product life greater by 9 days than that obtained with fruit not irradiated.

c. Dates

The availability and distribution of fresh dates are limited by their perishability. Fresh dates are harvested before they are fully ripe. They are consumed when they have attained the "soft" stage. Juiciness and tenderness are important indexes of quality. Extension of the market life of fresh dates can involve the use of irradiation to secure both delay of ripening and radurization (see Section I, B, 10, b).

The eating quality of fresh dates is not affected by doses as large as 5 kGy. The effect of irradiation on the ripening varies considerably with the variety and also with the development stage of the date at the time of irradiation. It is also dose dependent. Doses in the range 2.7–5.4 kGy reduce the time between irradiation and the attainment of the soft ripe stage, probably by shortening the induction period for the initiation of softening. The length of the softening process generally can be increased by low doses in the range 0.1–0.3 kGy. The amount of increase varies greatly with variety, from none for the 'Lelwi' variety to 75% or greater for 'Zahdi' and 'Tabarzel' varieties. Due to the fact that the picked fruits are not of uniform development, the fruit-to-fruit variation within a given quantity of dates can be large.

d. Guava

Limited studies have shown that doses of about 0.3 kGy delay the ripening of guavas 3–5 days when stored at 25 to 32°C. Stored irradiated fruits are firmer than nonirradiated. Doses greater than about 0.45 kGy cause browning of the skin.

e. Mangoes

Irradiation causes an immediate increase in the respiration rate of mangoes picked at the hard green stage and stored at 25°C. This increase is followed by a decline to about the level of nonirradiated fruit. The climacteric respiration peak occurs subsequently at about the same time for both irradiated and nonirradiated fruit. At the climacteric peak irradiated mangoes have a substantially smaller respiration rate than nonirradiated ones. This pattern of respiration changes is shown in Fig. 8.5.

Doses of approximately 0.25 kGy cause a delay of ripening of the order of 7 days with fruit stored at 25°C. Irradiation inhibits the disappearance of chlorophyll and also the formation of carotenoid pigments. In some varieties (e.g., 'Alphonso' and 'Bangalora') irradiation increases the formation of anthocyanin pigments in the skin, provided an anthocyanin precursor is present at the time of irradiation. Doses in the range 0.25–0.75 kGy suppress the rate of ethylene production, the effect increasing progressively with dose. The firmness of mangoes irradiated with a dose of 0.75 kGy is greater than that of nonirradiated fruit at the fully ripe stage. The color typical of fully ripe

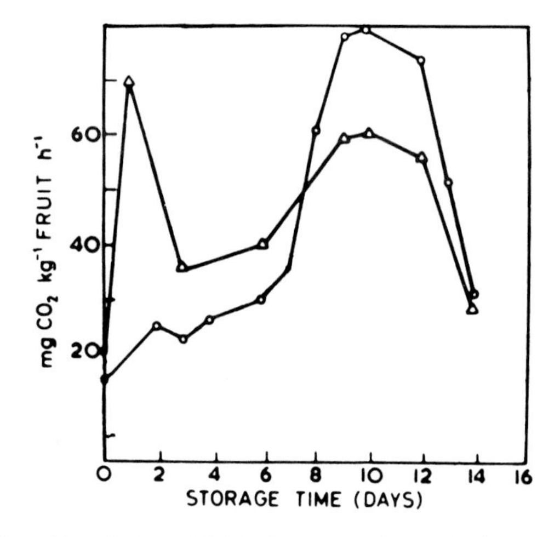

Figure 8.5 Effect of irradiation with 250 Gy (△) on the respiration pattern of mangoes. ○, Nonirradiated. From A. Sreenivasan, P. Thomas, and S. D. Dharkar, Physiological effects of gamma radiation on some tropical fruits. *In* "Disinfestation of Fruit by Irradiation." Int. Atomic Energy Agency, Vienna, 1971.

fruit may not develop. Doses greater than 1 kGy cause tissue darkening due to increased activity of the enzyme polyphenol oxidase.

Storage at temperatures in the range 7–15°C considerably extends the life of both irradiated and nonirradiated mangoes. Ripening occurs when such fruit is moved to higher temperatures. Low-temperature storage, however, impairs the development of the normal flavor and aroma in the ripened fruit. Soft brown rot also is favored by low-temperature storage.

The objectives for irradiating mangoes can be threefold: (1) insect disinfestation, (2) control of postharvest fungal decay, and (3) delay of ripening. The high dose requirements for control of postharvest fungal decay preclude the use of irradiation for this purpose. The use of a hot-water treatment instead has been advocated (see Section I, B, 10, g). The dose for insect disinfestation has been established to be about 0.75 kGy. This dose is sufficient to secure a significant delay of ripening. While doses in the range 0.25–2.0 kGy have been reported to be necessary to secure delay of ripening, it appears that a dose of 0.75 kGy is most likely to be used. With this dose, the product life extension obtained through ripening delay of fruit picked at the mature green stage and stored at about 11°C is of the order of a week. The total time from harvest can be 30 days or more.

f. Papayas

Preclimacteric papayas undergo a delay of ripening with doses in the range 0.50–0.75 kGy. Higher doses (e.g., 1 kGy) result in skin scald. The length of the ripening delay is related to the postirradiation storage temperature and at ambient temperatures is about 3 days. Retention of fruit firmness for a longer period is the principal effect of irradiation.

Papayas for export markets generally require insect disinfestation and also control of fungal decay. A combination treatment of the fruit at the quarter-ripe stage employing a dose of 0.75 kGy and a water dip at 48°C for 10 min (alternatively, 60°C for 20 sec) has been regarded as best suited to provide (1) delay of overripening, (2) insect disinfestation, and (3) control of fungal decay. With fruit so treated and stored at ambient temperatures the saleable life is about 10 days. This can be extended by storage at temperatures as low as 15.6°C and by use of a controlled atmosphere.

g. Persimmons

Irradiation of persimmons with 12 kGy at the stage of unripe green with a tinge of yellow and the use of hermetic packaging effectively preserves this fruit for more than 150 days. Ripening occurs on the seventh day. This compares with nonirradiated persimmons, which remain green and mold on the twenty-second day.

h. Pineapples

Although the pineapple is not considered to be a climacteric fruit, irradiation with 0.5 kGy delays the ripening 2–3 days when the fruit is picked fully mature, but green, and is stored at ambient temperatures. Doses of 1 to 3 kGy cause the fruit to be more susceptible to fungal decay. Surface browning, possibly due to loss of chlorophyll, limits the maximum dose tolerated to about 0.5 kGy.

i. Sapodilla (Sapota)

Sapodillas irradiated with a dose of 0.10 kGy exhibit a life extension of 3 to 5 days when stored at 25 to 32°C. Storage at 10°C increases the extension to 15 days.

j. Soursop

Doses as low as 0.1 kGy are reported to cause injury to the soursop. Irradiation at the preclimacteric stage markedly affects the respiration rate in an unusual and complex way. Failure of the fruit to soften when stored at 25°C and the development of a dark pulp color seem to preclude the use of irradiation with the soursop.

k. Tomatoes

Tomatoes are climacteric fruit, and irradiation applied at the preclimacteric stage delays ripening. Tomatoes picked at the mature green stage can be in the preclimacteric stage. Since tomatoes at the immature green stage do not ripen, determination of the degree of maturity is essential in using irradiation. This determination has occasioned problems. With holding temperatures in the vicinity of 15°C, ripening is delayed up to 1 week by irradiation with doses of 2 to 4 kGy. Although these doses do inactivate some species of fungi, in general, they have little fungicidal effect.

Doses greater than 5 kGy cause excessive injury to the fruit and cannot be used. At doses below 5 kGy the fruit is softened and this injury can cause severe transit damage, which accelerates fungal decay. Appropriate protective packaging during transit may be needed to offset fruit fragility. Irradiation intensifies chilling injury, which also can increase fungal decay. Color development may be uneven and result in mottling.

The use of irradiation to delay ripening of tomatoes appears to be impractical, mainly because of radiation-induced softening and the resultant fruit fragility. This also prevents the use of irradiation for fungal decay control.

D. Radappertization

1. Introduction

Only a small amount of information is available on the use of irradiation to provide fruit products which can be kept indefinitely without refrigeration. Fruits with a pH that is less than 4.5 require smaller doses for radappertization than those which do not have this acidity, for the reason that *Clostridium botulinum* does not grow and produce toxin at pH values lower than this. This situation parallels thermosterilization processes. While bacteria such as *Bacillus subtilis* are present, the principal spoilage organisms associated with fruits are yeasts and molds.

It is important to recognize that the information on radappertization of fruits and fruit products, generally was obtained without rigorous investigation of the microbiology. It would be unwise, therefore, to radappertize these foods without further check on their microbiological safety.

2. Fruits

Enzyme inactivation is an obvious requirement for long-term storage of fruits at ambient temperatures. This is best obtained by heating the fruits, as for example, to 70°C for 10 min. Mango and sapodilla (sapota) slices in 40% sucrose syrup in vacuumized cans are stable for at least 1 year when treated with 4 kGy. Such irradiated fruits have sensory qualities comparable with thermally sterilized counterparts. Mangoes are sensitive to O_2 and to high radiation doses. Sapodillas do not exhibit such sensitivity and can tolerate a dose of 12 kGy. Similar treatment of guavas and apples yields products considered superior to those obtained by thermal processing.

Peaches, whether heated for enzyme inactivation or not, packed in syrup and irradiated with doses of 15 to 25 kGy are acceptable from a sensory viewpoint. The same result is obtained whether the irradiation is carried out in the frozen state or not. A dose of 30 kGy produces unacceptable changes. Texture changes produced by irradiation are comparable with what is obtained by thermal sterilization.

3. Fruit Juices and Purees

Apparently there has been more interest in using irradiation to stabilize fruit juices than the fruits themselves. Apple juice has been studied extensively. The dose required to provide stability is partly determined by the solids content of the juice. A dose of 5 kGy does not stabilize 30% solids juice, but does so with a juice of 47% solids. Doses in excess of 10 kGy are required for 13% clear apple juice. Generally irradiation lightens the color of the juice. On storage, however, darkening occurs and stored irradiated apple juice shows little difference from the original color. Flavor changes caused by irradiation are slight. Analyses of the volatiles of irradiated apple juice confirm this observation. Irradiation in the frozen state with a dose of 15 kGy yields juice with less alteration of flavor than irradiation in the liquid state. The use of N_2 atmosphere during irradiation also improves flavor. γ Rays are reported to cause greater flavor changes than electrons.

The dose to obtain stability may be reduced by heating apple juice to 50°C prior to irradiation. In this way a dose of 3 kGy is adequate to secure stability at ambient temperatures for more than a year.

The radappertization of orange, grape, pear, and tomato juices follows patterns similar to that for apple juice. Table 8.11 shows the doses required to obtain stability.

Fruit purees prepared for remanufacturing can be stabilized for needed storage by irradiation. Guava and mango purees irradiated with doses in the range 5–10 kGy can be stored at 12°C. Stability at 0°C can be secured with a dose of 1 kGy. Higher doses produce some off-flavors.

E. Insect Disinfestation

In general there are two purposes for using radiation to carry out insect disinfestation of fresh fruits: (1) to prevent insect damage to the fruit and (2) to serve as a quarantine control measure. While death of the insect always meets the requirements for both purposes, lethality is not always an essential requirement. Irradiation can inhibit feeding, and although the insect may

Table 8.11
Doses to Secure Stability of Various Fruit Juices

	Dose (kGy)	
Juice	Irradiation only	Irradiation and heating
Apple	>10	3
Orange	8	4
Grape	15–20	10
Pear	>4	4
Tomato	5[a]	—

[a] Plus sorbic acid (40–50 mg/liter).

remain alive for a period, damage to the food does not occur. In quarantine control, inability of an insect to reproduce effectively conforms with the intended purpose.

While there is recognition that lethality may not always be needed, in general lethality has been sought in procedures other than irradiation, particularly in quarantine control measures. This has been reflected in the requirements of government regulations. Unless extremely large doses are employed, irradiation generally does not cause immediate insect death. Contemplated uses of irradiation for insect control are based upon smaller doses which do not provide an immediate kill, but instead cause "reproductive death." Apart from considerations of irradiation costs, which are higher with larger doses, the use of smaller doses often is dictated by the need to avoid or minimize radiation-induced phytotoxicity in the host fruit.

Irradiation of fruits for quarantine control purposes must meet government standards for effectiveness. In the United States these have required less than 32 survivors per 1 million insects, or a mortality of 99.9968%. Doses to meet this requirement must be provided. With irradiation used for quarantine control, reproductive death has been accepted as adequate in meeting the mortality requirement.

The insects carried by fresh fruits vary with the fruit and the location where it is grown. Various species of fruit flies exist in warm climates. These include the Oriental fruit fly, *Dacus dorsalis* Hendel; the Mexican fruit fly, *Anastrepha ludens* (Loew); the olive fruit fly, *Dacus oleae Gmelin*: the Queensland fruit fly, *Strumeta tryoni* (Froggatt); the Mediterranean fruit fly, *Ceratitis capitata* (Wiedemann); the mango fruit fly, *Cheatodacus ferrugienus* (F.); the Caribbean fruit fly, *Anastrepha suspensa* (Loew); and the melon fruit fly, *Dacus cucurbitae* (Coquillet). Other insects of interest in connection with irradiation are the codling moth, *Cydia pomonella* (L.); the mango weevil, *Cryptorhynchus (Sternochetus) mangiferae* (F.); and spider mites such as *Panonychus citri* (McGregor) and *Tetranychus telarius* (L.).

In most fruits the insect stages most likely to be encountered when irradiation is used are eggs, larvae, and possibly pupae. In the mango, however, adult forms constitute a substantial portion of the population. Mites also occur in adult forms.

The minimum dose for prevention of the development of eggs and larvae of the fruit flies into adults is about 0.21 kGy. That for the mango weevil is 0.33 kGy and for the codling moth about 0.23 kGy. Male and female mites require about 0.32 kGy for sterilization. All these doses generally are tolerated by bananas, guavas, longans, lychees, papayas, pineapples, rambutans, and tomatoes, although some varieties of each may suffer radiation injury. Avocados are unusually sensitive to radiation. They undergo pulp and skin darkening with doses less than 0.10 kGy, which is below that needed for disinfestation. Apples likewise undergo unacceptable changes with doses less than 0.25 kGy. Citrus fruits may exhibit some peel pitting, rind breakdown, and scald at doses needed for disinfestation. Mangoes may show dark, sunken areas at low doses, but generally tolerate disinfestation doses.

Since fresh fruits are living and metabolizing foods, the pre- and post-handling, as it can affect their continuing biological processes, can have important bearing on the effects of radiation in causing injury to the fruits. Among the important handling variables are age and maturity of the fruit at harvest, pre- and postirradiation holding time and temperature, and physical handling. Some injuries caused by irradiation become apparent only with the passage of time. Close control of the variables in handling sensitive fruits can be a requirement for success. For a number of commercially important fruits, nonetheless, radiation disinfestation is an effective and useful procedure.

II. VEGETABLES

A. Radurization

1. Introduction

The decay of fresh vegetables is due primarily to the growth of bacteria such as *Erwinia caratova*, which cause soft rot. In some cases, however, fungi and yeasts also are involved in vegetable spoilage.

While a substantial amount of research has been done on the radurization of fresh vegetables, no very promising applications have been uncovered. One explanation for this is that the doses needed to secure product life extension often are greater than can be tolerated by the vegetable. Much of the effort has been directed to the identification of those vegetables that have adequate radiation tolerance. As with fresh fruits, the principal problem area is the softening effect produced by radiation. Also important is the tendency for higher doses to increase the susceptibility to decay. Discolorations of various types and flavor changes also preclude the irradiation of some vegetables.

Probably because of discouraging difficulties that have been encountered with vegetables, much of the research was not done in sufficient depth as to provide information that can be regarded as adequate. Often the only information available is the maximum dose tolerated by the vegetable plus a description of injury that occurs with higher doses. Lacking is information on decay prevention. The infeasibility of irradiating certain vegetables has been based simply on injury caused by irradiation. There is need for more research, particularly in the area of approaches for avoiding radiation injury.

As with fruits, the great variation of the responses of individual vegetables to irradiation requires separate discussion of each.

2. Arracacha (Apio) (*Arracacia xanthorrhiza*)

A dose of 100 Gy about doubles the life at 22°C of this tuber. Doses greater than 200 Gy make it more susceptible to microbial attack.

3. Artichoke

Irradiation at doses as large as 4 kGy has no significant effect on *Botrytis* rot of artichokes. Stem pitting and internal discoloration increase with dose. External discoloration occurs with doses of 3 kGy or greater.

4. Asparagus

Control of *Phytophora* rot of asparagus requires doses in excess of 500 Gy. Injury to asparagus occurs with doses above 40 Gy and includes radial splitting of the butt end, slippery epidermis, cooked appearance, and darkened color.

5. Beans (String)

Mold control of green beans can be secured with doses in the range 2–3 kGy. Crispness decreases with dose, as does flavor acceptability.

6. Beets

Beets are softened by a dose of about 3 kGy, but otherwise are acceptable.

7. Broccoli

Doses as large as 4 kGy are tolerated by this vegetable. A slight decrease in flavor acceptability, related to dose, occurs.

8. Cabbage

Doses less than 3 kGy do not affect the quality of white cabbage. Higher doses cause severe texture loss and yellowing.

A dose of 1 kGy is tolerated by red cabbage, provided the postirradiation storage is in sealed packages. The use of vented packages leads to discolorations.

Irradiated sauerkraut exhibits purplish brown discoloration, possibly due to the action of H_2O_2.

9. Cantaloupe

A dose of about 4 kGy reduces stem scar molding. This dose has a minimal effect on quality. Doses about 4 kGy produce off-aromas, atypical flavors, and increased decay. Irradiation produces immediate softening. In storage, however, any difference from stored nonirradiated melons disappears.

10. Carrot

Doses of 50 to 100 Gy decrease rotting. A dose of 150 Gy increases it.

11. Cauliflower

Quality is adversely affected with a dose of 3 kGy. Softening and flavor changes occur.

12. Celery

A dose of approximately 500 Gy causes undesirable flavor and odor changes and also adversely affects texture.

13. Corn (Maize)

While doses of 3 to 5 kGy control molding of ear corn, off-flavors occur. "Denting" of the kernels, which usually is associated with aging, occurs in storage.

14. Cucumber

A dose of 3 kGy reduces the molding of stem scars. Irradiation increases yellowing and softening.

15. Cut Vegetables

Cut vegetables, prepared and packaged for kitchen uses such as soup greens, irradiated with doses of the order of 1 kGy, exhibit delay of spoilage for several days when stored at 10°C. Such spoilage is due mainly to bacterial growth. Red cabbage, onion, leek, rutabaga (swede), carrot, celery root, and cauliflower are adaptable to this procedure. Sealed packages yield better results than do perforated ones.

16. Endive

Dose up to 2 kGy do not reduce decay of endive. Spotting and leaf browning occur and increase with dose. A dose of 2 kGy inhibits the formation of green color.

17. Honeydew Melon

A dose of 4 kGy does not adversely affect commercially mature honeydew melons with respect to visual quality, ripeness, and skin and flesh color. Softening occurs, predisposes the melon to bruising and—differently from the case of bruising of nonirradiated melons—persists into the ripe stage. Doses above 4 kGy yield unmarketable melons.

18. Lettuce

Irradiation with doses of about 2 kGy reduces decay of head and leaf lettuce stored at 4°C, but causes unacceptable injury. Spotting, especially along the ribs occurs. The center of head lettuce exhibits browning, particularly along the leaf margins. Pink rib occurs, the severity of which increases with dose. Higher doses cause loss of "lettuce flavor" and turgidity.

19. Mushrooms

A dose of 1 kGy delays molding of mushrooms without causing sensory damage.

20. Parsley

Parsley irradiated with about 10 kGy has acceptable quality. Greater doses cause wilting and produce a sour grassy flavor.

21. Parsnips

Parsnips irradiated with approximately 5 kGy have acceptable sensory characteristics. Larger doses cause mushiness, darkening, and bitter flavors.

22. Peas

Peas tolerate doses of the order of 10 kGy without significant sensory changes.

23. Peppers

The control of decay of peppers by irradiation varies with the kind and variety of pepper. Doses as low as 1 kGy are beneficial with certain varieties of bell peppers ('Truhart' and 'California Wonder'). With other peppers doses as high as 5 kGy are not effective. Softening of the pericarp occurs and can limit the tolerable dose. Irradiation can affect red color formation, either accelerating or inhibiting its formation depending upon the dose employed.

24. Potatoes (White)

Although the principal interest in irradiating white potatoes has been to inhibit sprouting, spoilage due to microbial rot occurs with storage at high ambient temperatures (25–35°C), as may be the case in tropical regions. The dose for sprout inhibition is of the order of 100 Gy (see Section II,C), which is too low for controlling microbial spoilage. Reduction of microbial spoilage of potatoes irradiated for sprout control can be secured by treatment with water at 55°C for 10 min.

25. Radishes

Radishes are sensitive to doses greater than 1 kGy. Softening, bleaching, and the formation of a poor flavor occur.

26. Rutabaga (Swede)

Doses of approximately 1 kGy cause rutabagas to become gray or dark.

27. Spinach

Spinach blanched to inactivate enzymes, placed in sealed containers, irradiated with a dose of 15 kGy, and stored at 3 to 5°C exhibits extended storage life.

28. Squash

A dose of 3 kGy appreciably reduces decay of white bush, scallop, yellow crookneck, and zucchini squashes. A slight softening occurs. Some bleaching of the green surface of white bush squash occurs.

29. Turnips

A dose of 10 kGy is tolerated by turnips. Greater doses cause wateriness, softness, and color darkening.

B. Radappertization

1. Introduction

Investigation of the radappertization of vegetables was made only during the early work on food irradiation. Only a few vegetables were studied, and only with limited effort. In view of this, it cannot be said that the true potential for radappertization of vegetables is known.

While some picture of the effect of high doses on certain vegetables has been obtained, there has been no proper determination of the minimum radiation dose (MRD) for sterility. The MRD would be expected to vary with different vegetables, at least to the degree as to whether they are classified as acid (pH < 4.5) or not. For nonacid, low-salt vegetables, the dose must be based on the same considerations used in the radappertization of meats—that is, inactivation of the spores of the most radiation-resistant strain of *Clostridium botulinum* and an amount of radiation that provides a $12D_{10}$ reduction, or some appropriate modification of this level of treatment.

In view of this lack of accurate knowledge concerning the MRDs for vegetables, the doses so far determined must be viewed as only indicative of what is required for sterility and cannot be regarded as authoritative. Based on present knowledge, they should not be used in practicing radappertization of the vegetables discussed below.

It seems likely that development of a technology specifically appropriate for the radappertization of vegetables could lead to needed product improvements, such as has been obtained with meats, and provide comparable opportunities for radappertized vegetables.

2. Beans (String)

Doses in the range of about 20 to 30 kGy provide beans of acceptable quality and stability at ambient temperatures. Blanching is a requisite. Irradiation in the frozen state reduces chlorophyll degradation. Based on a study with spores of *Clostridium sporogenes* PA3679, an MRD of 45 kGy is indicated for string beans.

3. Carrot

Based on a study with spores of *Clostridium sporogenes* PA3679 an MRD of 25 kGy is indicated for carrots.

4. Peas

Blanched peas irradiated with a dose of approximately 20 kGy are stable at ambient temperatures. Sensory characteristics are satisfactory. Peas irradiated with doses in the range 8–10 kGy and heated to 100°C for 5 min are stable at ambient temperatures. Such peas have more acceptable sensory characteristics than peas sterilized by heating at 115°C for 40 min. Some softening may occur with irradiation. Irradiation in the frozen state improves flavor and color.

5. Pumpkin

Blanched pumpkin irradiated with doses in the range of about 15 to 25 kGy is stable at 22°C. Aroma is excellent and color good.

C. Inhibition of Ripening, Sprouting, and Senescence

As vegetables continue to metabolize after harvest, various kinds of changes occur. Virtually all of these changes have significance in the consumer acceptance of the vegetable and also, to a considerable degree, determine its market life. Changes of these kinds in vegetables which have been related to irradiation include

Color changes	Growth
Texture changes	Maturation
Flavor and aroma changes	Sprouting

Irradiation can affect the biological processes involved in these changes. As may be anticipated, such effects are dose dependent. Generally, relatively low doses may cause stimulation of a process, whereas higher doses lead to inhibition. For this reason the dose needed to obtain a particular result usually is quite specific. In addition, different vegetables can have particular dose requirements for a given kind of effect.

To a degree, there are similarities between fruits and vegetables in these effects, but intrinsic differences between these two groups of foods make for significant variations. For fruits delay of ripening generally has been the more important phenomenon. For vegetables inhibition of sprouting and delay of senescence have been of greater interest.

Some vegetables are harvested while green, and common custom dictates that they be consumed in this condition. Because irradiation may induce unwanted color changes in such foods, its utility is impaired. Doses of 1 or 2 kGy cause yellowing of green *cucumbers*, and this effect interferes with the use of irradiation to control microbial decay. Similarly, *bell peppers* yellow when irradiated with doses in the range 1.25–5 kGy. In connection with the after-ripening of red peppers, however, advantage can be taken of the radiation stimulation of color change to accelerate the increase of pigment formation, which is the purpose of the after-ripening process. Figure 8.6 shows the change of capsanthin content of peppers with time during the after-ripening process, as related to dose. It is seen that dose is critical in this process. Doses less than 100 Gy accelerate pigment increase, whereas larger doses inhibit pigment formation.

Endive is harvested while it is mostly white in color, and maintenance of this condition is important for consumer acceptance. Exposure to light after harvest, however, tends to turn endive green. A dose of 2 kGy retards greening. *White potatoes* undergo a similar greening, which also is inhibited by a dose of the order of 100 Gy (about the same as the dose for sprout inhibition). Since solanine production is associated with potato greening, this effect has added value. In the overall, peel losses associated with greening are reduced.

As can be seen from Fig. 8.7, the postharvest elongation of *asparagus* spears can be affected by irradiation. Irradiation, however, does not retard toughening. Doses greater than 0.5 kGy cause damage to the spears, including radial splitting of the butt end, slippery epidermis, a cooked appearance, and darkened color.

Irradiation with doses of 0.3 or 3 kGy does not improve moisture retention

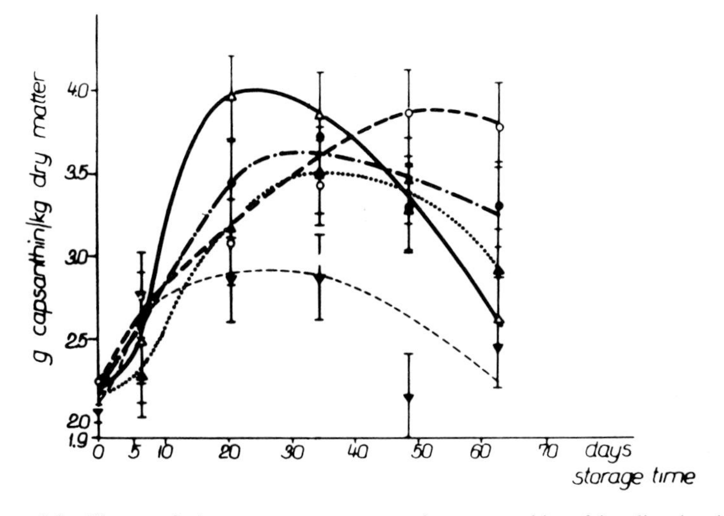

Figure 8.6 Change of pigment content, expressed as capsanthin, of irradiated red pepper (*Capsicum annuum*) with storage time. ○, Nonirradiated; ●, 80 rad; △, 2 krad; ▲, 10 krad; ▼, 100 krad. From J. Farkas, I. Kiss, and E. Andrassy, After ripening of red pepper (*Capsicum annuum* as affected by ionizing radiation. *In* "Food Irradiation, Proceedings of symposium, Karlsruhe." Int. Atomic Energy Agency, Vienna, 1966.

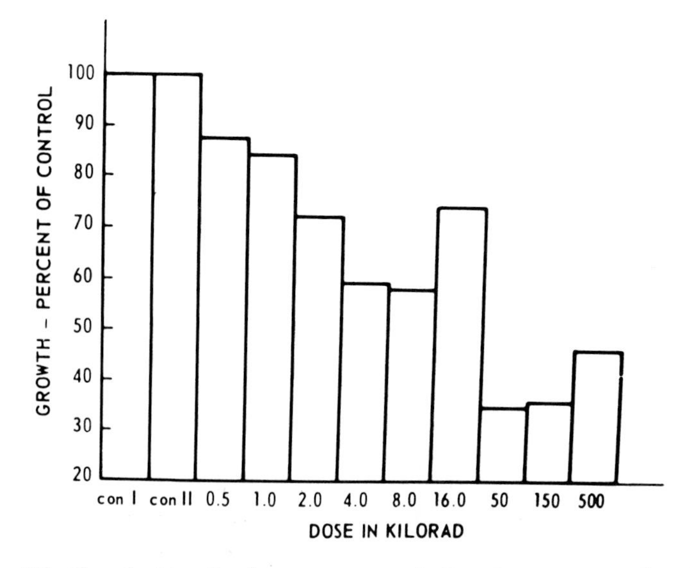

Figure 8.7 Growth of irradiated asparagus spears, indicated as percentage of nonirradiated control spears. From E. C. Maxie, N. F. Sommer, and D. S. Brown, "Radiation Technology in Conjunction with Postharvest Procedures as a Means of Extending the Shelf Life of Fruits and Vegetables," UCD-34-P80-6. U.S. Atomic Energy Comm., Washington, D.C., 1970.

of ear *sweet corn* over 14 days of storage at 4°C. Since moisture content is related to quality, in that higher moisture content indicates less conversion of sugar to starch, there appears to be no benefit derived from the irradiation of ear corn.

Mushrooms reach the market at an immature stage, with the cap or veil closed. Postharvest changes include opening and browning of the cap, darkening of the gills, and browning and elongation of the stem. These changes are regarded as undesirable by the consumer. Doses in the range 0.5–1 kGy strongly inhibit the postharvest changes and maintain a fresh-appearing condition. Appropriate packaging, mainly to retain moisture, is of value. Flavor is unaffected. The data of Table 8.12 support these findings.

Some vegetables are root parts of plants, and when stored they are apt to sprout. Changes as a result of sprouting can make these foods unacceptable. For this reason, sprouting can limit their storage life.

Irradiation can delay or prevent sprouting. For several reasons this use of

Table 8.12
Effects of Irradiation on Mushrooms Stored at Ambient Temperatures[a]

		Dose (Gy)			
Effect	Days	0	500	1000	3000
Cap opening rating[b]	1	4.2	4.9	4.9	4.9
	3	1.3	4.1	4.5	4.6
	6	0.9	4.0	4.5	4.8
Stem length (mm)	1	18	18	20	20
	3	30	25	20	20
	6	36	25	25	20
Skin color rating[c]	1	3.0	3.3	3.1	3.0
	3	1.6	3.1	2.9	3.1
	6	1.0	2.6	3.3	3.3
Flesh color rating[c]	1	4.3	4.1	3.8	4.2
	3	2.6	3.1	3.0	3.3
	6	1.8	2.2	2.6	3.0
Flavor when cooked rating[d]	1–6	1.1	3.1	3.0	3.0

[a] From R. A. Mercier and K. F. MacQueen, "Gamma-Irradiation to Extend Postharvest Life of Fruits and Vegetables," 1965 Report of the Horticultural Experiment Station and Products Laboratory. Ontario Dept. of Agriculture, Toronto, 1965.

[b] 0, Fully open; 5, closed.

[c] 0, Poor; 5, excellent.

[d] 1, Poorest; 5, best.

irradiation has attracted a great deal of interest worldwide:

1. It serves a very useful purpose in the preservation of certain important foods, such as white potatoes and onions.
2. The dose requirements are among the lowest employed in food irradiation, a fact that has significance in that radiation-induced damage to the food is minimal and also that irradiation costs are low.
3. It is dramatically effective.
4. It replaces chemicals used to control sprouting. While a number of root foods have been investigated in relation to sprout inhibition, the major effort has been concerned with white potatoes and onions.

While doses less than 10 Gy stimulate sprouting, *white potatoes* irradiated with doses in the range 30–150 Gy do not sprout when stored. Ordinarily the inhibition is irreversible, and unlike sprout inhibition obtained with chemicals, a single radiation treatment is sufficient, irrespective of postirradiation storage conditions. After sufficient storage time, radiation-induced sprout inhibition can be reversed by treating with gibberellic acid. This reversal is accomplished, however, by the formation of new buds.

Several mechanisms have been proposed for the sprout-inhibiting action of radiation, including (1) interference with the synthesis of nucleic acids in the meristematic tissue of the potato buds, (2) disturbance of the phosphorylation process, (3) inhibition of the formation of the auxin, indole acetic acid, and (4) chromosomal changes in the cells of the meristematic tissue which prevent normal cell division and growth. Deformed buds and tissue necrosis in growing points can be observed in stored irradiated potatoes. To be most effective, irradiation of white potatoes must be carried out during the natural postharvest dormancy period. This period varies substantially among potato varieties. In some locations this requirement is difficult to meet, since sprouting has already started at the time of harvest, as may occur with summer crops in warm climates. Such early sprouting can be overcome by the use of higher doses; for example, for potatoes of the 'Alpha' variety a dose of 140 Gy is needed.

Irradiation of potatoes affects more than sprouting. The harvesting and handling operations usually cause injuries such as cuts and bruises. Normally these will heal during postharvest storage. Irradiation, however, interferes with the healing process and unhealed lesions can become entry ports for invading spoilage microorganisms, which cause rotting of the potato. It is usually necessary, therefore, to delay irradiation for a period of several weeks after harvest that is sufficient to secure healing of the lesions. Any additional handling which would cause new injuries should be minimized.

Regardless of how long they are stored, irradiated potatoes can undergo blackening during or after cooking. This has been ascribed to an increase in the polyphenol content of the potato tissue and interaction of the polyphenols with ferrous salts present in the potato. A graying of processed potato products, such as prepeeled potatoes, precooked chips, and potato flakes, also occurs and is believed to be the same phenomenon. The intensity of the blackening has been related to (1) potato variety, (2) growing conditions, and (3) irradiation dose.

In fresh-cut raw potatoes a radiation-enhanced enzymatic browning can occur. Reduction of dose and the use of a holding period of 1 to 2 months provide a remedy for this kind of discoloration.

The content of the total sugars of potatoes increases immediately after irradiation, but soon drops to a level comparable with nonirradiated potatoes. It remains essentially unchanged thereafter up to 8 or 9 months of storage, when it again increases. Reducing sugars cause browning or darkening in heated potato products such as chips. The reducing-sugar content can be lowered by holding the potatoes at elevated temperatures ($\geq 15°C$) for a period of some weeks prior to processing.

The exact dose employed for sprout inhibition is critical for a given potato variety and for a given set of circumstances. In general it is desirable to use the smallest effective dose, since irradiation can cause changes in the potato which can reduce its quality and which are dose dependent. These changes include increased rotting and discoloration.

With some potato varieties (e.g., 'Ontario' and 'Pontiac'), the incidence of "black spot," a storage disorder, increases with does over the range 50–400Gy. Black spot originates with bruising and does not involve a surface cut. Since not all varieties are susceptible to this condition, the indicated remedy is selection of nonsusceptible varieties for irradiation. The use of the smallest dose otherwise effective also is indicated.

Postirradiation holding conditions have important effects on the quality and life of the potato. Low storage temperatures such as 5°C effect better overall preservation but increase the buildup of sugars. Higher storage temperatures do not affect the inhibition of sprouting but do favor rotting and loss of moisture. Rotting has been associated mainly with the bacteria of the *Erwinia* and *Micrococcus* species, although molds such as *Fusarium* spp. also occur. Storage humidities also are critical and affect both rotting and weight loss. Postirradiation holding conditions constitute variables of the total preservation process, which may be altered with circumstances and objectives.

In many locations climatic conditions allow only one potato crop per year. Even with good storage conditions, with this circumstance it is not possible to obtain a year-round supply of potatoes of local origin. Usually a time gap exists between the spoilage of last year's crop, primarily due to sprouting, and the availability of the new year's crop. While irradiation prevents sprouting, it alone does not accomplish potato preservation for a full 12-month period.

Processes other than sprouting deteriorate the potato and limit its life. Moisture losses due to evaporation and transpiration occur and can result in shriveling and loss of turgor. Handling during and after irradiation can cause new wounds which do not heal and which can lead to rotting. Such changes and others affecting chemical composition place a limit on the useful life of the stored potato, usually no greater than 8 or 9 months from harvest and shorter if the initial potato quality or storage conditions are not the best. Despite these limitations, sprout inhibition by irradiation is a useful process, which can be adjusted in regard to a number of variables to provide improved preservation of white potatoes. Fairly large-scale testing in a number of countries, plus experience with commercial operation over a number of years in Japan, support this view.

The use of irradiation to inhibit the sprouting of *onions* also has attracted a great deal of interest worldwide. The growth tissue of onions is in the interior of the bulb. Because they cannot reach the interior, chemical sprout inhibitors generally are not effective. Methylhydrazine applied to onions in the field at least 24 hr prior to rain can be used, but often such use is not practical.

The dormancy period of onions after harvest is short. When sprouting begins, growth progresses upward from the base, pushing through the scale leaves and emerging at the top of the bulb. If radiation is applied after the start of the growth, it is less effective and larger doses are needed. Additionally, if growth of the meristematic tissue has occurred, irradiation causes it to die, with consequent darkening. The amount of such tissue is related to the amount of growth that has occurred at the time of irradiation and the dose used. The significance of this necrotized tissue increases with its size. The dead tissue appears as a dry rot. The shoot dries and darkens. If the dry tissue extends into the neck region of the bulb, rot-causing organisms may gain entry through the neck and attack the bulb tissue. In the home use of onions a very small amount of necrotized tissue may be acceptable, whereas a larger amount usually is not. In onions used for processing into dehydrated flakes or powder, darkening to any substantial degree is objectionable.

In order to avoid problems associated with necrotized meristematic tissue, irradiation should be carried out shortly after harvest, probably within a period of 1 to 4 weeks. The dormancy period can be extended by storing the onions at low temperatures prior to irradiation. Doses in the range 20–150 Gy can prevent sprouting, but the practical range is considered to be 60–120 Gy. The lower end of the range can be used if irradiation is done very soon after harvest (e.g., 1 week), or if the preirradiation storage temperature is low (e.g., 14–16°C).

Irradiation at the doses for sprout inhibition also inhibits root growth.

Storage losses are the result of (1) sprouting, (2) rot, and (3) moisture loss. Organisms that have been associated with rotting include the bacteria

Erwinia, *Sarcina*, and *Micrococcus* spp., and the fungi *Fusarium* and *Aspergillus* spp. Postirradiation storage conditions greatly affect losses. A fairly dry storage is better than moist. Varietal differences apparently exist in relation to the best storage temperature over the range of about 5 to above 25°C.

The sensory characteristics of onions are little affected by irradiation in the dose range 20–150 Gy. With just a few varieties some decrease of pungency occurs. As with potatoes, irradiated onions should be stored only for periods which provide good retention of quality.

Not surprisingly, the irradiation of *garlic* for sprout inhibition follows a pattern similar to that of onions. The dormancy period of garlic following harvest is about the same as for onions, but comparable darkening of the meristematic tissue does not occur with irradiation, and the browning that does occur is not regarded as objectionable. Dehydrated garlic made from irradiated bulbs has a lighter color than that made from nonirradiated bulbs. In view of this, the time period after harvest before irradiation is less critical than it is with onions. Garlic bulbs irradiated with a dose of 100 to 150 Gy and stored at 0 to 5°C develop internal sprouting at 2 months, but do not exhibit external sprouting. Storage at temperatures in the range 24–34°C does not result in external sprouting but can cause dry rot, moisture loss, and shriveling, which are less important at lower storage temperatures. Bulbs irradiated with 125 Gy more than 3 months after harvest undergo some sprouting. Doses up to 175 Gy do not cause rot or other damage to garlic bulbs, including loss of pungency.

The doses needed to prevent sprouting of other root foods for which there is information are listed in Table 8.13.

Table 8.13

Irradiation Doses for Sprout Inhibition of Various Root Foods

	Dose (Gy)
Beet	50–150
Carrot	190
Ginger	10
Jerusalem artichoke	50–150
Parsnip	190
Shallot	60–120
Sweet potato	60–500
Turnip	50–190
Yam	75–200

D. Insect Disinfestation

The use of irradiation for insect disinfestation of vegetables has not received the same attention as has been given to the radiation disinfestation of fruits. Yet some of the same possibilities exist with vegetables as with fruits, both in the area of quarantine applications and in the prevention of food damage. Some vegetables, such as green bell peppers, are presumed hosts for certain insects, including the Mediterranean fruit fly, and require disinfestation before they can be shipped into the United States from a quarantined area. The potato tuber moth *Pthorimaea operculella* (Zeller) causes substantial losses in warm countries. As a consequence of the lack of attention to this use of irradiation, only very limited information is available on insect disinfestation of fresh vegetables.

A dose of 100 Gy prevents emergence of the adult tuber moth *Pthorimaea operculella* (Zeller) on white potatoes, and larval burrowing or castings are absent. A dose of 100 Gy also can inhibit sprouting of potatoes, and there is therefore a double effect of the applied radiation. Weevil growth in garlic is inhibited by doses in the range 140–200 Gy. Doses of 200 to 300 Gy eliminate 70–80% of the population of the nematode *Scutellonema bradys* Adsiyan in white yams.

III. DRIED FRUITS AND VEGETABLES

The low moisture content of dried fruits and vegetables ordinarily provides the needed preservative action. In cases in which the moisture content is not sufficiently low to preserve these foods, radurization or radappertization may be useful. Most commonly, however, the principal use for irradiation of dried fruits and vegetables has been to secure insect disinfestation.

Typical insects that infest dried fruits include the saw-toothed grain beetle, *Oryzaephilus surinamensis*; the fig moth, *Ephesia cautella*; and the Indian meal moth, *Plodia interpunctella*. The moth *Nemapogon granellus* (L.) has been identified in dried mushrooms. Other insects may occur in various dried fruits. As would be expected, insect forms other than adult require smaller doses. For example, the eggs of *P. interpunctella* treated with 200 Gy do not hatch. Adult forms may maintain activity for a large number of days after irradiation with doses as large as 1 kGy and present problems with quarantine and other requirements. The use of higher doses (e.g., 2 kGy) can effect a quicker kill. One procedure that may be used to hasten the death of adult *O. surinamensis* is to use a lower dose such as 350 Gy combined with the holding of the fruit, such as dates, for a period at 40°C. The data of Table 8.14 exhibit this combination effect. Complete disinfestation of dry dates results, since eggs or other early

Table 8.14

Estimated Mean Period in Days Required to Produce 50% (LT_{50}) or 95% (LT_{95}) Mortality of Irradiated *Oryzaephilus surinamensis* Adults[a]

Dose (Gy)	Days at 25°C		Days at 40°C	
	LT_{50}	LT_{95}	LT_{50}	LT_{95}
0	34.0	87.0	10.5	28.3
350	7.6	17.0	3.4	9.6
700	6.9	16.5	3.5	9.0
1050	5.5	16.1	1.5	8.0

[a] From M. S. H. Ahmed, Z. S. Al-Hakkak, S. K. Al-Maliky, A. A. Khadhum, and S. B. Lamooza, Irradiation disinfestation of dry dates and the possibility of using combination treatments. *In* "Combination Processes in Food Irradiation." Int. Atomic Energy Agency, Vienna, 1981.

stages of the insect are prevented from developing as a consequence of irradiation.

Unlike fresh fruits and vegetables, the dried products have little sensitivity to radiation and generally withstand doses greater than those needed for insect disinfestation without causing undesirable sensory changes. Some darkening may occur as the dried fruits are stored. A small amount of softening due to conversion of pectin substances can occur. Generally, rehydration, swelling, and cooking properties are improved. If there is a need, improvement of sensory characteristics can be obtained by irradiating at low temperatures in the range −80 to −180°C.

The principal vegetable of interest has been dried mushrooms. Dried fruits of interest are apples, apricots, currants, dates, figs, peaches, pears, prunes, quinces, and raisins. Insect disinfestation doses generally are less than 1 kGy.

IV. NUTS

Nuts may be irradiated in order to secure various kinds of results, including (1) inactivation of pathogenic bacteria such as salmonellae and molds which can produce aflatoxin, (2) insect disinfestation, and (3) sprout inhibition. Since nuts may contain substantial amounts of oils, undesired flavors may develop from radiation-accelerated lipid oxidation. Almonds, chestnuts, chalghoza, and peanuts (ground nuts) exhibit no change of sensory characteristics with doses up to 1 kGy. Walnuts, however, develop an unacceptable flavor when stored for 6 months at 25 to 40°C.

Shelled peanuts irradiated with doses of 1 to 1.5 kGy remain free of aflatoxin during storage for 10 months. Doses of about 1 kGy are adequate for insect control with almonds, chalghoza, peanuts, and walnuts. Doses in the range 120–600 Gy inhibit sprouting of chestnuts when irradiated 2 months after harvest and stored at 18 to 20°C. After 3 months of storage, rotting, however, is not controlled by these doses.

SOURCES OF ADDITIONAL INFORMATION

Abdel-Kader, A. S., and Maxie, E. C., "Radiation Pasteurization of Fruits and Vegetables. A Bibliography," ORNL-IIC-11. U.S. Atomic Energy Agency, Oak Ridge, Tennessee, 1967.

Akamine, E. K., and Moy, J. H., Delay in postharvest ripening and senescence of fruits. *In* "Preservation of Food by Ionizing Radiation," (E. S. Josephson and M. S. Peterson, eds.), Vol. III. CRC Press, Boca Raton, Florida, 1983.

Anonymous, "Application of Food Irradiation in Developing Countries," Tech. Rep. Series No. 54. Int. Atomic Energy Agency, Vienna, 1966.

Anonymous, "Preservation of Fruits and Vegetables by Irradiation, Proceedings of a Panel." Int. Atomic Energy Agency, Vienna, 1968.

Anonymous, "Disinfestation of Fruit by Irradiation, Proceedings of a Panel." Int. Atomic Energy Agency, Vienna, 1971.

Anonymous, "Irradiation of Plant Products." Council for Agricultural Science and Technology, Ames, Iowa, 1984.

Anonymous, "Use of Irradiation as a Quarantine Treatment of Agricultural Commodities," IAEA-TECDOC-326. Int. Atomic Energy Agency, Vienna, 1985.

Bramlage, W. J., and Covey, H. M., "Gamma Radiation of Fruits to Extend Market Life," Marketing Res. Rept. 717. U.S. Dept. of Agriculture, Washington, D.C., 1965.

Bramlage, W. J., and Lipton, W. J., "Gamma Radiation of Vegetables to Extend Market Life," Marketing Res. Rept. 703. U.S. Dept. of Agriculture, Washington, D.C., 1965.

Burditt, A. K., Jr., Food irradiation: A quarantine treatment of fruits. *Food Technol.* (*Chicago*) **36,** 51 (1982).

Kader, A. A., and Heintz, C. M., "Gamma Irradiation of Fresh Fruits and Vegetables," An Indexed Reference List (1965–1982). University of California, Davis, 1983.

Kahan, R. S., "Radiation Preservation of Foodstuffs. A Bibliography." Israel Atomic Energy Comm., Yavne, 1967.

McKinney, F. E. "Radiation Pasteurization of Fruits and Vegetables. A Bibliography," ORNL-IIC (Rev.). U.S. Atomic Energy Comm., Oak Ridge, Tennessee, 1977.

Matsuyama, A., and Umeda, K., Sprout inhibition in tubers and bulbs. *In* "Preservation of Food by Ionizing Radiation," (E. S. Josephson and M. S. Peterson, eds.), Vol. III. CRC Press, Boca Raton, Florida, 1983.

Maxie, E. C., and Abdel-Kader, A., Food irradiation—physiology of fruits as related to feasibility of the technology. *Adv. Food Res.* **15,** 105 (1966).

Mercier, R. G., and MacQueen, K. F., "Gamma-Irradiation to Extend Postharvest Life of Fruits and Vegetables," 1965 Report of the Horticultural Expt. Station and Products Lab. Ontario Dept. of Agriculture, Toronto, 1965.

Metlitskii, L. V., Rogachev, V. N., and Krushchev, V. G., "Radiation Processing of Food Products," ORNL-IIC-14. Translation of Russian publication. U.S. Atomic Energy Comm., Oak Ridge, Tennessee, 1968.

Moy, J. H., Potential for gamma irradiation of fruits—a review. *J. Food Technol.* **12,** 449 (1977).

Moy, J. H., Radurization and radicidation: fruits and vegetables. *In* "Preservation of Food by Ionizing Radiation" (E. S. Josephson and M. S. Peterson, eds.), Vol. III. CRC Press, Boca Raton, Florida, 1983.

Moy, J. H., ed., "Radiation Disinfestation of Food and Agricultural Products." University of Hawaii Press, Honolulu, 1985.

Romani, R. J., Radiobiological parameters in the irradiation of fruits and vegetables. *Adv. Food. Res.* **15,** 57 (1966).

Salunke, D. K., Gamma radiation effects on fruits and vegetables. *Econ. Bot.* **15,** 28 (1961).

Sommer, N. F., and Fortlage, R. J., Ionizing radiation for control of postharvest diseases of fruits and vegetables. *Adv. Food Res.* **15,** 147 (1966).

Thomas, P., Radiation preservation of foods of plant origin. Part 1. Potatoes and other tuber crops. *CRC Crit. Rev. Food Sci. Nutr.* **19**(4), 327 (1983).

Thomas, P., Radiation preservation of foods of plant origin. Part 2. Onions and other bulb crops. *CRC Crit. Rev. Food Sci. Nutr.* **21**(4), 95 (1984).

Thomas, P., Radiation preservation of foods of plant origin. III. Tropical fruits: Bananas, mangoes, and papayas. *CRC Crit. Rev. Food Sci. Nutr.* **23**(2), 147 (1986).

Thomas, P., "Efficacy of Irradiation Treatment for Shelf-life Improvement of Tropical Fruits." Int. Atomic Energy Agency, Vienna, in press.

Thomas, P., "A Code of Technological Practice for Sprout Inhibition of Root Crops by Irradiation." Int. Atomic Energy Agency, Vienna, in press.

CHAPTER 9

Cereal Grains, Legumes, Baked Goods, and Dry Food Substances

I. CEREAL GRAINS

Cereal grains, one of humanity's oldest foods, generally are preserved by one of the oldest methods, namely, drying. Drying to a sufficient degree, coupled with maintenance of an adequately low moisture level during storage, ordinarily effectively prevents spoilage due to microbial action. Partly for this reason there has been no strong interest in radurization of cereal grains. Dry grains, however, are subject to insect infestation and, as a consequence, can undergo insect feeding damage, which can be severe. Additionally, insect metabolic processes can increase the moisture content of grain and raise it to a level that allows microbial growth, which can add to the grain damage. In view of these phenomena, the principal interest in grain irradiation has been to secure insect disinfestation. Cereal grains that have been examined from the standpoint of irradiation include wheat, rice, maize, milo, and barley.

Broadly speaking, insects of two orders, the Coleoptera (beetles) and Lepidoptera (moths), infest grains. In addition certain arachnids (mites) may be present. Radiation resistance varies with the species of each; generally the

Lepidoptera are more resistant than the Coleoptera. The grain mite *Acarus siro* has intermediate resistance (see Tables 4.7 and 4.8).

While standard practices for insect disinfestation by chemical insecticides usually yield lethality within 24 hr, disinfestation by irradiation will not provide lethality within such a short time period, unless very large doses, in the range 3–5 kGy, are used. Some cereal grains are subject not only to what may be termed the usual considerations regarding sensory characteristics (e.g., flavor and color) as affected by dose, but also require retention of certain functional properties related to their common food usage. Wheat flour, for example, in the making of fermentation-leavened bread must yield certain critical rheological characteristics to the dough, which are dependent upon the presence of specific starch and protein components. Large doses may affect these functional properties and may, therefore, not be usable with some cereal grains. The decrease in loaf volume of bread made with irradiated wheat flour as a function of dose over the range 2.5–10 kGy (0.25–1 Mrad) is shown in Fig. 9.1.

Partly to avoid radiation damage to grains and partly to secure lower irradiation costs, disinfestation of cereal grains is based on a time-delayed effect on insects. The data of Tables 4.7 and 4.8 indicate that doses in the range

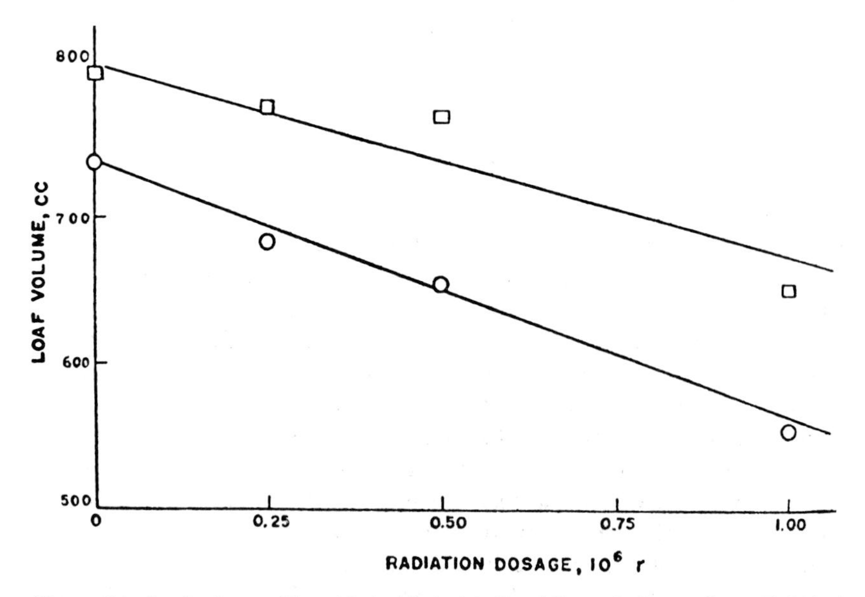

Figure 9.1 Loaf volume of bread baked from irradiated flour. ○, Patent flour; □, baker's flour. The radiation dosage is shown in 10^6 r (or rep); 1 rep ≈ 0.01 Gy (see Note on p. 341). From C. C. Lee, *Cereal Chem.* **36,** 70 (1959).

130–250 Gy permit some development of eggs and larvae but not to the adult stage. Doses in the range 400–1000 Gy prevent development to the next stage, that is, from eggs, larvae, or pupae. For beetles doses of 130 to 250 Gy produce sterility; for moths 250 to 400 Gy are needed, and the same range applies to the mite *Acarus siro*. For immediate lethality 3–5 kGy are needed. If lethality within a few days is acceptable, a dose of 1 kGy is sufficient. A 500-Gy dose produces sterility in survivors and death within a few weeks.

For the disinfestation of cereal grains a dose of 500 Gy generally is considered sufficient. This will control beetles and immature stages of moths. Any progeny of moths would be sterile. Insects irradiated with this dose do not feed normally and may undergo as much as 97% reduction of this activity. This dose does not adversely affect sensory or functional properties of the grains.

One aspect of commercial grain irradiation can be somewhat different from that of other foods. Grains can be treated as a fluid and can be made to "flow" past a radiation source. By using an air stream to accelerate the grain to a high velocity and by impacting the grain in an appropriate manner at the end of the pathway through the irradiation, physical destruction of adults can be accomplished. This manner of destruction of the most radiation-resistant portion of the insect population significantly adds to the radiation effect. A schematic diagram of an electron beam grain irradiator is shown in Fig. 9.2.

The embryo of grains can be killed by radiation. The dose to accomplish this varies with the kind of grain and its moisture content. Significant reduction of viability occurs at relatively low doses, as may be seen from the data of Table 9.1, in which the viability of maize grains is related to dose. From the standpoint of the storage of grains, reduced seed viability results in better maintenance of a low moisture content of the grain, since grain respiration is reduced. This effect, in combination with the radiation-reduced respiration of any insects present is beneficial in avoiding moisture buildup sufficient to permit mold growth.

Radurization doses for several cereal grains are given in Table 9.2. These doses are high enough to result in changes which may affect the functional characteristics of the grains.

II. LEGUMES

The seeds of various leguminous plants, generally called legumes or pulses, constitute important foods in many parts of the world. They are normally preserved by drying. Generally, they are subject to insect infestation and in some situations also to molding. While there can be other reasons for irradiating these foods (see Chapter 10), the chief interest has centered on insect disinfestation and to a lesser extent on radurization.

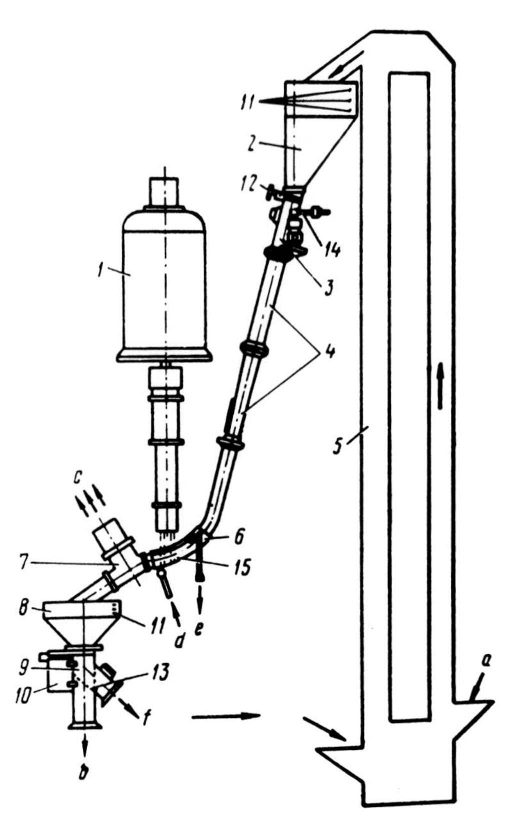

Figure 9.2 Commercial electron irradiation facility for disinfestation of cereal grains. a, Infested grain delivered for processing; b, processed grain; c, suction ducts; d, inlet for cooling water; e, outlet for cooling water; f, grain fed for recycling; 1, electron accelerator; 2, feeding bin; 3, grain flow distribution unit; 4, grain speedingup duct; 5, bucket elevator; 6, irradiation chamber cooling radiator; 7, suction chamber; 8, damping bin; 9, grain flow redistribution chamber; 10, automatic device for maintaining grain level in damping bin; 11, grain level sensors; 12, grain flow control valve; 13, selector valve; 14, fast-action valve; 15, irradiation chamber. From High-efficiency Industrial Plant for Radiation Disinfestation of Grain. Techsnabexport, Moscow, 1984.

Molds identified with some legumes (e.g., cowpeas) include *Aspergillus*, *Pencillium*, and *Mucor* spp. If the moisture content of the legume is kept sufficiently low, molding does not occur. During storage, however, moisture levels can rise due to respiration of infesting insects and lead to molding. Mold control, therefore, can be assisted by insect disinfestation.

A principal insect found in legumes is *Callosobruchus subinnotatus* Pic. It can cause severe damage and make pulses unfit for consumption. A dose of 50 Gy sterilizes adult males and females. A dose of 200 Gy inactivates pupae and

Table 9.1

Percentage Seed Viability of Irradiated Maize Grains Incubated for 2 Weeks after Irradiation[a]

Dose (Gy)	Viability (%)	Dose (Gy)	Viability (%)
0	100	400	3
100	90	800	0
200	45	1600	0

[a] From B. Amoakoatta, Simulated radiation disinfestation of infested maize in Ghana. *In* "Combination Processes in Food Irradiation." Int. Atomic Energy Agency, Vienna, 1981.

Table 9.2

Radurization Doses for Cereal Grains

	Dose (kGy)	
Cereal grain	Bacteria	Yeasts and molds
Wheat	1.5	2.5
Maize	15.5	6.0
Milo	5.4	5.4
Rice		
Korean	—	4.0
Japanese	2.0	2.0
Spanish	3.0	3.0
Thai	3.0	3.0

earlier stages of this insect. These observations suggest that a dose of 200 Gy can be used for insect disinfestation of legumes.

III. BAKED GOODS

Irradiation can be used to advantage with baked goods through two pathways: (1) irradiation of materials used as ingredients and (2) irradiation of the finished baked product. The objectives can be radurization or insect disinfestation.

Radurization of ingredients can be directed to inactivation of bacteria and fungi. The bacterium *Bacillus subtilis* occurs in baked goods as a contaminant of wheat flour, and its spores, being heat resistant, can survive the baking process. Germination and growth, which can occur as bread is held after baking, can cause spoilage. A dose of 0.75 kGy applied to wheat flour yields bread which has a shelf-life at 30°C that is about 50% greater than normal.

Bread made with flour irradiated with a dose of 500 Gy has increased firmness. The staling rate increases with dose. Low doses such as 50 Gy produce little or no off-odors and/or off-flavors, but higher doses do. Storage of flour for a period after irradiation may reduce off-flavors in bread.

Irradiation of bread with a dose of 5 kGy suppresses mold growth for many weeks. By irradiating the bread at 65°C, or by heating before or after irradiation, the dose can be reduced to 0.5 kGy. Chapatties irradiated with a dose of 10 kGy are free of mold for more than 6 months. Generally these doses do not reduce consumer acceptance, although larger ones may.

It is for baked goods that radiation-induced impairment of functional characteristics of cereal grains is of greatest concern. Most of the investigations relating to functional characteristics have been carried out on wheat flour as used in fermentation-leavened bread. It has been observed that loaf volume decreases when bread is made from irradiated flour (see Fig. 9.1). This has been ascribed to changes in the starch and protein constituents of the flour.

At moderate doses it has been observed that loaf volume may increase, as may be seen from Fig. 9.3. This has been explained as the result of radiation-induced starch degradation to form sugars which increase gas production in the dough and cause larger loaves. When less than adequate amounts of sugar

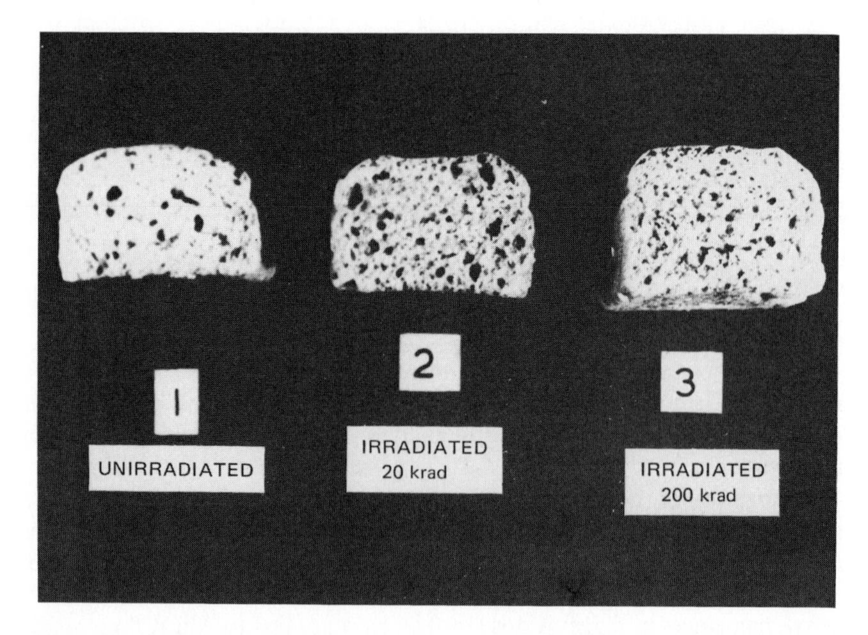

Figure 9.3 Breads prepared from nonirradiated and irradiated wheat using a lean dough formulation (1% sugar). From A. Sreenivasan, Compositional and quality changes in some irradiated foods. *In* "Improvement of Food Quality by Irradiation." Int. Atomic Energy Agency, Vienna, 1974.

are used in formulating the dough or when bromate is not used, the radiation-formed sugars serve in their places and increase loaf volume. Larger doses applied to the flour result in reduced loaf volume, due to the magnitude of the changes produced which cannot be offset by the radiation-increased sugar content.

Increased loaf volumes do not occur with "rich" formulations. Baking powder biscuits made of flour irradiated with a dose of 1 kGy or less show no increase in volume. White cakes made of soft wheat and formulated with large proportions of sugar ordinarily are not leavened by fermentation of sugar, as is done with bread. When made of wheat flour irradiated with a dose of 100 Gy, white cakes display no change in volume. With greater doses a decrease in volume occurs, as well as texture changes.

Noodles made from wheat flour irradiated in the range 200–1000 Gy have greater than normal cooking losses. The increase in cooking loss is ascribed to starch degradation.

It is clear that there are limitations on the irradiation of at least some of the cereal grains used for preparing baked goods. Although not all cereal grains have been studied with regard to the effects of irradiation on the functional properties required for their common usage, it is possible that undesired changes may occur as a result of irradiation. For this reason there is a need to carry out a thorough evaluation of each irradiated cereal grain in relation to its intended use, particularly in baked goods.

IV. DRY FOOD SUBSTANCES

A. Introduction

In the production of many food products, dry ingredients or intermediates are used. Some of these materials can be contaminated with microorganisms which inoculate the food product of which they become a component and can cause subsequent spoilage. Alternatively, for some food products for which the contaminated dry material are components, it is necessary to increase processing, such as heating, in order to secure a final acceptable microbial population. In such cases the necessary increased processing may reduce the quality of the food product. Since often there is no way to obtain these ingredient materials free of contaminating microorganisms, it has become customary to treat them prior to use in some manner so as to inactivate the microorganisms.

Irradiation provides an effective means to accomplish the needed inactivation. Since the materials are dry, the action of radiation is principally direct and chemical alteration is minimized. On the other hand, the low moisture

Table 9.3
Effect of Irradiation on Microbial Counts of Spices[a]

Spice and irradiation level (kGy)	Number of microorganisms per gram		
	Bacteria	Yeasts	Molds
Allspice			
0	2.28×10^6	<10	0
10	<10	<10	0
Greek oregano			
0	1.21×10^6	4×10^4	9×10^3
10	<10	<10	<10
Black pepper			
0	3.2×10^7	0	0
10	60	0	0
Garlic powder			
0	4.14×10^5	<10	7.8×10^3
10	700	<10	<10
Egyptian basil			
0	3×10^6	$>3 \times 10^4$	$>1.1 \times 10^4$
10	1×10^3	<10	0
Thyme			
0	1.5×10^5	0	300
10	40	0	<10
Mexican oregano			
0	1.5×10^6	3×10^4	5×10^3
10	30	<10	10
Domestic paprika			
0	1×10^6	0	0
6.5	100	0	0
Spanish paprika			
0	2.2×10^6	0	0
6.5	260	0	0
Celery seed			
0	4.4×10^5	1.5×10^3	200
10	<10	<10	<10
Crushed red pepper			
0	1.31×10^5	<10	0
6.5	<10	<10	0

[a] From M. I. Eiss, Irradiation of spices and herbs. *In* "Ionizing Energy Treatment of Foods, Proceedings of National Symposium." ISBNO 85856-0534. Sydney, 1983.

content results in a requirement for fairly large doses for microbial inactivation.

Since the objective of this use of irradiation usually is not the preservation of the irradiated material per se (it is of low moisture content), and since pathogens are not necessarily involved, the use of the terms radurization, radappertization, or radicidation is not appropriate. Instead terms such as decontamination or hygienization seem more correct and meaningful.

In addition to microbial contamination, some dry materials may also have insect contamination. The use of irradiation in such cases also is effective.

B. Spices and Vegetable Seasonings

Spices and vegetable seasonings such as dried onion and garlic powder usually are contaminated with both bacteria and molds. In some cases insects also may be present.

Table 9.3 provides microbiological data for a number of spices before and after irradiation with a dose of 6.5 or 10 kGy. A wide range of bacterial and fungal organisms as well as yeasts occurs. Bacteria that may be present include anaerobic spoilage bacteria, spore-forming bacteria, and proteolytic and gas-forming bacteria. Aerobic spore formers constitute a large proportion of the bacteria present, especially *Bacillus* spp. Molds that may be present include *Penicillium* spp., *Aspergillus flavus*, and *Aspergillus glaucus*. Pathogens such as *Bacillus cereus*, *Clostridium perfringens*, and salmonellae also occur. The level of contamination usually is in the range 10^3–10^8 organisms per gram. Spices normally are used at the level of 0.1 to 1% in meat products, and untreated spices can cause a product contamination level of as many as 10^5 to 10^6 bacteria per gram.

Figure 9.4 shows the relationship between dose and microorganism survival for a number of spices. The data also demonstrate a postirradiation effect which increases the inactivation beyond that which occurs immediately following irradiation.

As noted, the low moisture content of spices and seasonings limits chemical change resulting from irradiation. As might be expected, the effects of radiation are different for different spices. The threshold doses required to produce a detectable change in flavors in several spices are given in Table 9.4. That changes are dose dependent may be seen from the data of Table 9.5 for the essential-oil content of cardamom, as determined by gas chromatographic analysis. Gas chromatograms of the essential oils of juniper for doses over the range 5–30 kGy are shown in Fig. 9.5. The 13 peaks observed can be separated into those with a short retention time due to terpene hydrocarbons and those with a longer retention time due to a group of oxygen compounds. The oxygen compounds undergo greater changes as dose is increased.

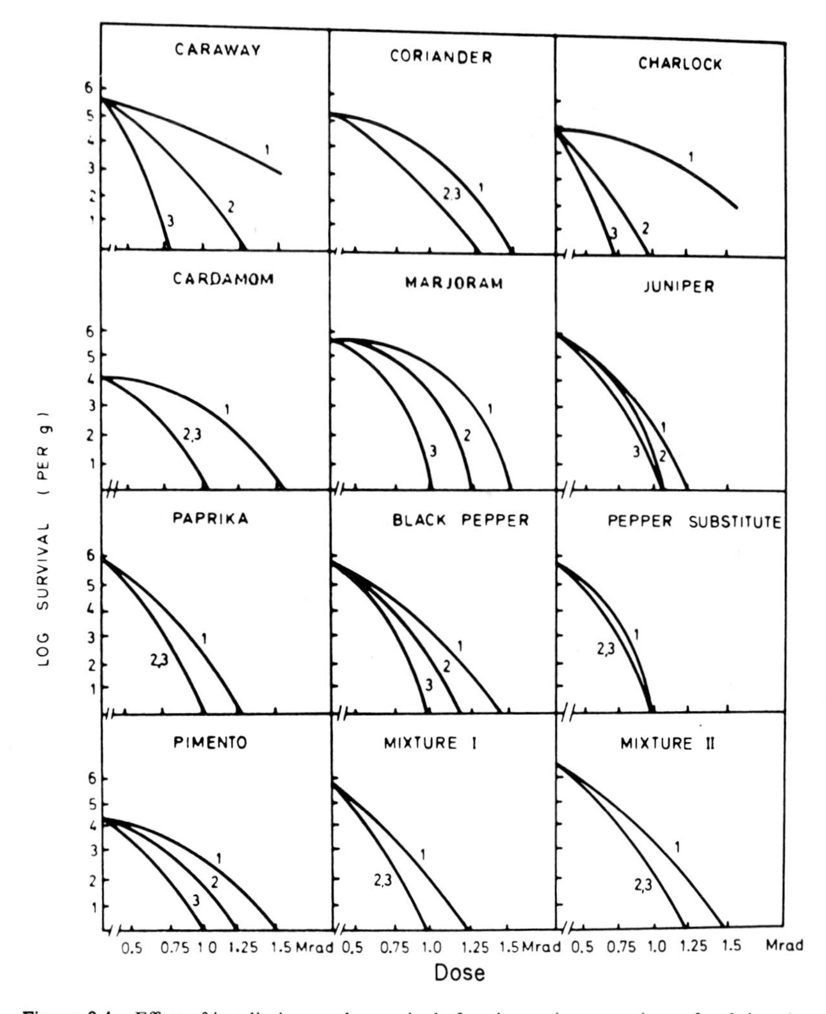

Figure 9.4 Effect of irradiation on the survival of various microorganisms after 3 days (curve 1), 2 weeks (curve 2), and 6 weeks (curve 3). From S. Bachman and J. Gieszczynska, Studies on some microbiological and chemical aspects of irradiated spices. *In* "Aspects of the Introduction of Food Irradiation in Developing Countries." Int. Atomic Energy Agency, Vienna, 1973.

Typical chromatograms obtained for the volatiles present in onion powder irradiated with 0 and 40 kGy are shown in Fig. 9.6. While there is great similarity between the two chromatograms, there are some differences.

The yield of essential oils from spices, important in spice quality, can be changed by irradiation. Reduction of the yield of essential oils as dose is

Table 9.4
Threshold Doses Required to Produce Detectable Changes in Flavor of Several Spices[a]

Spice	Dose (kGy)
Caraway	12.5
Coriander	7.5
Cardamom	7.5
Charlock	10.0
Juniper	>15.0
Marjoram	7.5–12.5
Black pepper	12.5
Pimento	15.0

[a] From S. Bachman and J. Gieszczynska, Studies on some microbiological and chemical aspects of irradiated spices. *In* "Aspects of the Introduction of Food Irradiation in Developing Countries." Int. Atomic Energy Agency, Vienna, 1973.

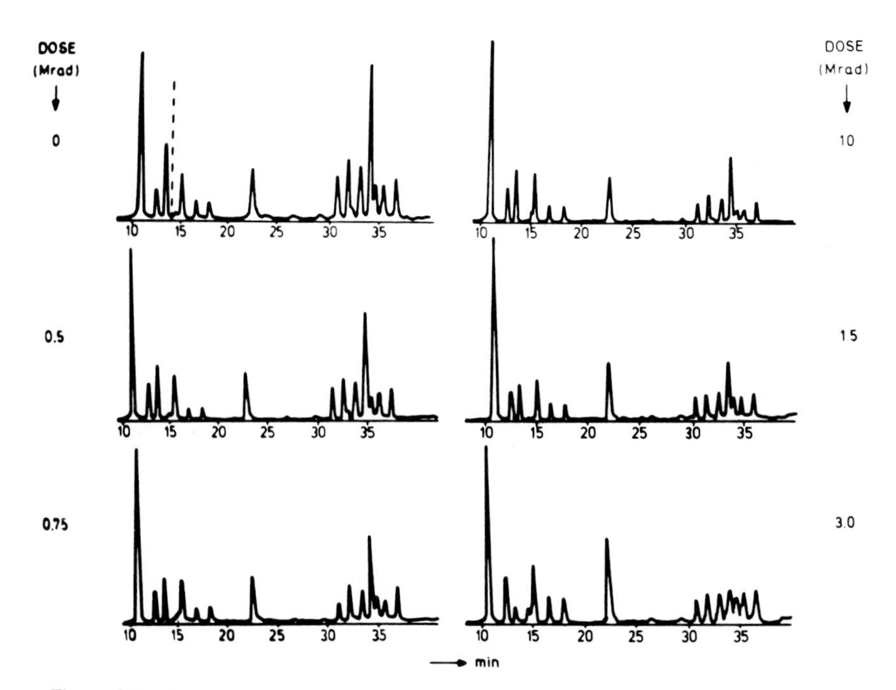

Figure 9.5 Gas chromatograms of essential oils of juniper before and after irradiation. From S. Bachman and J. Gieszczynska, Studies on some microbiological and chemical aspects of irradiated spices. *In* "Aspects of the Introduction of Food Irradiation in Developing Countries." Int. Atomic Energy Agency, Vienna, 1973.

Table 9.5

Effect of Irradiation on Percentages of Essential Oils of Cardamom[a]

Component	Percentage at dose (kGy)			
	0	5	10	15
α-Pinene	1.33	1.22	1.28	1.21
β-Pinene + sabinene	3.56	2.7	2.86	2.71
	0.4	0.36	0.32	0.39
	0.64	0.52	0.75	0.68
1,8-Cineole (eucalyptol)	37.8	38.02	34.57	38.04
	0.34	0.32	0.43	0.35
	0.26	0.38	0.6	0.23
	0.17	0.13	0.29	0.21
Linalool	2.34	2.03	2.01	2.09
	0.24	0.2	0.31	0.25
	0.22	0.24	0.20	0.23
	1.08	0.95	0.94	0.93
Terpinenol-4	1.27	1.01	1.11	1.03
	0.08	0.24	0.16	0.08
α-Terpineol	1.86	2.59	2.83	2.34
	0.18	0.16	0.18	0.33
	0.12	0.12	0.11	0.2
Terpinyl acetate	46.77	47.31	49.79	47.34
	0.33	0.45	0.13	0.22
	0.31	0.34	0.34	0.34
	0.07	0.1	0.09	0.11
	0.54	0.57	0.63	0.61

[a] From S. Bachman, S. W. Witkowski, and A. Zegota, Some chemical changes in irradiated spices (caraway and cardamom). *In* "Food Preservation by Irradiation, Vol. 1, Proceedings of Symposium, Wageningen." Int. Atomic Energy Agency, Vienna, 1978.

increased is indicated by the data of Figs. 9.7 and 9.8 for black pepper and juniper. For the essential oil of marjoram, however, yield is increased, as may be seen from the data of Fig. 9.9.

For paprika color is an important quality factor. As can be seen from Fig. 9.10, irradiation only slightly affects the content of the benzene-soluble pigment (carotenoid content). The effect of storage temperature on color retention of paprika is shown in Fig. 9.11. Color retention is slightly better with irradiated paprika. The contents of vitamin C, linolic acid, linolenic acid, capsaicin (pungency component), and reducing sugar are unaffected by irradiation.

The essential-oil compositions of black and white peppers irradiated in the dose range 2–9 kGy do not change in storage.

Irradiation accomplishes the needed reduction of microbial content of spices and vegetable seasonings without causing chemical changes which can significantly affect their normal sensory characteristics and uses. Panel tests

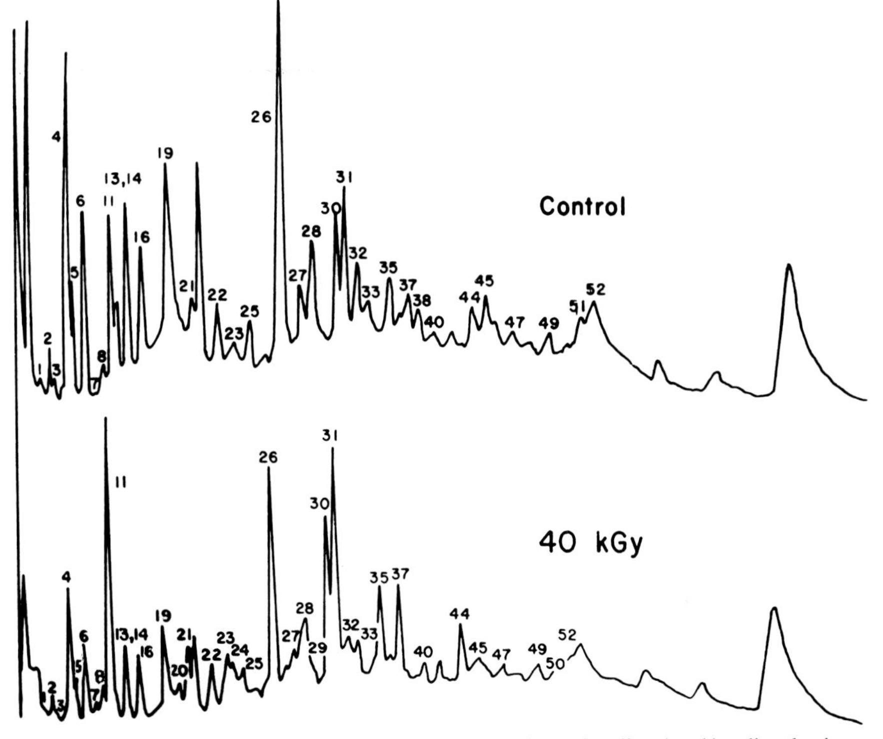

Figure 9. 6 Typical chromatograms of volatiles present in nonirradiated and irradiated onion powder. From W. Galleto, J. Kahan, M. Eiss, J. Welbourn, A. Bednarczyk, and O. Silberstein, *J. Food Sci.* **44,** 591 (1979). Copyright © by Institute of Food Technologists.

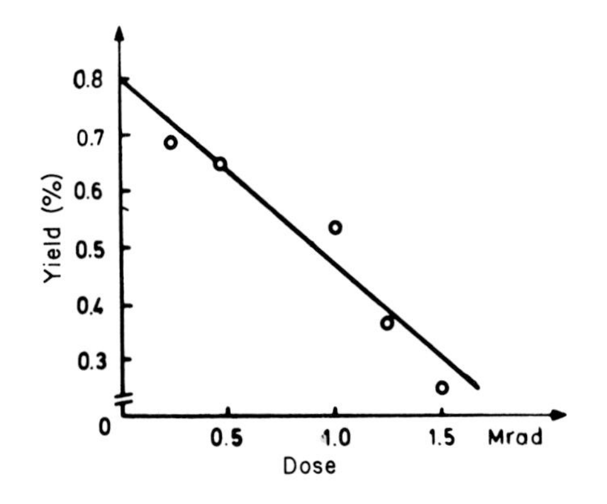

Figure 9.7 Changes in the yield of essential oils of black pepper as a function of radiation dose. From S. Bachman and J. Gieszczynska, Studies on some microbiological and chemical aspects of irradiated spices. *In* "Aspects of the Introduction of Food Irradiation in Developing Countries." Int. Atomic Energy Agency, Vienna, 1973.

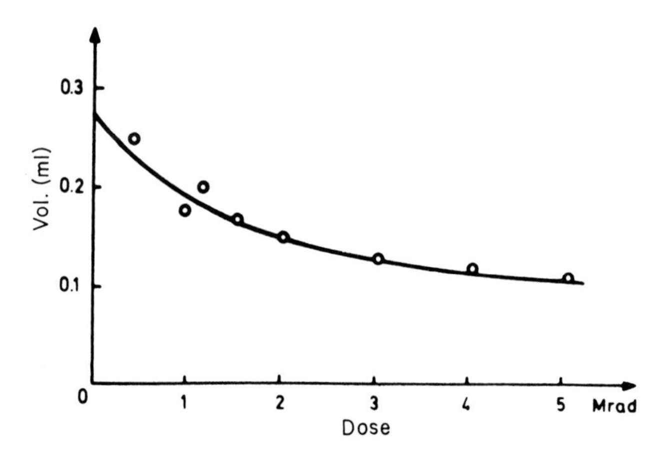

Figure 9.8 Changes in the yield of essential oils of juniper as a function of radiation dose. From S. Bachman and J. Gieszczynska, Studies on some microbiological and chemical aspects of irradiated spices. *In* "Aspects of the Introduction of Food Irradiation in Developing Countries." Int. Atomic Energy Agency, Vienna, 1973.

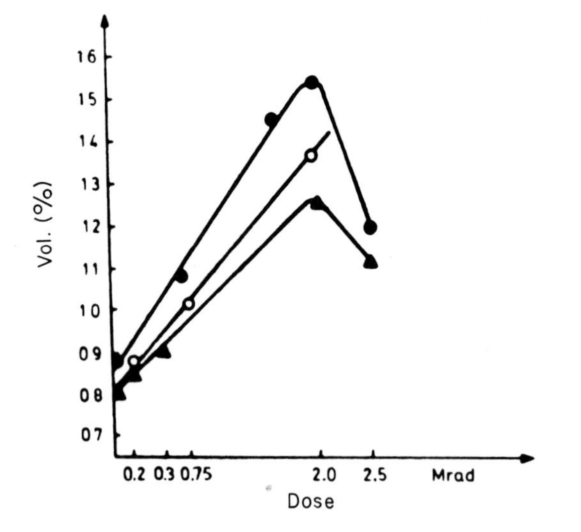

Figure 9.9 Changes in the yield of essential oils of marjoram as a result of irradiation in different atmospheres: ●, O_2; ○, air; ▲, N_2. From S. Bachman and J. Gieszczynska, Studies on some microbiological and chemical aspects of irradiated spices. *In* "Aspects of the Introduction of Food Irradiation in Developing Countries." Int. Atomic Energy Agency, Vienna, 1973.

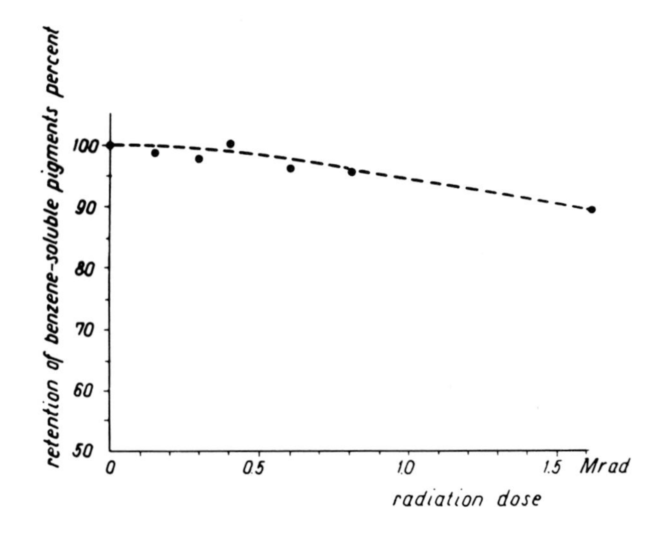

Effect of irradiation on the retention of the benzene-soluble pigment content of paprika powder. From J. Farkas, J. Beczner, and K. Incze, Feasibility of irradiation of spices with special reference to paprika. *In* "Radiation Preservation of Food, Proceedings of Symposium, Bombay." Int. Atomic Energy Agency, Vienna, 1973.

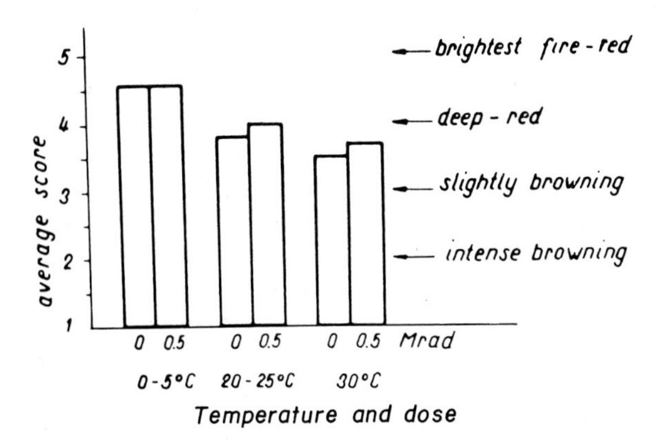

Figure 9.11 Visual color scores of ground paprika after storage for 120 days in closed aluminum tubes as affected by irradiation and by storage temperature. From J. Farkas, J. Beczner, and K. Incze, Feasibility of irradiation of spices with special reference to paprika. *In* "Radiation Preservation of Food, Proceedings of Symposium, Bombay." Int. Atomic Energy Agency, Vienna, 1973.

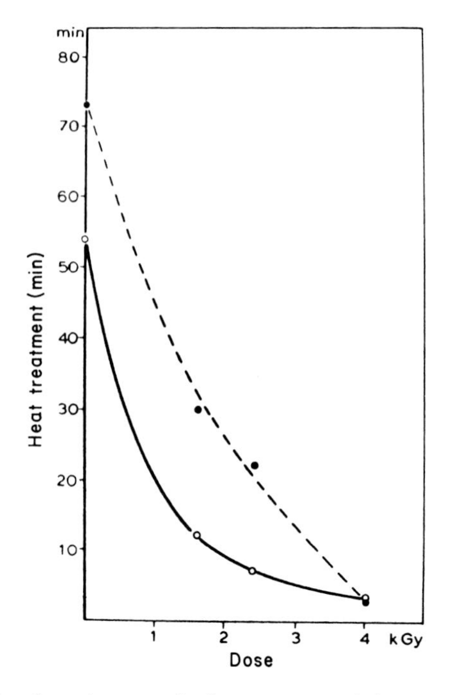

Figure 9.12 Additional requirements for heat treatment (minutes at 95°C) as related to radiation dose for achieving a combined reduction of initial viable cell counts of black pepper (●) and paprika (○) by 4 log cycles. From I. Kiss and J. Farkas, Combined effect of gamma radiation and heat treatment of microflora of spices. *In* "Combination Processes in Food Irradiation, Proceedings of Symposium, Columbo." Int. Atomic Anergy Agency, Vienna, 1981.

confirm this conclusion. Doses employed for decontamination can be 30 kGy or greater.* It is to be noted that the doses needed for microbial inactivation greatly exceed those required for insect disinfestation (~500 Gy) and, therefore insect disinfestation, if needed, also is accomplished.

A reduction of dose for microbial decontamination can be obtained by employing irradiation and heat in combination. This is illustrated in Fig. 9.12. The reduction of the microbial load in spiced or seasoned meats through the use of irradiated spices significantly reduces the heat-processing requirement to obtain a satisfactory end-product microbial level. The heat process for canned luncheon meat needed to produce commercial sterility (≤ 100 bacteria per gram) can be reduced from an F_0 of 4.7 to 1.1 if irradiated spices are used. This reduction in heating yields a demonstrable product quality improvement.

* U.S. Food and Drug Administration regulation limits the dose to a maximum of 30 kGy.

Table 9.6

Effects of Ethylene Oxide and Irradiation on Total Bacterial Counts of Selected Spices[a, b]

Spice	Untreated	Ethylene oxide	Irradiated	
			Count	Dose (kGy)
Black pepper	4.0×10^6	1.5×10^3	0	16
Paprika	9.9×10^6	0	0	10
Oregano	3.3×10^4	0	0	6
Allspice	1.7×10^6	42	0	10
Celery seed	3.7×10^5	8	0	10
Garlic	4.6×10^4	1.4×10^4	0	8

[a] From M. Vadji and R. R. Peireira, *J. Food Sci.* **38,** 893 (1973). Copyright © by Institute of Food Technologists.

[b] Expressed as number of organisms per gram.

Comparable results can be secured with heat-processed, though not sterilized, meat products such as brawn (cooked pork product).

The irradiation of spices and dried vegetable seasonings, primarily for the purpose of reducing microbial content, is carried out commercially in Europe and the United States. Irradiation has proved to be superior to chemical sterilization with ethylene oxide. Data for irradiation and ethylene oxide sterilization of several spices are shown in Table 9.6. Irradiation inactivates molds more effectively than does ethylene oxide. Additionally, it is easier to use in that spices in sealed containers can be penetrated with radiation and treatment times can be significantly shorter. No questionable residues result from irradiation. The safety of operating personnel is not a problem with irradiation. Adequate product volume, however, is essential for satisfactory economic feasibility.

C. Food Starches

Dry starches such as those of potato and maize are ingredients of many prepared foods. Since specification of upper limits for microbial content is common, a treatment to reduce an initial contamination level may be necessary. Table 9.7 shows the kind and level of microbial contamination typical of maize starch. At the level of contamination indicated, a dose of 3 kGy or less accomplishes a 90–95% reduction of microbial content and brings the count down to an acceptable level without significantly affecting the technological properties of the starch. Higher contamination levels may require larger doses but may also impair the technological properties of the starch.

Table 9.7

Microbial Contamination of Maize Starch[a]

Microorganism	Average number of organisms per gram
Mold spores	100
Spores of sulfite-reducing *Clostridia*	71
Spores of thermophilic *Bacillus*	1055
Spores of mesophilic *Bacillus* (10 min at 80°C)	559
Spores of mesophilic *Bacillus* (3 min at 100°C)	442
Mesophilic aerobic bacteria	610

[a] From L. Saint-lebe, A. Mucchieli, P. Leroy, and H. Beerens, Preliminary studies on the microflora of maize starch before and after irradiation. *In* "Radiation Preservation of Food, Proceedings of Symposium, Bombay." Int. Atomic Energy Agency, Vienna, 1973.

Table 9.8

Effect of Irradiation on Total Bacterial Count of Dry (11–14% Moisture) Gelatin[a]

Days after irradiation	Dose (kGy)	Total number of bacteria per gram
3	0	18,720
	5.0	1,600
	7.5	800
	10.0	0
	15.0	0
14	0	21,000
	5.0	1,000
	7.5	480
	10.0	0
	15.0	0
42	0	20,000
	5.0	800
	7.5	500
	10.0	0
	15.0	0

[a] From S. Bachman, S. Galant, Z. Gasyna, S. Witkowski, and H. Zegota, Effects of ionizing radiation on gelatin in the solid state. *In* "Improvement of Food Quality by Irradiation." Int. Atomic Energy Agency, Vienna, 1974.

D. Gelatin

Gelatin is used in a number of foods as an ingredient, and microbial contamination therefore is a matter of concern. Food-grade gelatin which has an initial relatively low bacterial count undergoes a sufficient reduction to an acceptable level with doses in the range 5–10 kGy, as indicated by the data of Table 9.8. A further reduction of count occurs during storage.

At doses above 5 kGy a peptone odor is produced. Small changes in viscosity and gel strength occur in the dose range 5–10 kGy, as may be seen from the data of Figs. 3.4 and 3.5. On the whole, however, the radiation-induced changes in the range 5–10 kGy do not seriously impair the food-ingredient uses of gelatin.

E. Gums

Gums usually are natural polysaccharides. They are available as dry products and used as food ingredients primarily for their rheological properties. Due to their origin, they may be contaminated with microorganisms and consequently require treatment prior to use. Irradiation may be used to reduce the contamination, but as may be anticipated, it can cause changes which affect the usefulness of the gum. This is illustrated by the data of Fig. 9.13, in which is shown the effect of a dose of 10 kGy applied to three

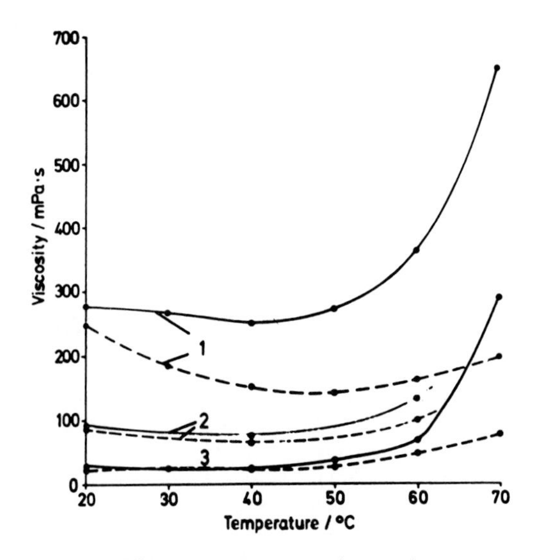

Figure 9.13 Viscosity at different temperatures of several gums suspended in water as measured with a rotating viscosimeter at a shear rate of 225 per second. 1, Carob gum; 2, Xanthan 4288; 3, Danagel.——Control; ---- 10 kGy. From T. Grünewald, Electron irradiation of dry food products. *Radiat. Physics Chem.* **22,** 733 (1983). Copyright (1983) Pergamon Press Ltd.

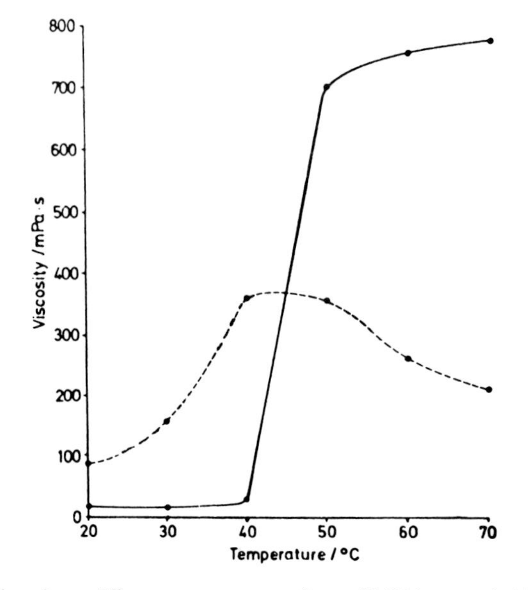

Figure 9.14 Viscosity at different temperatures of gum SS4324 suspended in water (5 g/dl) as measured with a rotating viscosimeter at a shear rate of 214 per second.——Control; ----- 10 kGy. From T. Grünewald, Electron irradiation of dry food products. *Radiat. Phys. Chem.* **22,** 733 (1983). Copyright (1983) Pergamon Press Ltd.

dry commercial gums on the viscosity of water suspensions of the gums over the indicated temperature range. The reduction in viscosity constitutes an undesirable change. The data of Fig. 9.14, however, indicate that a reverse effect may occur. With this gum (a commercial mixture of several gums) an increase in viscosity occurs with irradiation at temperatures below 40°C, although at higher temperatures, the viscosity is reduced.

Irradiation also may cause color changes in gums.

SOURCES OF ADDITIONAL INFORMATION

Anonymous, "Irradiation of Grain and Grain Products for Insect Control." Council for Agricultural Science and Technology, Ames, Iowa, 1984.

Cornwell, P. B., "The Entomology of Radiation Disinfestation of Grain—A Collection of Original Research Papers." Pergamon, Oxford, 1966.

Farkas, J., Radurization and radicidation: spices. *In* "Preservation of Food by Ionizing Radiation," (E. S. Josephson and M. S. Peterson, eds.), Vol. III. CRC Press, Boca Raton, Florida, 1983.

Giddings, G. G., Sterilization of spices: Irradiation versus gaseous sterilization. *Act. Rep. Res. Dev. Assoc. Mil. Food Packag. Syst.* **36** (2), 20 (1984).

Gottschalk, H. M., "A review on spices," Food Irradiation Information No. 7. Int. Project in the

Field of Food Irradiation, Karlsruhe, West Germany, 1977.

Lorenz, K., Irradiation of cereal grains and cereal grain products. *CRC Crit. Rev. Food Sci. Nutr.* **6** (1), 317 (1975).

Tilton, E. W., and Burditt, A. K., Jr., Insect disinfestation of grain and fruit. *In* "Preservation of Food by Ionizing Radiation," (E. S. Josephson and M. S. Peterson, eds.). Vol. III. CRC Press, Boca Raton, Florida, 1983.

CHAPTER 10

Miscellaneous Foods. Useful Food Modifications

I. DAIRY PRODUCTS

Dairy products may develop objectionable changes in flavor, odor, and color when irradiated, even with doses as small as 500 Gy. A dose of 45 kGy applied to fluid milk at 5°C produces a brown color and a strong caramelized flavor. Irradiation at temperatures in the range −80 to −185°C eliminates the brown discoloration and the caramelized flavor but causes the occurrence of an extremely bitter flavor. There is no significant difference between whole and skim milk in these effects.

With doses up to approximately 20 kGy, concurrent irradiation and vacuum distillation yield milk of acceptable flavor. Such milk, however, develops unacceptable browning on storage. Gelation also occurs at ambient temperatures.

Table 10.1
Effects of Sorbic Acid and Irradiation on Product Life of Khoa Cheese[a]

Treatment	Number of days for mold to appear	
	35–40°C	10–12°C
Nonirradiated	1–2	7–10
Irradiated only (kGy)		
1	3–4	8–9
2	4–6	9–12
5	6–7	13–20
Sorbic acid-coated wrappers[b]	2–4	8–12
Sorbic acid-coated wrappers plus irradiation (kGy)		
1	5–7	10–15
2	6–8	15–20
5	12–15	20–25
Sorbic acid-coated wrappers + binding substances[c]	4–6	10–14
Sorbic acid-coated wrappers + binding substances[c] plus irradiation (kGy)		
2	8–11	17–21
5	15–17	22–27

[a] From D. R. Bongiwar and U. S. Kumta, *Food Irradiat.* (*Saclay*) **8**(1,2), 16 (1967).
[b] Sorbic acid concentration used was 0.02 g per 100 cm^2 of cellophane wrapper.
[c] Gum-ghatti 0.1%, agar 0.1%, and carboxymethyl cellulose 0.1% were used as binding substances for sorbic acid-coated wrappers.

Skim-milk powder of moisture content of about 5% irradiated with doses in the range 2–16 kGy and reconstituted in the normal manner exhibits the typical "irradiated" flavor. Irradiation increases free, masked, and total –SH groups. The addition of ascorbic acid, and especially ascorbyl palmitate, which presumably act as free-radical scavengers, decreases the amount of –SH groups in milk powder.

Khoa, a milk product prepared by concentrating whole milk in open pans to a moisture content of about 35%, has acceptable flavor with doses up to 5 kGy. Combining irradiation at 5 kGy with the use of a wrapper treated with sorbic acid prevents mold growth for extended time periods, as may be seen from the data of Table 10.1.

Turkish kashar cheese (similar to cheddar) and plain yogurt have acceptable flavor and color when irradiated with doses up to 1.5 kGy. Larger doses cause off-flavors and pronounced fading of color. The effect of irradiation on the bacterial count of kashar cheese (in colony-forming units per gram) stored at ambient temperatures is shown in Fig. 10.1. Inhibition of mold growth also

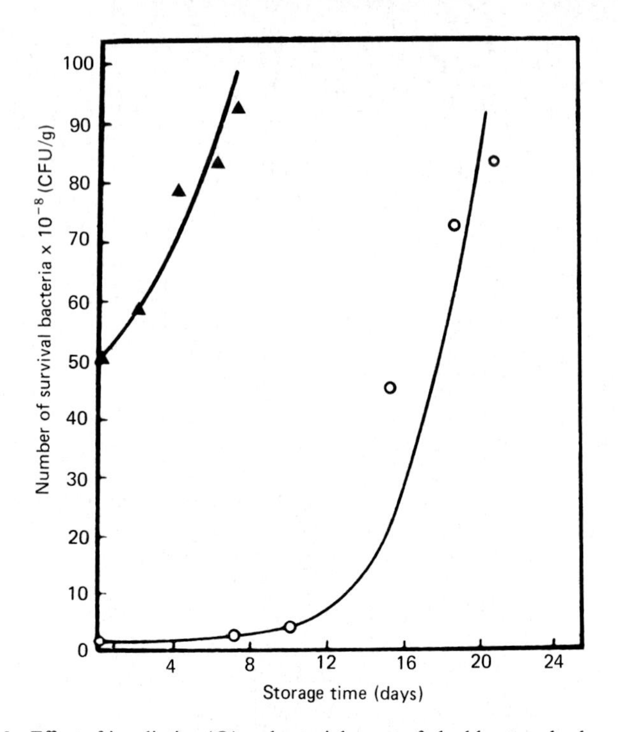

Figure 10.1 Effect of irradiation (○) on bacterial count of cheddar-type kashar cheese stored at ambient temperatures. ▲, Nonirradiated. From S. Yüceer and G. Gündüz, *J. Food Prot.* **43,** 114 (1980).

occurs. Doses of about 400 Gy extend the storage life of kashar cheese and plain yogurt by four to five times the normal.

II. EGGS AND EGG PRODUCTS

Irradiation of shell eggs is not regarded as feasible, since it thins the thick white and weakens the yolk membrane, both changes giving the egg the characteristics of lack of freshness.

Early interest in the irradiation of frozen and dried eggs and egg products was directed to reduction of the salmonellae content. Table 10.2 lists the doses for a 7-log cycle reduction of the more resistant strains of salmonellae in various egg products. Whether there is need for such a large dose has been debated. The value of using irradiation to reduce the *Salmonella* content of eggs and egg products has been diminished by the availability of heating procedures for reducing the count and by improvements in egg production.

Table 10.2

Doses Required for a 7-Log Cycle Reduction of the More Resistant Strains of Salmonellae in Various Egg Products[a]

Product	Strain	Dose (kGy)
Whole egg (liquid)	*Salmonella typhimurium*	2–4.4
Whole egg (frozen)	*S. typhimurium*	4.8
Whole egg (dried)	*S. typhimurium* + *S. senftenberg*	3.7
Egg yolk (frozen)	*S. typhimurium* + *S. senftenberg*	3.2
Egg yolk (dried)	*S. typhimurium* + *S. senftenberg*	5.7
Egg white (liquid)	*S. typhimurium*	2.6
Egg white (frozen)	*S. typhimurium*	2.1

[a] From M. J. Thornley, Microbiological aspects of the use of radiation for the elimination of salmonellae from foods and feeding stuffs. *In* "Radiation Control of Salmonellae in Food and Feed Products," Tech. Rept. Series No. 22. Int. Atomic Energy Agency, Vienna, 1963.

Slightly reduced volumes are obtained with cakes made with irradiated eggs. The stability of egg white foam is lessened by irradiation. Irradiation can develop off-flavors in egg products.

III. MOLASSES

As produced, edible molasses may contain several different types of microorganisms, which ordinarily do not grow in molasses due to its high solids content. In some uses, however, molasses is diluted and growth can occur. In such cases, rope-forming thermophilic bacteria, including *Bacillus subtilis*, can cause difficulty. Due to its action to cause quality damage, heat cannot be used to inactivate such bacteria in molasses. Doses in the range 5–15 Gy effect a 95–99% reduction of bacterial count without loss of quality.

The irradiation of fermentation media which are composed principally of molasses increases the yield of ethanol. For this, doses in the bacterial radurization range are sufficient.

IV. POI

Poi, a viscous dark-gray paste with a moisture content in the range 64–82%, is made from cooked taro corms. Lactic acid bacteria cause a drop in pH and in time render this staple food of Oceania inedible. Irradiation with doses of 10 kGy extends the refrigerated life of poi for a period greater than 7 days without adverse effects.

V. BEVERAGE MATERIALS AND BEVERAGES

A. Cocoa Beans

In countries which have warm and humid climates, cocoa beans are subject to molding and insect infestation. Molds present include *Aspergillus* spp., and *Penicillium* spp. Several kinds of insects may be present, the most frequent

Table 10.3

Effect of Heat Treatment and/or Irradiation on Fungal Infection of Stored Cocoa Beans[a]

Air heat treatment (30 min)	Dose (kGy)	Percentage of moldy beans at 28 days storage at 28°C	
		75% R.H.	90% R.H.
20°C, <40% R.H.	0	24.5	100
	1.5	17.5	100
	3.5	17.3	100
20°C, >85% R.H.	0	25.0	100
	1.5	13.0	56.9
	3.5	8.6	36.8
60°C, <40% R.H.	0	8.6	100
	1.5	0	40.0
	3.5	0	19.3
60°C, >85% R.H.	0	3.7	100
	1.5	0	21.5
	3.5	0	0
70°C, <40% R.H.	0	0	95.0
	1.5	0	16.1
	3.5	0	0
70°C, >85% R.H.	0	0	94.4
	1.5	0	15.1
	3.5	0	0
80°C, <40% R.H.	0	0	100
	1.5	0	17.0
	3.5	0	0
80°C, >85% R.H.	0	0	53.7
	1.5	0	0
	3.5	0	0

[a] From V. Appiah, G. T. Odamtten, and D. Is. Langerak, *Food Irradiat. Newsl.* **5**(2), 17 (1981).

being *Lasioderma serricorne*, the tobacco beetle. These contaminants not only damage the beans but also can taint products made from them.

A dose of 3.5 kGy does not alter the fatty acid content. The maximum dose, however, which can be used without causing unacceptable changes has not been well established.

The dose required to prevent molding is determined partly by the humidity encountered during storage. The data of Table 10.3 indicate this. As may be seen from these data, the most effective treatment is the combined use of heated, humidified air and irradiation. Heating the beans for 30 min with air at 80°C and 85% R.H. combined with irradiation with doses in the range 1.5–3.5 kGy is effective in controlling molds such as *Aspergillus flavus*, even when very high ambient humidities exist during storage.

This treatment exceeds what is needed for insect disinfestation.

B. Coffee Beans

Insect infestation of coffee beans can cause significant losses. Chemical fumigants are not sufficiently effective against insect eggs and can leave residues which change taste and aroma. The irradiation of green coffee beans with doses as high as approximately 10 kGy does not cause detectable flavor change in beverage coffee brewed from such beans after conventional roasting.

A dose of 500 Gy has been suggested as adequate for insect disinfestation of green coffee beans.

C. Herbal Teas

Herbal teas such as camomile, mint, linden, and dog-rose hip can contain substantial numbers of bacteria and molds. Bacteria include sulfite-reducing *Clostridia*, *Proteus* spp., *Escherichia coli*, *Enterobacter*, and *Staphylococcus aureus* (coagulase positive). A dose of 5 kGy reduces the aerobic bacteria content to about 100 per gram. This does not change the amount of ethereal oil extractible from dried camomile flowers, nor does it alter its composition. Doses up to 10 kGy appear feasible.

D. Alcoholic Beverages

1. Beer

The radurization dose for beer is about 500 Gy. This dose causes unacceptable changes, including loss of color and the development of a burnt flavor and objectionable odors.

Table 10.4

Chemical Analyses of Irradiated Wines[a]

Type of wine and dose (kGy)	pH	Tannin[b]	Titratable acid[c]	Extract[d]	Volatiles[e]			
					Ethanol (% v/v)	Esters (mg EtOAc/liter)	Aldehydes (mg AcH/liter)	Acid (g AcOH/dl)
Red dry								
Control	3.57	684	0.533	1.7	10.2	110	11.2	0.013
1	3.48	692	0.561	2.0	10.6	114	8.9	0.018
5	3.47	666	0.537	2.0	10.2	111	19.3	0.013
10	3.68	656	0.528	2.1	10.3	111	64.8	0.012
Red sweet								
Control	3.76	555	0.403	11.7	18.9	24	56.2	0.004
1	3.77	545	0.406	12.0	18.9	24	47.7	0.003
5	3.79	535	0.399	11.6	18.7	25	82.9	0.003
10	3.83	565	0.392	12.1	18.1	28	103.9	0.004
White sweet								
Control	3.83	233	0.359	12.7	19.8	74	118.3	0.007
1	3.89	227	0.360	12.6	19.3	72	113.3	0.008
5	3.91	219	0.356	12.5	19.3	73	147.6	0.006
10	3.92	231	0.355	12.7	19.7	73	138.2	0.006

Shermat								
Control	3.59	154	0.418	1.8	20.4	143	30.2	0.013
1	3.52	147	0.428	1.8	20.1	128	42.5	0.008
5	3.57	147	0.435	1.8	10.5	132	69.8	0.008
10	3.54	180	0.409	1.8	20.5	136	68.9	0.010
White dry								
Control	3.60	162	0.497	2.1	11.0	127	65.4	0.016
1	3.54	168	0.496	2.2	10.9	123	51.4	0.017
5	3.52	168	0.493	2.2	10.9	127	77.8	0.017
10	3.60	173	0.493	2.4	10.8	124	89.3	0.016

[a] From V. L. Singleton, *Food Technol.* (*Chicago*) **18,** 790 (1963). Copyright © by Institute of Food Technologists.
[b] Expressed as milligrams of gallic acid per liter.
[c] Expressed as grams of tartaric acid per 100 ml.
[d] Expressed as grams of dissolved solids per 100 g.
[e] Abbreviations: EtOAc, ethyl acetate; AcH, acetaldehyde; AcOH, acetic acid.

2. Wines and Brandy

There can be several reasons for irradiating wines: (1) as a sterilization measure to terminate fermentation and provide stability, (2) to change a wine's normal characteristics, and (3) to accelerate aging. The latter can apply also to brandies.

A dose of 6 to 7 kGy is needed to inactivate yeasts and bacteria present in wines after fermentation. The responses of different wines to doses of these magnitudes vary with the type of wine. Dry wines have been reported to maintain acceptable sensory characteristics. Semisweet wines, however, develop a bitter taste. Bleaching of color occurs. Chemical changes observed in various California wines as a result of irradiation in the dose range 1–10 kGy are given in Table 10.4. Table 10.5 gives the results of sensory analyses of the same California wines. Large changes are produced with doses of 5 to 10 kGy. A significant increase in "oxidized" flavor occurs, which may be associated with increased aldehyde content. At doses less than 0.1 kGy no significant differences due to irradiation have been observed.

For the Korean wines *tackjoo* and *yackjoo*, made by fermenting rice and wheat flours and consumed without sterilization, a dose of 2.4 kGy plus heating at 70°C for 10 min yields stability at 33°C for 10 days and does not alter sensory characteristics.

In the aging process for brandies, improved flavor is associated with extraction of material from oak containers. Irradiation of oak wood shavings permits easier extraction and improves brandy flavor.

Irradiation of brandy with a dose of 0.1 kGy is reported to produce a flavor equivalent to that obtained with 3 years of normal aging. Higher doses are damaging to flavor.

VI. HOSPITAL DIETS

Certain kinds of hospital patients require extraordinary protection against infection. As a part of such care, they may be given only sterilized foods. Irradiation can be used for such sterilization. It offers advantages of versatility and of foods with better patient acceptance. It is applicable to both wet and dry foods. Foods that have been irradiated for this purpose include meats, fish, vegetables, breakfast cereals, breads, rolls, crackers, cookies, pastries, condiments, complete meal items, nutritional supplements, snacks, candies, nuts, gum, and beverage powders. Any needed cooking is done prior to irradiation. The foods are irradiated while frozen, for example, at −40°C. Doses of 25 to 35 kGy are employed. These doses are used in order to obtain the needed low bacterial count and not for preservation purposes. Moist irradiated foods are kept at subfreezing temperatures until used.

Table 10.5

Sensory Analyses of Irradiated Wines: Mean Score by Treatment[a,b]

Overall quality score (max, 20)				
Red dry	C 13.5	x 12.9	y 10.9	z 9.2
Red sweet	x 12.8	C 12.7	y 10.0	z 7.9
White sweet	x 14.7	C 13.8	y 13.0	z 12.3
Shermat	x 12.2	C 12.1	y 10.3	z 8.2
White dry	C 13.3	x 12.8	y 10.8	z 8.8
Combined data	C 13.08	x 13.07	y 11.00	z 9.28
Complexity, richness (max, 10)				
Red dry	x 3.4	C 3.1	y 2.7	z 2.0
Red sweet	x 4.2	C 4.0	y 2.7	z 1.8
White sweet	x 5.3	C 4.9	y 3.8	z 3.6
Shermat	C 3.8	y 3.4	x 3.2	z 2.4
White dry	y 3.7	x 3.4	C 3.3	z 2.9
Combined data	x 4.00	C 3.87	y 3.27	z 2.57
Color changes related to aging (max, 10)				
Red dry	z 6.4	y 4.8	x 2.8	C 1.9
Red sweet	z 5.8	y 4.9	x 4.1	C 3.0
White sweet	C 5.8	x 4.8	y 2.8	z 2.4
Shermat	C 3.6	x 2.5	z 2.1	y 1.8
White dry	y 2.5	C 2.2	z 2.2	x 1.8
Combined red data	z 6.08	y 4.80	x 3.48	C 2.96
Combined white data	C 3.80	x 3.00	y 2.34	z 2.23
Oxidized flavor (max, 10)				
Red dry	z 4.1	y 3.1	x 1.8	C 0.7
Red sweet	z 3.7	y 3.0	x 2.6	C 2.2
White sweet	y 3.4	z 3.3	C 2.8	x 2.4
Shermat	z 3.8	y 3.2	x 2.5	C 2.1
White dry	z 3.8	y 2.8	x 1.8	C 1.6
Combined data	z 3.77	y 3.12	x 2.27	C 1.92
Bottle bouquet (max, 10)				
Red dry	C 3.0	x 2.5	z 2.3	y 2.2
Red sweet	C 2.6	x 2.5	y 2.5	z 2.5
White sweet	C 3.2	x 3.2	z 2.8	y 2.6
Shermat	x 2.5	C 2.1	z 2.1	y 1.5
White dry	C 2.8	y 2.5	x 2.2	z 1.7
Combined data	C 2.72	x 2.58	z 2.28	y 2.25
Hotness (max, 10)				
Red dry	x 1.1	y 1.0	z 1.0	C 0.8
Red sweet	C 3.8	x 3.5	z 3.5	y 3.1
White sweet	x 3.8	y 3.8	z 3.6	C 3.4
Shermat	z 3.9	x 3.8	y 3.2	C 3.1
White dry	y 1.1	z 1.1	C 1.0	x 1.0
Combined data	x 2.72	z 2.72	y 2.54	C 2.51

(continued)

Table 10.5 (*Continued*)

Bitterness (max, 10)				
Red dry	C 2.3	x 2.3	z 2.3	y 2.0
Red sweet	C 3.0	z 2.8	y 2.6	x 2.6
White sweet	z 2.2	C 1.9	x 1.9	y 1.8
Shermat	z 3.4	y 2.7	C 2.6	x 2.4
White dry	y 1.7	C 1.6	x 1.6	z 1.6
Combined data	z 2.47	C 2.32	x 2.19	y 2.19
Astringency (max, 10)				
Red dry	C 4.1	z 3.5	y 3.3	x 3.2
Red sweet	C 3.5	z 3.2	x 3.1	y 2.8
White sweet	x 2.3	y 2.2	C 2.0	z 1.9
Shermat	z 2.9	y 2.2	C 2.1	x 2.1
White dry	y 1.4	z 1.2	x 1.1	C 1.0
Combined data	C 2.55	z 2.53	y 2.42	x 2.37
Grape aroma (max, 10)				
Red dry	C 5.7	x 4.8	y 2.9	z 1.7
Red sweet	C 4.2	x 3.3	y 2.3	z 1.2
White sweet	x 5.2	C 4.7	y 3.3	z 3.2
Shermat	C 3.8	x 2.5	z 1.5	y 1.4
White dry	C 4.4	x 4.3	y 2.6	z 1.5
Combined data	C 4.54	x 4.02	y 2.51	z 1.82

[a] From V. L. Singleton, *Food Technol.* (*Chicago*) **18,** 790 (1963). Copyright © by Institute of Food Technologists.

[b] Arranged in decreasing order of mean scores: C, control, x, 1 kGy; y, 5 kGy; z, 10 kGy.

VII. ANIMAL FEEDS

A. Laboratory Animal Diets

Irradiation offers an effective method of sterilizing diets prepared for laboratory animals which are raised to be specified-pathogen free or germ free (gnotobiotic). Such irradiated diets are nutritionally superior to diets sterilized with heat or ethylene oxide. Doses of the order of 25 kGy are employed. It is stated that irradiated diets are more convenient and that they are better accepted by the animals.

B. Stock Feeds

It is generally accepted that the principal sources of human salmonellosis are various foods of animal origin, such as poultry and red meats, eggs, and dairy products. The farm animals which are the sources of these foods receive their infections from the feeds they consume. Enterobacteriaceae can be found in many feed ingredients, but especially in those that are primarily protein sources, such as fish meal, bone meal, and blood meal. Typical counts of Enterobacteriaceae are of the order of 10^5 or 10^6 microorganisms per gram.

Terminal sterilization of the mixed feeds fed to farm animals long has been regarded as the most effective approach to the elimination of human salmonellosis and other diseases of similar origin. Unlike treating the food products, this approach would break the infection sequence. A dose of 7.5 kGy has been suggested as sufficient.

C. Pet Foods

Frozen raw meat (e.g., horse meat) used for dog feeding can be freed of salmonellae with irradiation with doses in the range 10–15 kGy. The dose requirement is halved by irradiation of the meat in the unfrozen state.

Radappertization of moist pet foods appears feasible.

VIII. USEFUL MODIFICATIONS OF FOODS

A. Introduction

Foods can be altered by irradiation in ways that yield beneficial effects. Often such effects are impossible to obtain by other means, and such uses of irradiation challenge the ingenuity of investigators.

Most beneficial alterations involve degradation of protein or carbohydrate components of foods obtained through chemical changes. A second way foods may be altered is by action of radiation on a biological system of a living food, which consequently causes chemical modification of the food. Beneficial effects of these two kinds can be in the nature of either product or process improvements.

B. Product Improvements

The time to rehydrate *dehydrated vegetables* varies with different vegetables and often is so long as to constitute an inconvenience. Irradiation, in proportion to dose, reduces the time for rehydration. By selection of a particular dose for each vegetable of a mixture, the same rehydration time for all can be secured. In this manner a soup mix can be quickly and uniformly rehydrated. Table 10.6 gives the doses which substantially reduce the rehydration times for the listed vegetables.

Table 10.6

Doses to Reduce Rehydration Time of Dehydrated Vegetables from 10 to 20 min to 1 to 2 min[a]

Vegetable	Dose (kGy)	Vegetable	Dose (kGy)
White onion flakes	3	Cabbage flakes	30
Tomato flakes	6	Green lima beans	30
Potato dice	10	Celery flakes	33
Carrot dice	20	Cut green beans	40
Fresh peas	20	Okra pieces	40
Leek	24	Beet cubes	>40
Bell pepper dice	25		

[a] From "Training Manual on Food Irradiation Technology and Techniques," 2nd Ed., Tech. Rept. Series No. 114. Int. Atomic Energy Agency, Vienna, 1982.

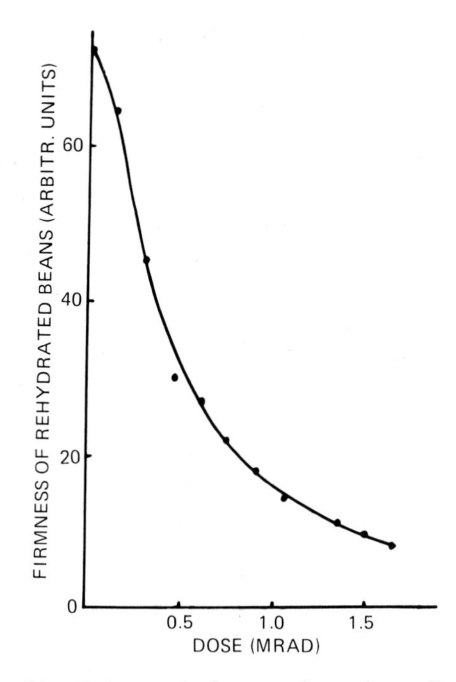

Figure 10.2 Effect of irradiation on the firmness of navy beans. From P. Markakis, R. C. Nicholas, and B. S. Schweigert, *In* "Irradiation Preservation of Fresh-water Fish and Inland Fruits and Vegetables," C00-1283-27. U.S. Atomic Energy Comm., Washington, D.C., 1965.

Irradiation of dried *legumes* such as red gram or white navy beans causes them to be softer when rehydrated. This effect is shown in Fig. 10.2. Such radiation-induced softening can reduce cooking or processing time. A similar reduction of cooking time occurs with dried *curly kale* irradiated with a dose of 10 kGy.

Meats such as beef are significantly tendered by radappertization doses, presumably through depolymerization of collagen in the connective tissue (see Chapter 6).

Certain *legumes* cause flatulence, which is ascribed to the presence of oligosaccharides such as raffinose, verbascose, and stachyose. In the process of germination of the seed, the oligosaccharides are metabolized. Irradiation may be used to control the germination of soybeans, broad beans, and red gram. Soybeans, for example, may be steeped and then incubated for 24 h, irradiated with a dose of 2.5 kGy, and incubated an additional 3 days. This procedure results in a reduction of about 80% of the content of raffinose and stachyose of the bean. As may be seen from the data of Fig. 10.3, irradiation serves to limit growth. A similar reduction of oligosaccharides in green gram can be obtained by this procedure as may be seen from the data of Fig. 10.4.

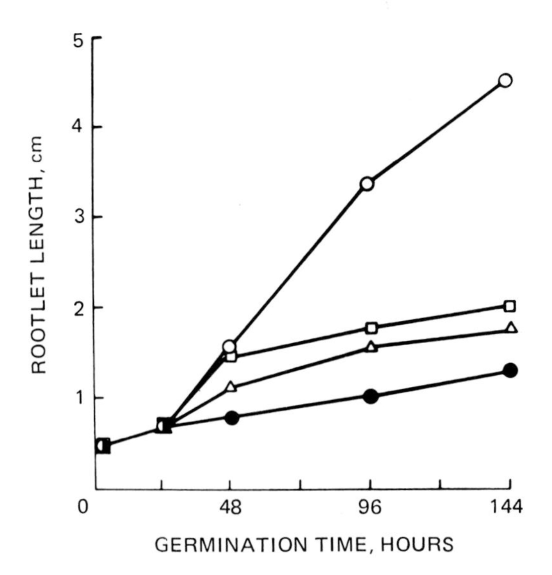

Figure 10.3 Average length of rootlets of steeped soybeans irradiated with 2.5 kGy at the end of various 30°C incubation periods. ○, Not irradiated; ●, irradiated after 2 hr; △, irradiated after 24 hr; □, irradiated after 48 hr. From Y. Hasegawa and J. H. Moy, Reducing oligosaccharides in soybeans by gamma-radiation controlled germination. *In* "Radiation Preservation of Food, Proceedings of Symposium, Bombay." Int. Atomic Energy Agency, Vienna, 1973.

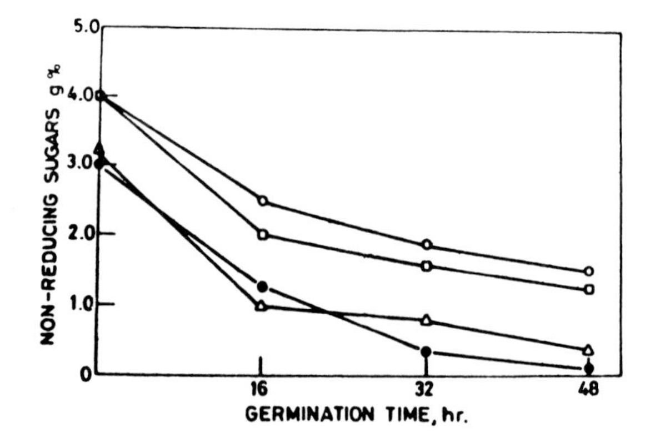

Figure 10.4 Effect of irradiation on total oligosaccharide (nonreducing sugars) content of green gram during early germination. ○, Not irradiated; □, 2.5 kGy; △, 5 kGy; ●, 10 kGy. From V. S. Rao and U. K. Vakil, *J. Food Sci.* **48,** 1791 (1983). Copyright © by Institute of Food Technologists.

Irradiated *white potatoes* stored at ambient or somewhat lower temperatures (e.g., 10–15°C) yield better quality fried chips and fresh fries as compared with those made from nonirradiated potatoes that have been cold stored (3–4°C). Lower amounts of reducing sugars present in the irradiated potatoes account for this better quality.

The *oyster mushroom, Pleurotus ostreatus,* releases quantities of spores during postharvest transport and can cause severe allergy diseases among personnel involved in its handling. Doses greater than 1 kGy inhibit spore production and thereby overcome an obstacle to the marketing of this desirable mushroom.

C. Process Improvements

The *rate of dehydration* of prunes is increased as irradiation dose is increased. This effect may be seen from the data of Fig. 10.5. Blanching by dipping in water at 93°C for 30 sec after irradiation is necessary in order to secure acceleration of the dehydration rate. Skin breakage during blanching occurs. Irradiation causes greater retention of anthocyanin pigments.

The *juice yield* of fruits is increased by irradiation prior to pressing. This is attributed to one or more of the following radiation-induced changes: (1) increase of the permeability of the walls of the skins, (2) weakening of the skin, and (3) breakdown of protopectin and pectin of the intermediate layers, which weakens the bonds between the cells, in particular the bonds between the skin and the parenchymatous cells and between the cells of the flesh.

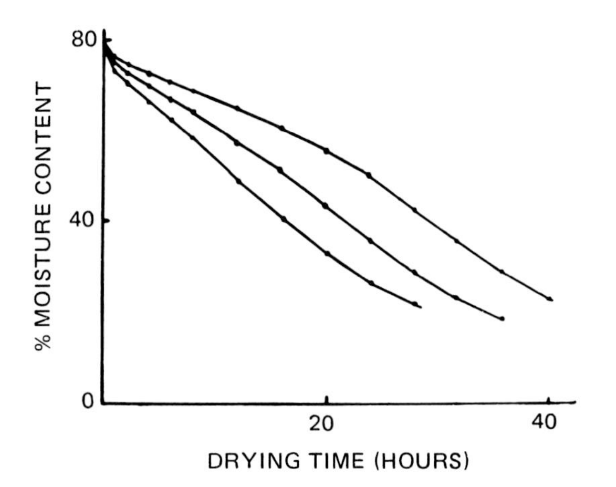

Figure 10.5 Effect of irradiation on drying rate of 'Stanley' prunes. Upper curve, non-irradiated; middle curve, 2 kGy; lower curve, 4 kGy. From P. Markakis, R. C. Nicholas, and B. S. Schweigert, *In* "Irradiation Preservation of Fresh-water Fish and Inland Fruits and Vegetables," C00-1283-27. U.S. Atomic Energy Comm., Washington, D.C., 1965.

Irradiation of grated coconut meat with doses up to 0.5 kGy improves the efficiency of pressing to secure coconut milk. While the oil of the milk remains white, the carbohydrate portion is browned, as is the press residue.

The yield of juice from grapes is increased by irradiation. The data of Table 10.7 show this for several varieties of grapes. At doses of 8 to 16 kGy the grape skin is thin and discolored, the stem browns, and the texture is altered. The grape cells are damaged. Even with higher doses fermentation of the juice is normal. The composition of wine made from grapes given 16 kGy has higher

Table 10.7

Percentage Increase in Juice Yield of Several Hungarian Grape Varieties as a Function of Radiation Dose[a]

Radiation dose (kGy)	Egri csillagok	Olasz-rizling	Kocsis Irma	Kövidinka	Harslevelü 1	Harslevelü 2	Harslevelü 3
0.5	7.1	2.1	4.0	2.3	0	2.6	2.5
2.0	9.1	5.0	7.6	11.5	1.9	10.2	5.9
8.0	16.5	4.0	14.9	18.8	13.3	13.3	20.8
16.0	20.9	4.4	25.0	28.6	18.5	25.8	28.1

[a] From I. Kiss, J. Farkas, S. Ferenczi, B. Kalman, and J. Beczner, Effects of irradiation on the technological and hygienic qualities of several food products. *In* "Improvement of Food Quality by Irradiation." Int. Atomic Energy Agency, Vienna, 1974.

Table 10.8
Effect of Irradiation on Dibis (Syrup) Yield of Two Date Varieties[a]

Dose (kGy)	Dibis yield[b] Lelwi	Tabarzel
0.375	—	161
0.5	132	—
0.75	—	186
1.0	143	—
1.5	—	151
2.0	127	—

[a] From J. Farkas, F. Al-Charchafchy, M. H. Al-Shaikhaly, J. Mirjan, and H. Auda, *Acta Aliment. Acad. Sci. Hung.* **3**(2), 151 (1974).
[b] Expressed as percentage of nonirradiated yield.

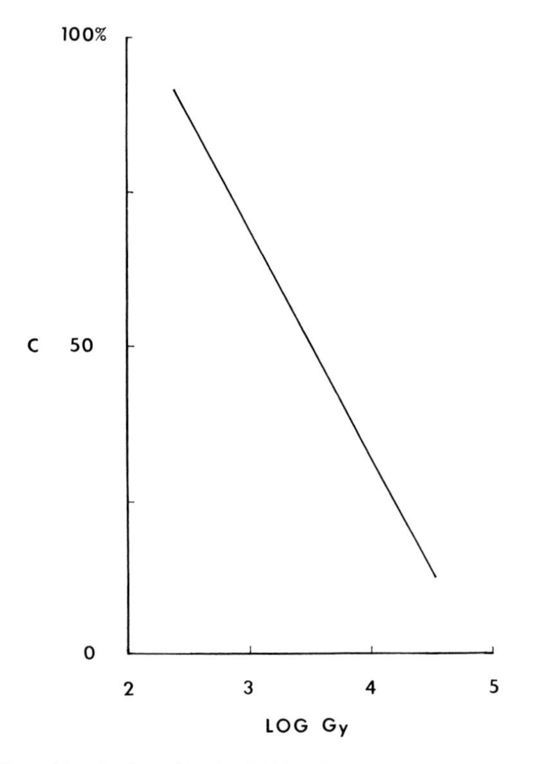

Figure 10.6 Effect of irradiation of barley (11% moisture) upon sprout length. For irradiated barley, C is the percentage of the sprout length of nonirradiated barley obtained on the fourth day of germination. From J. Farkas, I. Kiss, Z. Razga, and K. Vas, *Keki-Közlemenyek* **I-II,** 19 (1963).

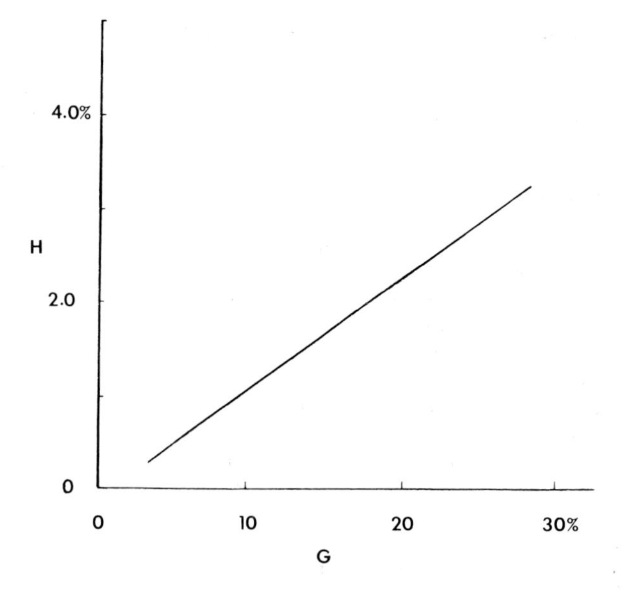

Figure 10.7 Relation between decrease of barley sprout length by irradiation and yield of malt. For irradiated barley, G is the percentage of the sprout length of nonirradiated barley obtained on the fourth day of germination. H is the percentage increase of yield of malt. From J. Farkas, I. Kiss, Z. Razga, and K. Vas, *Keki-Közlemenyek* **I-II,** 19 (1963).

sugar-free extract, ash content, and pH, but otherwise is normal. Color intensity of the wine increases with dose. Wines made from grapes irradiated with doses below about 5 kGy have normal sensory characteristics. At such doses an increase in juice yield of 10 to 12% may be obtained.

The yield of dibis (date syrup) from fully ripe fresh dates is increased by irradiation. The data of Table 10.8 show this effect for two varieties of dates. The yield peaks with doses of about 0.75 kGy.

By *control of germination and sprouting*, irradiation reduces malting losses with barley. Figure 10.6 shows the relationship between relative sprout length of germinated barley and dose. Figure 10.7 shows the correlation between the radiation-induced reduction of sprout length and increase of the yield of malt obtained. Malting losses may be reduced 1–2% by irradiation of the dry barley with doses of 0.5 to 8 kGy. Quality characteristics of the malt produced are unchanged by irradiation.

Reduction of the growth of barley by irradiation also permits malting to proceed satisfactorily at higher than normal temperatures, and in this manner to extend operations into periods ordinarily too warm for malting.

SOURCES OF ADDITIONAL INFORMATION

Adamiker, D., "A Comparison of Various Methods for Treating Feedstuffs for Laboratory Animals—A review," Food Irradiation Information No. 5. Int. Project in the Field of Food Irradiation, Karlsruhe, West Germany, 1975.

Akers, S. N., On the cutting edge of science. (Irradiated hospital diets) *Nutr. Today* **19** (4), 24 (1984).

Anonymous, "Radiation Control of *Salmonellae* in Food and Feed Products." Int. Atomic Energy Agency, Vienna, 1963.

Anonymous, "Application of Food Irradiation in Developing Countries." Int. Atomic Energy Agency, Vienna, 1966.

Anonymous, "Elimination of Harmful Organisms from Food and Feed by Irradiation." Int. Atomic Energy Agency, Vienna, 1968.

Anonymous, "Improvement of Food Quality by Irradiation." Int. Atomic Energy Agency, Vienna, 1974.

Anonymous, "Decontamination of Animal Feeds by Irradiation." Int. Atomic Energy Agency, Vienna, 1979.

Bregvadze, U. D., "Action of Gamma Radiation on Nonalcoholic Drinks, Wine and Brandy," ORNL-IIC-28. Translation of Russian publication. U.S. Atomic Energy Comm., Oak Ridge, Tennessee, 1972.

Gerrard, M., "Milk Preservation by Ionizing Radiation—Literature Review," ORNL-IIC-26. U.S. Atomic Energy Comm., Oak Ridge, Tennessee, 1969.

Ley, F. J., Bleby, J., Coates, M. E., and Paterson, J. S., Sterilization of laboratory animal diets using gamma radiation. *Lab. Anim.* **3,** 321 (1969).

CHAPTER 11

Combination Processes

I. INTRODUCTION

It is clear that a desired effect relating to the control of microbiological spoilage, insect disinfestation, decontamination of a food of bacteria, yeasts, molds, and parasites, and other uses of food irradiation involving living organisms always can be secured if a sufficient quantity of radiation is employed. From the information provided in other sections of this book, it is clear also that often there are limits to the amount of radiation that can be used, stemming from radiation-induced changes in foods. In some cases the dose to produce such changes is less than the dose needed to produce the desired technical effect. For example, some fruits cannot tolerate doses needed for inactivation of fungi.

One approach for securing the desired effect without incurring radiation damage in a food is to use a smaller quantity of radiation in combination with some other agent in such a way that the two acting in concert produce the desired effect.

Reference to the section on the economics of food irradiation (see Chapter 15) indicates that a principal factor in the cost of irradiation is the dose employed. The use of a combination process can reduce a treatment cost by lowering the amount of radiation needed.

In addition to the two advantages described above in using irradiation in combination with another agent, both of which are directed to dose reduction, there is a third reason for combining irradiation with another measure. The pathways of spoilage of some foods involve more than microbial spoilage, and, consequently, control of the spoilage of such foods is only partially secured through irradiation. This situation calls for irradiation to be used in combination with other measures which are effective in controlling the spoilage not controlled by irradiation. The use of refrigeration in combination with radurization for meats and fishery products is an example of such a necessary combination.

Agents or measures that can be considered for use in combination with irradiation may be classified as either physical or chemical in nature.

II. PHYSICAL PROCESSES

Since irradiation is a process involving energy input, the combination of radiation with another energy form for the purpose of reducing the amount of radiation is logical. The broad experience with the use of *heat* in treating foods points to the likelihood of the usefulness of this energy form in this connection.

As has been noted, the actual target in the irradiation of a living organism is its DNA. Synergistic effects of ionizing radiation and other agents can occur if both produce the same type of DNA damage, or if both similarly impair the organism's enzymatic mechanism for repairing its DNA. Heat and radiation appear to have this commonality.

If applied separately, the order in which heat and radiation are applied is significant in the effect of the combination. Figure 11.1 shows that heating the spores of *Clostridium botulinum* before their irradiation causes no change in the radiation sensitivity. The data of Fig. 11.2, however, demonstrate that

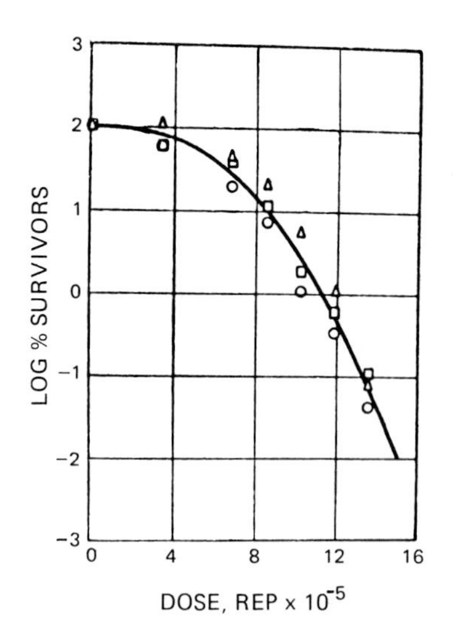

Figure 11.1 Effect of preliminary heating at 99°C on subsequent radiation resistance of spores of *Clostridium botulinum* 213B suspended in 10% gelatin at pH 7.0. ○, Control; □, heated 8 min; △, heated 25 min. From L. L. Kempe, *In* "The Preservation of Foods by Ionizing Radiations." Mass. Inst. Tech., Cambridge, 1959.

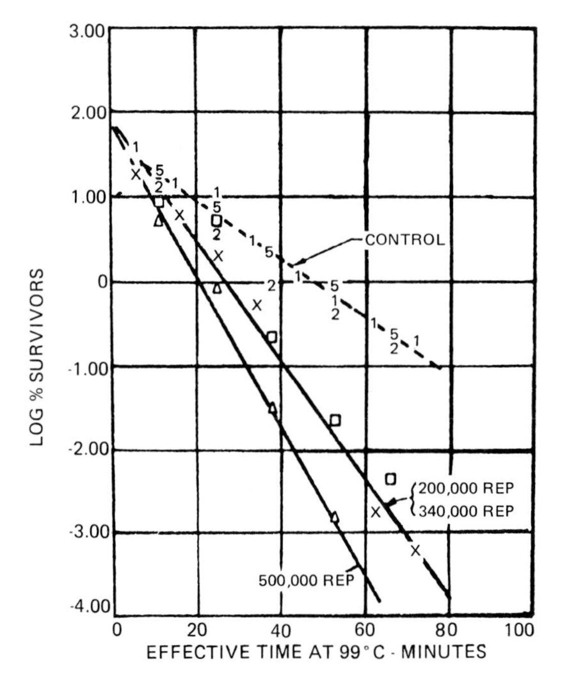

Figure 11.2 Effect of irradiation on subsequent heat resistance of spores of *Clostridium botulinum* 213B suspended in broth at pH 6.7. From L. L. Kempe, *In* "The Preservation of Foods by Ionizing Radiations." Mass. Inst. Tech., Cambridge, 1959.

there is a reduction in the heat resistance when the order is irradiation followed by heating.

That the combined action of irradiation and sufficient heating (in that order) involves synergism may be seen from the data of Table 11.1 for spores of *Bacillus subtilis* 168. The observed decimal reduction in spore numbers by the combination process exceeds the arithmetic sum of the reductions obtained from the individual actions.

While the preferred order is first irradiation and then heating, it may not be possible in all cases to have this order. A combination effect can be obtained, nonetheless, using the reverse order.

The combination of heat and radiation can be used to reduce radappertization doses. Ham, for example, heated to 65 to 70°C and subsequently irradiated with only 5 kGy has been reported to be sterile. Similarly, goose liver given a heat treatment of F_0 of 1.6 min and irradiated with a dose of 5 kGy has been found to be sterile. Fruits heated to 70°C and irradiated with a dose of 4 kGy are stable. Orange juice heated to 50°C and irradiated with 4 kGy is reported to be sterile.

Table 11.1
Synergistic Effect of Radiation and Heat on Spores of *Bacillus subtilis* 168[a]

		Decimal reductions in spore numbers			
			Combined effect		
Treatment	Amount	Single treatments	Simple addition	Actually observed	Log difference due to synergism
Radiation	0.5 kGy	0.76	0.82	0.82	0.0
Heat	10 min at 90°C	0.06			
Radiation	0.5 kGy	0.76	2.58	3.80	1.22
Heat	30 min at 90°C	1.82			
Radiation	1 kGy	0.83	0.89	0.81	−0.08
Heat	10 min at 90°C	0.06			
Radiation	1 kGy	0.83	2.65	8.37	5.72
Heat	30 min at 90°C	1.82			

[a] From N. Grecz, G. Bruszer, and I. Amin, Effect of radiation and heat on bacterial spore DNA. *In* "Combination Processes in Food Irradiation." Int. Atomic Energy Agency, Vienna, 1981.

Heat and radiation can be combined to treat fruits to control postharvest fungal infections. Papayas, for example, are damaged by doses in the range 2.5–6 kGy, which are needed to inactivate fungi present. A combination process of a dose of 0.75 kGy to accomplish insect disinfestation with a 20-min dip in water at 48°C (alternatively, 20 sec at 60°C) to inactivate fungi can be used to avoid fruit damage. The combined process provides what is needed in order to disinfest and preserve the fruit.

As has been noted, the doses required for indigenous enzyme inactivation are very high and are above doses employed in radappertization. In such cases, needed enzyme inactivation can be secured by heating the food. In radappertized meats, for example, enzyme inactivation is accomplished by low-temperature (77°C) treatment prior to irradiation.

A combination of heating with air at 40°C and irradiation with a dose of 35.0 Gy disinfests dry dates of insects by increasing the killing effect and ensuring inhibition of reproduction. Without heat, a dose of 700 Gy is needed and the killing effect is delayed.

The temperature during irradiation can have important significance in end results. In Fig. 4.6 is shown the change in D_{10} value for spores of *Clostridium botulinum* 33A. The precipitous drop in the D_{10} value at temperatures near 100°C was noted (see Chapter 4, Section II,C,2,b,i), and suggests that irradiation in this temperature region can be a means of reducing dose.

As noted in the discussion of radappertized meats (see Chapter 6), irradiation is carried out at subfreezing temperatures in order to reduce the indirect action of radiation and in this way to minimize radiation-induced flavor change.

Hydrostatic compression during or following irradiation increases the sensitivity of spores of bacteria to radiation by affecting germination. This can be a means for reduction of dose to effect inactivation. This effect is shown in Fig. 11.3 for spores of *Bacillus pumilis.*

The *temperature of storage* after irradiation can have significance in the prevention of spoilage of a food. For many foods refrigeration must be combined with radurization in order to obtain meaningful product life extension. Refrigeration not only helps delay microbial spoilage but also may

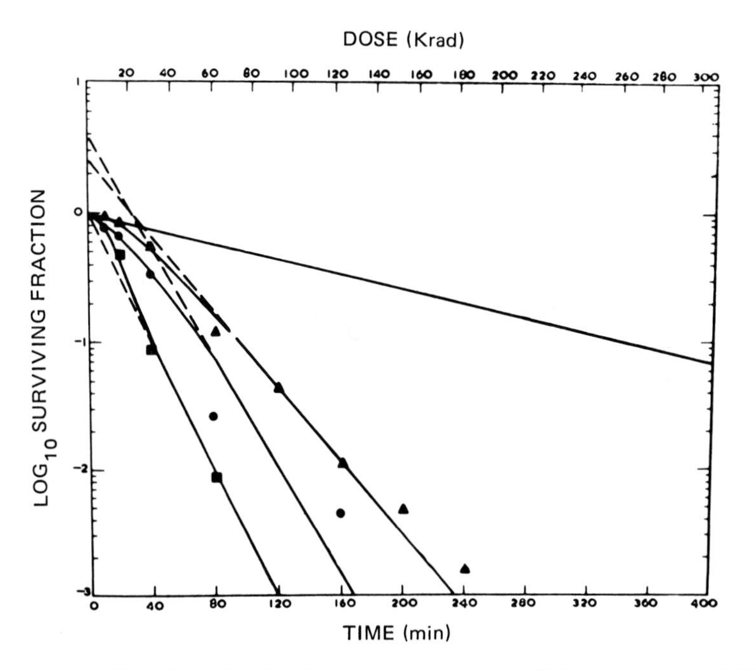

Figure 11.3 Effect of combined hydrostatic compression–radiation treatment at 25°C on survival of spores of *Bacillus pumilus*. Top curve, 1 atm; ▲, 820 atm; ●, 950 atm; ■, 1085 atm. From P. A. Wills, J. G. Clouston, and N. L. Gerraty, Microbiological and entomological aspects of the food irradiation program in Australia. *In* "Radiation Preservation of Food, Proceedings of Symposium, Bombay." Int. Atomic Energy Agency, Vienna, 1973.

be the means of preventing toxigenesis due to organisms such as *Clostridium botulinum*, which would not be inactivated by radurization doses. Control of chemical degradation of a food also may require the use of refrigeration temperatures. In the case of living foods (e.g., fruits), the postharvest physiological processes, which could lead to undesirable changes, can call for the use of particular storage temperatures.

Packaging of the irradiated food, appropriate to the food and the objectives of the treatment, may be needed, and unless combined with irradiation, the absence of such packaging may cause the intended effects of irradiation to be nullified. The package may constitute a physical barrier to recontamination of the food with spoilage microorganisms, insects, etc. Similarly, the package may exclude atmospheric O_2, whose action on the food may be damaging. Likewise it can serve as a moisture barrier to prevent either loss or gain of water. Packaging may be needed to protect the food from damage during handling. This may be a special requirement if irradiation alters the food (e.g., causes softening).

III. CHEMICAL PROCESSES

Theoretically, the use of additives with particular actions can be the means of obtaining combination effects and in this way securing reduction of dose. Practical problems of obtaining effective additives that are acceptable for use in foods and in addition can be conveniently incorporated in foods very much limit this approach.

The use of nitrites in combination with irradiation in the preparation of bacon has been discussed (see Chapter 6). This combination treatment is directed primarily to the reduction of the nitrite content of bacon.

Sodium chloride can be used in moderate amounts to lower dose requirements. This combination is illustrated in the radappertized meat products listed in Table 6.8. The addition of 100 ppm of *nisaplin* or 1 ppm of *tylosin lactate* to a pea preparation reduces the dose for microbiological stability to one-fourth that needed without these additives. *2-(2-Furyl)-3-(5-nitro-2-furyl) acrylamide* and *potassium sorbate* reduce dose requirements for the radurization of pork sausage. Potassium sorbate at the 0.1% level combined with irradiation at a dose of 4 kGy retards mold growth on salted chub mackerel, boiled chub mackerel, and smoked milk fish. The combination is more effective in extending the life of these fish products than irradiation or potassium sorbate used individually at the same indicated levels.

Lowering the *water content*, or A_w (water activity), can reduce the dose requirements for microbial stability. Shrimp, for example, dried to 40% moisture, heated to 80°C for 5 min, and irradiated with 2.5 kGy are stable.

Table 11.2

Effect of Radiation and CO_2 on Bacterial Content of Enzyme-Inactivated Beef Stored 24 Days at 38°C[a]

	Bacterial count per gram			
	No CO_2		CO_2	
Dose (kGy)	Aerobic	Anaerobic	Aerobic	Anaerobic
0	1×10^4	1×10^6	6×10^6	1
0	2.6×10^5	1×10^6	1.4×10^5	1
10	1×10^2	1×10^6	1×10^2	1
10	1×10^2	1×10^3	1×10^2	1

[a] From W. M. Urbain, J. L. Shank, and F. L. Kauffman, Irradiation with CO_2 under pressure. U.S. Patent No. 3,483,005 (1969).

Lowering the *pH* can reduce the dose requirements. *Clostridium botulinum* does not grow and produce toxin at pH values less than 4.5. If a food has a pH less than 4.5, the radappertization dose can be based on considerations other than radiation inactivation of the spores of *C. botulinum*. This combination effect can be obtained with beef in a closed container to which has been added sufficient CO_2 to yield the needed low pH (e.g., 8 atm). The data of Table 11.2 illustrate this. Viable anaerobic spores are absent in the irradiated beef to which CO_2 has been added.

SOURCES OF ADDITIONAL INFORMATION

Anonymous, "Combination Processes in Food Irradiation." Int. Atomic Energy Agency, Vienna, 1981.

Tilton, F. W. and Brower, J. H., Supplemental treatments for increasing the mortality of insects during irradiation of grain. *Food Technol.* (*Chicago*) **39**(12) 75 (1985).

CHAPTER 12

Packaging

I. INTRODUCTION

In general the same objectives apply to the packaging of both irradiated and nonirradiated foods. In certain instances there may be a particular packaging requirement in order to make irradiation effective. In most cases the food will be placed in the container before irradiation and, as a consequence, the container will be irradiated along with the food. For this reason material used in making the container must be such as not to undergo radiation-related changes which interfere with the intended function of the container. There is also the requirement that packaging material in contact with the food must not impart to the food any toxic or otherwise undesirable material as a result of irradiation.

Principal classes of materials used in making food packages are given below along with a discussion of the effects of irradiation on them.

A. Cellulose

Packaging materials composed largely of cellulose are paper, paperboard, fiberboard, and wood. Cellulose is a natural high molecular weight polymer which is crystalline in character. A number of chemical derivatives of cellulose also are available, including cellophane, rayon, and cellulose acetate. Cellulose polymers undergo chain scission and other changes when irradiated. Such changes can cause significant losses in strength properties but generally allow the normal usage of these materials in food packaging.

B. Glass

Free electrons, present in glass as a result of irradiation, can be trapped and can form color centers. With large amounts of radiation this causes a brown discoloration of the glass. This discoloration is essentially permanent, although heating the glass causes it to disappear. The addition of cerium to glass prevents radiation-induced browning.

C. Metals

Irradiation increases the mobility of electrons of metals. This energy degrades to heat, usually of negligible amount. There is no other effect of irradiation on metals.

D. Organic Polymers

Free radicals may be formed in polymeric packaging materials and result in chemical changes including chain scission and cross-linking. Scission, which decreases chain length, can decrease tensile and flexural strength. Cross-linking has the reverse effect, but can decrease elongation, crystallinity, and solubility. The overall effect of irradiation on polymers is determined by the predominant reaction.

Polymers as used in food packaging often are components of multi-ply laminates. The effects of irradiation on such laminates depend upon the nature of the individual components and the adhesives used to bond the various plies.

Because polymeric materials used in food packaging are likely to contact the food, consideration must be given to effects of radiation on changes in the polymers as they might affect the sensory and wholesomeness characteristics of the food.

E. Coloring Materials

Inks and coloring materials, either incorporated in the body or applied to the surface of polymeric films, papers, or other materials may be subject to changes by irradiation. Of such materials, organic dyes are likely to be the most sensitive to radiation. Generally, however, selection of suitable radiation-resistant coloring materials for food packages is not a major problem.

F. Adhesives and Waxes

In order to avoid failure due to irradiation, consideration must be given to adhesives and waxes employed as functional components of packages. This aspect of the packaging of irradiated foods, however, appears not to be a major problem.

II. RIGID CONTAINERS

The tinplate can, commonly used for thermally processed foods, can be used also for radappertized foods. As may be anticipated, however, certain features of this container require consideration of particular aspects of the irradiation process. These features include composition of the inside enamel, if any, composition of the end-sealing compound, and container integrity in relation to irradiation at subfreezing temperatures and to radiation-produced head-space gases. In addition, in general other considerations, normal to the use of metal containers for purposes other than irradiation, apply. Table 12.1 shows specifications for metal cans which have performed satisfactorily in the radappertization of beef.

In the use of metal cans for radappertized foods, provision must be made for

Table 12.1
Specifications for Tinplate Cans for Use with Radappertized Meats[a,b]

Components	Requirements
1. Tinplate	
a. 400 × 400 × 1110	100# MRT2–25# Electrolytic tinplate
b. 404 × 700	95# LTU–25# Electrolytic tinplate
2. Side seam	Conventional 2-98 solder
3. Enamel	a. Epoxy-phenolic with aluminum pigment
	b. Epoxy type with aluminum pigment and wax
4. End-sealing compound	a. Blend of cured and uncured butyl rubber
	b. Blend of neoprene and butadiene-styrene

[a] From E. Wierbicki, Technological feasibility of preserving meat, poultry and fish products by using a combination of conventional additives, mild heat treatment and irradiation. *In* "Combination Processes in Food Irradiation." Int. Atomic Energy Agency, Vienna, 1981.

[b] Dose: 30–75 kGy. Irradiation temperatures: 5, −30, −90°C.

Table 12.2

Head-Space Gases in Cans of Enzyme-Inactivated, Nonirradiated and Irradiated Beef[a]

Product	Percentage				
	O_2	N_2	H_2	CO_2	CH_4
Frozen control	2.7	80.9	0.0	14.0	0.0
Irradiated[b]	0.9	45.5	36.3	16.4	0.7

[a] From E. Wierbicki, Technological feasibility of preserving meat, poultry and fish products by using a combination of conventional additives, mild heat treatment and irradiation. *In* "Combination Processes in Food Irradiation." Int. Atomic Energy Agency, Vienna, 1981.

[b] Irradiated with 47 kGy at $-30 \pm 10°C$.

head-space gases formed by irradiation. At closure, a vacuum of at least 63 cm (25 in.) must be secured. With this vacuum, a final value of about 38 cm (15 in.) is obtained after irradiation. Table 12.2 provides information on the composition of the head-space gases formed in cans of radappertized beef.

Cans made of aluminum have less radiation absorption loss due to the lower density of aluminum compared with that of steel tinplate. Containers made of the usual glasses available for use with foods generally are unacceptable due to a radiation-induced browning. Rigid containers made of cellulose materials or of plastics generally are little changed by irradiation and ordinarily are satisfactory.

III. FLEXIBLE CONTAINERS

Various types of flexible containers are commonly used with foods and generally would be expected to be suitable for use with irradiated foods. The following materials have been found to be satisfactory for use with radurized haddock fillets:

Nylon-11
Saran-coated nylon-11
Polyethylene-coated polyester
Laminated paper/aluminum polyethylene
Laminated paper/aluminum/polyolefin-coated polyester
Laminated nylon-11/aluminum
Laminated saran/polyethylene/nylon

Polyethylene, polypropylene, cellophane, nylon-6, and rubber hydrochloride are not satisfactory for radurized fish.

For radurized fresh meats the following oxygen-permeable films are satisfactory:

Polyvinyl chloride
Cellophane (fresh-meat type)
Polyethylene

Oxygen-impermeable films that are suitable for radurized meats include the following:

Polyvinylidine chloride (saran)
Laminated polyvinylidine chloride, polyester, and polyethylene

For radappertized meats a laminate of 25 nylon-6 (outside), 9 aluminum foil (middle), and 62 intermolecularly bonded polyethylene terephthalate medium-density polyethylene (inside), used to form a heat-sealable flexible pouch is satisfactory. The adhesive between the outer layer and the aluminum ply is an epoxy-modified polyester. Between the inside layer and the aluminum ply the adhesive is an ethylene–acrylic acid copolymer.

As a general comment, it may be said that practically all of the technology ordinarily employed in food packaging is applicable to irradiated foods.

SOURCES OF ADDITIONAL INFORMATION

Anonymous, "Determination of Effect of Packaging Materials on the Properties of Irradiated Foods," No. DA19-129-QM-752. U.S. Army Quartermaster Corps, Chicago, 1958.

Anonymous, "Research Study on Can Enamels as Part of the Radiation Preservation of Foods Program," No. DA19-129-QM-968. U.S. Army Quartermaster Corps, Chicago, 1960.

Anonymous, "Survey of Packaging Requirements for Radiation Pasteurized Foods," TID-15-144. U.S. Atomic Energy Comm., Washington, D.C., 1962.

Anonymous, "Training Manual on Food Irradiation Technology and Techniques," 2nd Ed. Int. Atomic Energy Agency, Vienna, 1982.

Highland, H. A., Packaging–radiation disinfestation relationships, *In* "Radiation Disinfestation of Food and Agricultural Products," (J. H. Moy, ed.). Univ. of Hawaii Press, in press.

Killoran, J. J., Chemical and physical changes in food packaging materials exposed to ionizing radiation. *Radiat. Res. Rev.* **3,** 369 (1972).

Killoran, J. J., Packaging irradiated foods. *In* "Preservation of Food by Ionizing Radiation," (E. S. Josephson and M. S. Peterson, eds.), Vol. II. CRC Press, Boca Raton, Florida, 1983.

Killoran, J. J., Agarwal, S. R., and Burke, P. T., "Effect of Ionizing Radiation on Physical and Chemical Properties of Fiberboard and Paperboard," Tech. Rept. 74-6-GP. U.S. Army Natick Labs., Natick, Massachusetts, 1972.

Killoran, J. J., Cohen, J. J., and Wierbicki, E., Reliability of flexible packaging of radappertized beef under production conditions. *J. Food Process. Preserv.* **3,** 25 (1979).

Killoran, J. J., Howker, J. J., and Wierbicki, E., Reliability of the tinplate can for packaging of radappertized beef under production conditions. *J. Food Process. Preserv.* **3,** 111 (1979).

Pratt, G. B., Kneeland, L. E., Heiligman, F., and Killoran, J. J., Irradiation-induced gases in packaged foods. 1. Identification and measurements. *J. Food Sci.* **32,** 200 (1967).

CHAPTER 13

Wholesomeness of Irradiated Foods

I. INTRODUCTION

Much of the processing that is applied to foods today had its origin in the distant past. With few exceptions, these historical processes were used for long periods with little or no knowledge about associated health hazards for consumers of such processed foods. Ultimately, usage established safety, and unless it did, the process did not survive. Today, however, in an era of extensive scientific knowledge, a new food process is evaluated as to its safety before it is used. Serious interest in food irradiation developed in such circumstances, and the need to obtain adequate proof that irradiated foods are wholesome was recognized early and led to a great deal of investigational work. An important part of this activity has been the role of government regulations (see Chapter 14), which establish requirements for suitable evidence of wholesomeness and which control important aspects of the usage of food irradiation.

A wholesome food is one that has satisfactory nutritional quality and is toxicologically and microbiologically safe for human consumption. Areas of concern about the wholesomeness of irradiated foods include (1) added or induced radioactivity, (2) the formation of one or more new chemical substances, termed radiolytic products, in the food which in some manner are toxic to the consumer, (3) significant impairment of the nutritional value of the food, and (4) health hazards of a microbiological nature or origin. Each of these areas is treated separately.

II. INDUCED RADIOACTIVITY

Due to the presence of naturally occurring radionuclides all foods have some radioactivity. It has been accepted without challenge that irradiation should not cause an increase in radioactivity of a food. To this end, limitations on the energy levels and the kinds of ionizing radiations employed in food irradiations have been established. As a result, there is no added radioactivity in an irradiated food. This subject is discussed in detail in Chapter 1.

III. TOXICOLOGICAL ASPECTS

In the early period of the development of food irradiation there was little understanding of the mechanisms by which it operated, nor was there much knowledge of the nature of chemical changes it produced in foods. In accord with the practices of that period for dealing with possible toxicological hazards associated with a food, animal feeding was employed to evaluate the wholesomeness of irradiated foods. This approach had the value that one did not need to know the identity of the possible hazard or in what manner it might be unhealthful. Wholesomeness, or the lack of it, was indicated by the performance of the test animals.

Animal-testing procedures had been developed for the evaluation of chemical food additives. In such procedures an additive was added to foods in different concentrations and this provided a means of determining at what level, if any, it showed toxic effects. Commonly it is a practice to accept a substance as safe as a food additive if satisfactory animal performance is secured at 100 times the level of intended use.

Application of this approach to the evaluation of the wholesomeness of irradiated foods presents a serious problem, in that the test material is not a food additive but a food. In using this approach, the irradiated food is treated as a discrete chemical entity, which it is not. Any radiolytic product present might be so regarded, but its concentration is tied to the food of which it is a component and there is no way to increase its level for the evaluation test. The proportion of the diet which can be composed of the irradiated food is limited by nutritional and other requirements of the test animal. It is clear that the aspect of the food additive-testing procedure which allows testing at exaggerated levels is not applicable when the test material is a food.

Exaggeration of dose as a means of paralleling exaggeration of the concentration of a food additive is regarded as invalid, since very high doses can alter the normal characteristics of the food.

Despite this difficulty with animal testing, for many years this was the only approach to the evaluation of the toxicological aspects of the wholesomeness of irradiated foods that was considered possible. As time progressed protocols for this kind of testing increased in complexity and involved a number of ancillary determinations to provide certain specific types of information. Generally the animal-feeding tests employed mammals and often were multigeneration, multispecies tests in which various indexes of animal health and performance were determined, such as growth, feed consumption and efficiency, reproduction, longevity, gross and microscopic pathology, urology, hematology, and hepatic microsomal enzyme function. The numbers of test animals employed were sufficient to enable judgment of statistical significance of the findings. Findings regarding teratology were obtained by mating test animals fed irradiated foods and examining uteri and fetuses for malformations. Carcinogenicity was evaluated through separate appropriate animal tests.

Tests for cytotoxicity were carried out by feeding mammals the irradiated food and subsequently making an examination of the chromosomes of the cells of specific tissues, such as peripheral lymphocytes and bone marrow. This permits detection of gross chromosomal damage, but not effects at the gene level.

The use of mammalian test animals to investigate mutagenic effects requires very large numbers of animals and extended time periods to obtain meaningful data. In order to avoid the use of such cumbersome procedures, other organisms such as bacteria, plant and animal cell cultures, and higher life forms such as *Drosophila* were employed. The use of such procedures, however, is open to question, since they involve extrapolation from relatively simple life forms to the human, which is not necessarily valid.

Several kinds of tests which can be performed *in vivo* in mammals have been employed for the purpose of improving the extrapolation. The dominant lethal test reveals chromosomal mutations of the type responsible for death at the stage of the zygote or during embryogenesis. The host-mediated assay involves the use of an indicator bacterium such as *Salmonella typhimurium*, or conidia such as *Neurospora*. The host mammalian animal (e.g., a mouse) is fed the test food and the indicator microorganism is injected into the peritoneal cavity. Later the microoragnism is withdrawn and the number of mutants formed is determined.

Initially there was agreement that the wholesomeness testing of irradiated foods could be done by testing a single food which was representative of a particular class of foods and by extrapolating the tested food to the other foods of the same class. In this manner, at one time, 21 foods were selected and as a group were considered to represent all foods. Testing was to be limited to these 21 representative foods. In time, this approach was not accepted by the U.S. Food and Drug Administration and the requirement that each and every irradiated food be tested separately was imposed. Additionally, different conditions for the irradiation of a given food required separate evaluations.

The combination of the complexity of the testing procedures for evaluating the wholesomeness of irradiated foods plus the requirement for separate testing of individual foods became an overwhelming obstacle to the use of food irradiation. This was so despite an accumulation of a great deal of information secured from various studies, including animal-feeding experiments, all of which, when properly conducted, revealed no effect attributable to the irradiation of a food that could be regarded as a health hazard to the consumer. This impasse existed for more than 10 years.

The principal difficulty in assessing the toxicological wholesomeness of an irradiated food lay in the ignorance of the chemical changes in the food caused by irradiation. This ignorance resulted in the use of "shotgun" techniques which did not require such knowledge. Fortunately, studies were made on the radiation chemistry of foods (see Chapter 3), originally not so much for the purpose of assessing wholesomeness, but to investigate other problems such as the "irradiation flavor" that develops in some foods such as meats.

These studies provided an understanding of the chemical changes in a food produced by irradiation. With regard to toxicological aspects of wholesomeness, the following findings in the field of radiation chemistry are pertinent and significant:

1. No radiolytic product so far identified has significant toxic characteristics.
2. The amount of each identified radiolytic product is small, even with doses as large as about 50 kGy. This low-level incidence is significant in limiting any potential toxicity.
3. The kind and amount of radiolytic products correlate with the initial composition of the food; foods of similar initial composition yield similar radiolytic products.
4. The amount of each radiolytic product is a function of the dose and is related to the conditions of irradiation, primarily as they may alter the indirect action of radiation.
5. The radiolytic products so far identified can be found in nonirradiated foods and are not unique to irradiated foods.

These findings of the radiation chemistry of foods led to the following conclusions regarding the toxicological wholesomeness of irradiated foods:

1. From the standpoint of toxicology, foods irradiated at the doses studied (a maximum of $\sim$50 kGy) are safe for human consumption.
2. Findings relating to a given food can be extrapolated to other foods of similar composition.
3. For a given food, findings relating to a particular radiation dose and the particular irradiation conditions can be extrapolated to other doses and to other irradiation conditions.

The findings of radiation chemistry also indicate that much of the approach for evaluating the wholesomeness of irradiated foods by animal testing is an inappropriate procedure, largely due to the small amounts of radiolytic products formed and which are inadequate to produce a detectable response in the test animals.

The word *chemiclearance* has been devised to designate the concept that the toxicological wholesomeness of an irradiated food can be judged on the basis of the nature and amounts of the radiolytic products present after irradiation. This concept has received broad acceptance.

IV. NUTRITIVE VALUE

The low levels of radiolytic products in irradiated foods, plus the related fact of the small amount of energy employed in the irradiation process and available to produce chemical change (see Chapter 3, Section III, A), provide an explanation for findings that the macronutrients of a food undergo little change, and, as a consequence, the normal nutritional values of proteins, lipids, and carbohydrates of foods are maintained. This conclusion has been affirmed by animal-feeding tests, as may be seen from the data of Table 13.1.

Of the micronutrients, concern generally involves minerals and vitamins. Irradiation does not alter the mineral content of a food. It could, however, alter the nutritional availability of minerals. Animal-feeding studies have not revealed a problem in this area. It has been observed that phosphorus in wheat is made more available by irradiation.

Vitamins, on the other hand, can be destroyed by irradiation, as noted previously (see Chapter 3). In most cases the losses are not large and of little significance nutritionally. In cases such as vitamin B_1 (thiamine), in which

Table 13.1

Average Nutrient Availability through Four Generations of Rats Fed a Nine-Component Irradiated Diet[a]

	Availability (%)		
	Protein	Fat	Carbohydrate
Control[b]	87.7	93.6	90.6
Irradiated[c]	88.5	94.1	90.4

[a] From M. S. Read, H. F. Kraybill, G. J. Isaac, and N. F. Witt, *Toxicol. Appl. Pharmacol.* **3**, 153 (1961).
[b] Stored frozen.
[c] Dose, 55.8 kGy; stored at ambient temperature.

substantial loss can occur, the use of techniques such as irradiation of meats in the frozen state greatly reduces the loss. In general vitamin losses, if they occur, are not greater than losses obtained with other types of food processing and often are less.

Based on consideration of both macronutrients and micronutrients, the nutritive value of foods is not changed in any significant way by irradiation.

V. MICROBIOLOGICAL SAFETY

Assurance of safety from potential microbiological hazards associated with irradiated foods largely depends upon proper specification of certain important pertinent parameters of the irradiation process, of which the most critical is correct dose. In some cases irradiation may be used in combination with another agent, such as heat. In these situations proper specification of the combination effect is required.

In particular irradiation processes for specific foods, so far the required specification of the dose for microbiological safety generally has been secured. There are findings, however, that have been reported, especially with early work on food irradiation where this may not be true and further work to assure safety may be needed. Any future work must recognize the need to determine and use a proper dose which assures microbiological safety.

Concern for a potential microbiological hazard associated with the possibility of irradiation acting to cause mutations of bacteria has existed. This hazard would be a possibility in applications in which not all of a population of microorganisms is inactivated and there are survivors (e.g., radurization). All evidence available to date indicates that radiation-induced mutants of bacteria are not a problem in food irradiation (see Chapter 4). Any changes in attributes of bacteria that have been observed as a result of radiation-induced mutations have been in the direction of reduction of possible health hazards. Also, such changes have not interfered with standard procedures for the identification of the bacteria.

In applications such as radurization, the postirradiation outgrowth flora may differ as to kind from what is the typical outgrowth of bacteria on the same food not irradiated. Depending upon the flora and the kind of food involved, plus other circumstances such as postirradiation holding temperatures, there could be a microbiological hazard in such a circumstance. One such possibility has been identified with marine and freshwater animal foods, wherein radurization may favor growth and toxin formation by *Clostridium botulinum* type E (see Chapter 7).

One other area of potential microbiological hazard involves the effect of radurization to increase mycotoxin formation during postirradiation growth of certain molds, notably *Aspergillus flavus* and *A. parasiticus*. This effect may involve mutation.

VI. GENERAL COMMENT ON WHOLESOMENESS

From the above discussion it is clear that the subject of the wholesomeness of irradiated foods has received a great deal of attention. Unlike the development of most other food processes, irradiation has had to establish its safety prior to use. As a consequence, it can be stated that no other food process has been scrutinized with regard to safety as intensively as has food irradiation. The wealth of scientifically sound information that has been obtained establishes the wholesomeness of irradiated foods and provides a proper basis for the appropriate use of food irradiation. The actions of international organizations and national governments relative to irradiated foods in view of this evidence of their wholesomeness are reviewed in Chapter 14.

SOURCES OF ADDITIONAL INFORMATION

Anonymous, "Recommendations for Evaluating the Safety of Irradiated Foods, "Rep. of the Irradiated Foods Committee. U.S. Food and Drug Administration, Washington, D.C., 1980.

Anonymous, "Fourth Activity Report." Int. Project in the Field of Food Irradiation, Karlsruhe, West Germany, 1981.

Anonymous, "Is Radiation a Food Additive?" Council for Agricultural Science and Technology, Ames, Iowa, 1984.

Basson, R.A., Chemiclearance. *Nucl. Act.* **17,** 3(1977).

Chauhan, P., "Assessment of irradiated foods for toxicological safety—Newer methods," "Food Irradiation Information No. 3. Int. Project in the Field of Food Irradiation, Karlsruhe, West Germany, 1974.

Diehl, J. F., Effects of combination processes on the nutritive value of food. *In* "Combination Processes in Food Irradiation." Int. Atomic Energy Agency, Vienna, 1981.

Hickman, J. R., The problem of wholesomeness of irradiated food. *In* "Radiation Preservation of Food, Proceedings of Symposium, Bombay." Int. Atomic Energy Agency, Vienna, 1973.

Josephson, E. S., Thomas, M. H., and Calhoun, W. F., Nutritional Aspects of Food Irradiation: An Overview. *J.* Food *Process. Preserv.* **2,** 299(1978).

Murray, T. K., "Nutritional Aspects of Food Irradiation," Food Irradiation Information No. 11. Int. Project in the Field of Food Irradiation, Karlsruhe, West Germany, 1981.

Raica, N., Jr., Scott, J., and Nielsen, W., The nutritional quality of irradiated foods. *Radiat. Res. Rev.* **3,** 447(1972).

Taub, I. A., Angelini, P., and Merritt, C., Jr., Irradiated food: Validity of extrapolating wholesomeness data. *J.* Food. *Sci.* **41,** 942(1976).

Teufel, P. "Microbiological Implications of the Food Irradiation Process," Food Irradiation No. 11. Int. Project in the Field of Food Irradiation, Karlsruhe, West Germany, 1981.

CHAPTER 14

Government Regulation of Irradiated Foods

I. INTRODUCTION

Virtually all national governments have imposed regulatory control on irradiated foods. The general approach has been to ban irradiated foods except as provided for by specific regulations which authorize production, importation, and sale in accord with designated limitations. The issuance of such regulations follows procedures as established by statutes. Details of the regulations usually are determined by the government agency which has statutory jurisdiction.

Various segments of the United Nations are concerned with the regulatory aspects of food irradiation. These include the World Health Organization (WHO), the Food and Agriculture Organization (FAO), the International Atomic Energy Agency (IAEA), and the Codex Alimentarius Commission. Basically, these groups serve member nations (of the United Nations) in an advisory capacity. Their leadership, however, often has a strong influence on the actions of individual nations regarding regulations for foods, and this has been the case with irradiated foods.

In the United States the primary federal government agency having jurisdiction over irradiated foods in the civilian sector is the Food and Drug Administration (FDA). It operates under the Federal Food, Drug and Cosmetic Act. Other federal agencies which work within the FDA regulations and which also have certain jurisdictions in relation to irradiated foods include the Department of Agriculture, which regulates meat, poultry, and plant foods; the Department of Commerce, which regulates marine and freshwater animal foods; the Department of Energy, which regulates irradiation facilities and radionuclide materials; and the Department of Labor, which regulates occupational safety and health.

II. WORLD HEALTH ORGANIZATION, GENEVA

"The World Health Organization is a specialized agency of the United Nations with primary responsibility for international health matters and public health."* WHO, working jointly with FAO and IAEA, investigated the wholesomeness of irradiated foods and in 1981 published Technical Report Series 659, "Wholesomeness of Irradiated Food." The following quotations are excerpted from that report:

10. Conclusions on the Acceptability of Irradiated Food

10.1 Toxicological acceptability of irradiated food

The Committee, having reviewed new evidence, was able to formulate a recommendation on the acceptability of food irradiated up to an overall average dose of 10 kGy (see sections 2 and 3). This development follows logically from the approaches to the assessment of the wholesomeness of irradiated food adopted in the past by previous Joint Expert Committees, as described in the Introduction. The following considerations led to this development:

(a) All the toxicological studies carried out on a large number of individual foods (from almost every type of food commodity) have produced no evidence of adverse effects as a result of irradiation.

(b) Radiation chemistry studies have now shown that the radiolytic products of major food components are identical, regardless of the food from which they are derived. Moreover, for major food components, most of these radiolytic products have also been identified in foods subjected to other, accepted types of food processing. Knowledge of the nature and concentration of these radiolytic products indicates that there is no evidence of a toxicological hazard.

(c) Supporting evidence is provided by the absence of any adverse effects resulting from the feeding of irradiated diets to laboratory animals, the use of irradiated feeds in livestock production, and the practice of maintaining immunologically incompetent patients on irradiated diets.

The Committee therefore concluded that the irradiation of any food commodity up to an overall average dose of 10 kGy presents no toxicological hazard; hence, toxicological testing of foods so treated is no longer required.

10.2 Microbiological and nutritional acceptability of irradiated food

The Committee considered that the irradiation of food up to an overall average dose of 10 kGy introduces no special nutritional or microbiological problems. However, the Committee emphasized that attention should be given to the significance of any changes in relation to each particular irradiated food and to its role in the diet.

* "Wholesomeness of Irradiated Foods," Tech. Rep. Series No. 659. World Health Organization, Geneva, 1981.

10.3 High-dose irradiation

The Committee recognized that higher doses of radiation were needed for the treatment of certain foods but did not consider the toxicological evaluation and wholesomeness assessment of foods so treated because the available data are insufficient for this purpose. Further studies in this area are therefore needed.

III. CODEX ALIMENTARIUS COMMISSION

The Codex Alimentarius Commission develops standards and codes relative to foods, which are submitted to member nations for acceptance. Member nations may accept these standards and codes fully or partly, or they may not accept them. The standards and codes have substantial importance in international trade.

The Commission has issued (1) International General Standard for Irradiated Foods and (2) International Code of Practice for the Operation of Radiation Facilities used for the Treatment of Foods. The principal parts of these are reproduced below with the permission of the Food and Agriculture Organization (Rome) and the World Health Organization (Geneva) (both units of the United Nations).

CODEX GENERAL STANDARD FOR IRRADIATED FOODS*
(World-wide Standard)

1. SCOPE

This standard applies to foods processed by irradiation. It does not apply to foods exposed to doses imparted by measuring instruments used for inspection purposes.

2. GENERAL REQUIREMENTS FOR THE PROCESS

2.1. Radiation Sources

The following types of ionizing radiation may be used:

(a) Gamma rays from the radionuclides ^{60}Co or ^{137}Cs;

(b) X-rays generated from machine sources operated at or below an energy level of 5 MeV;

(c) Electrons generated from machine sources operated at or below an energy level of 10 MeV.

2.2. Absorbed Dose

The overall average dose absorbed by a food subjected to radiation processing should not exceed 10 kGy.[1,2]

2.3. Facilities and Control of the Process

2.3.1. Radiation treatment of foods shall be carried out in facilities licensed and registered for this purpose by the competent national authority.

2.3.2. The facilities shall be designed to meet the requirements of safety, efficacy and good hygienic practices of food processing.

* Revised version identified as: Codex Stan 106-1983.

2.3.3. The facilities shall be staffed by adequate, trained, and competent personnel.

2.3.4. Control of the process within the facility shall include the keeping of adequate records including quantitative dosimetry.

2.3.5. Premises and records shall be open to inspection by appropriate national authorities.

2.3.6. Control should be carried out in accordance with the Recommended International Code of Practice for the Operation of Radiation Facilities used for the Treatment of Foods (CAC/RCP 19-1979, Rev. 1).

3. HYGIENE OF IRRADIATED FOODS

3.1. The food should comply with the provisions of the Recommended International Code of Practice—General Principles of Food Hygiene (Ref. No. CAC/RCP 1-1969, Rev. 1, 1979) and, where appropriate, with the Recommended International Code of Hygienic Practice of the Codex Alimentarius relative to a particular food.

3.2. Any relevant national public health requirement affecting microbiological safety and nutritional adequacy applicable in the country in which the food is sold should be observed.

4. TECHNOLOGICAL REQUIREMENTS

4.1. Conditions for Irradiation

The irradiation of food is justified only when it fulfils a technological need or where it serves a food hygiene purpose[3] and should not be used as a substitute for good manufacturing practices.

4.2. Food Quality and Packaging Requirements

The doses applied shall be commensurate with the technological and public health purposes to be achieved and shall be in accordance with good radiation processing practice. Foods to be irradiated and their packaging materials shall be of suitable quality, acceptable hygienic condition, and appropriate for this purpose and shall be handled, before and after irradiation, according to good manufacturing practices taking into account the particular requirements of the technology of the process.

5. RE-IRRADIATION

5.1. Except for foods with low moisture content (cereals, pulses, dehydrated foods and other such commodities) irradiated for the purpose of controlling insect reinfestation, foods irradiated in accordance with sections 2 and 4 of this standard shall not be re-irradiated.

5.2. For the purpose of this standard food is not considered as having been re-irradiated when: (a) the food prepared from materials which have been irradiated at low dose levels e.g. about ~ 1 kGy, is irradiated for another technological purpose; (b) the food, containing less than 5% of irradiated ingredient, is irradiated, or when (c) the full dose of ionizing radiation required to achieve the desired effect is applied to the food in more than one instalment as part of processing for a specific technological purpose.

5.3. The cumulative overall average dose absorbed should not exceed 10 kGy as a result of re-irradiation.

6. LABELLING

6.1. Inventory Control

For irradiated foods, whether prepackaged or not, the relevant shipping documents shall give appropriate information to identify the registered facility which has irradiated the food, the date(s) of treatment and lot identification.

6.2. Prepackaged Foods Intended for Direct Consumption
The labelling of prepackaged irradiated foods shall be in accordance with the relevant provisions of the Codex General Standard for the Labelling of Prepackaged Foods.[4]

6.3. Foods in Bulk Containers
The declaration of the fact of irradiation shall be made clear on the relevant shipping documents.

[1] For measurement and calculation of overall average dose absorbed see Annex A of the Recommended International Code of Practice for the Operation of Radiation Facilities used for Treatment of Foods (CAC/RCP 19-1979, Rev. 1).

[2] The wholesomeness of foods, irradiated so as to have absorbed an overall average dose of up to 10 kGy, is not impaired. In this context the term "wholesomeness" refers to safety for consumption of irradiated foods from the toxicological point of view. The irradiation of foods up to an overall average dose of 10 kGy introduces no special nutritional or microbiological problems (Wholesomeness of Irradiated Foods, Report of a Joint FAO/IAEA/WHO Expert Committee, Technical Report Series 659, WHO, Geneva, 1981).

[3] The utility of the irradiation process has been demonstrated for a number of food items listed in Annex B to the Recommended International Code of Practice for the Operation of Radiation Facilities used for the Treatment of Foods.

[4] Under revision by the Codex Committee on Food Labelling.

RECOMMENDED INTERNATIONAL CODE OF PRACTICE FOR THE OPERATION OF IRRADIATION FACILITIES USED FOR THE TREATMENT OF FOODS*

1. INTRODUCTION

This code refers to the operation of irradiation facilities based on the use of either a radionuclide source (^{60}Co or ^{137}Cs) or X-rays and electrons generated from machine sources. The irradiation facility may be of two designs, either "continuous" or "batch" type. Control of the food irradiation process in all types of facility involves the use of accepted methods of measuring the absorbed radiation dose and of the monitoring of the physical parameters of the process. The operation of these facilities for the irradiation of food must comply with the Codex recommendations on food hygiene.

2. IRRADIATION PLANTS

2.1. Parameters
For all types of facility the doses absorbed by the product depend on the radiation parameter, the dwell time or the transportation speed of the product, and the bulk density of the material to be irradiated. Source–product geometry, especially distance of the product from the source and measures to increase the efficiency of radiation utilization, will influence the absorbed dose and the homogeneity of dose distribution.

2.1.1. Radionuclide Sources
Radionuclides used for food irradiation emit photons of characteristic energies. The statement of the source material completely determines the penetration of the emitted radiation. The source activity is measured in Becquerel (Bq) and should be stated by the supplying organisation. The actual activity of the source (as well as any return or

replenishment of radionuclide material) shall be recorded. The recorded activity should take into account the natural decay rate of the source and should be accompanied by a record of the date of measurement or recalculation. Radionuclide irradiators will usually have a well separated and shielded depository for the source elements and a treatment area which can be entered when the source is in the safe position. There should be a positive indication of the correct operational and of the correct safe position of the source which should be interlocked with the product movement system.

2.1.2. Machine Sources

A beam of electrons generated by a suitable accelerator, or after being converted to X-rays, can be used. The penetration of the radiation is governed by the energy of the electrons. Average beam power shall be adequately recorded. There should be a positive indication of the correct setting of all machine parameters which should be interlocked with the product movement system. Usually a beam scanner or a scattering device (e.g., the converting target) is incorporated in a machine source to obtain an even distribution of the radiation over the surface of the product. The product movement, the width and speed of the scan and the beam pulse frequency (if applicable) should be adjusted to ensure a uniform surface dose.

2.2. Dosimetry and Process Control

Prior to the irradiation of any foodstuff certain dosimetry measurements[1] should be made, which demonstrate that the process will satisfy the regulatory requirements. Various techniques for dosimetry pertinent to radionuclide and machine sources are available for measuring absorbed dose in a quantitative manner.[2]

Dosimetry commissioning measurements should be made for each new food, irradiation process and whenever modifications are made to source strength or type and to the source–product geometry.

Routine dosimetry should be made during operation and records kept of such measurement. In addition, regular measurements of facility parameters governing the process, such as transportation speed, dwell time, source exposure time, machine beam parameters, can be made during the facility operation. The records of these measurements can be used as supporting evidence that the process satisfies the regulatory requirements.

3. GOOD RADIATION PROCESSING PRACTICE

Facility design should attempt to optimalize the dose uniformity ratio, to ensure appropriate dose rates and, where necessary, to permit temperature control during irradiation (e.g. for the treatment of frozen food) and also control of the atmosphere. It is also often necessary to minimize mechanical damage to the product during transportation, irradiation and storage, and desirable to ensure the maximum efficiency in the use of the irradiator. Where the food to be irradiated is subject to special standards for hygiene or temperature control, the facility must permit compliance with these standards.

4. PRODUCT AND INVENTORY CONTROL

4.1. The incoming product should be physically separated from the outgoing irradiated products.

4.2. Where appropriate, a visual colour change radiation indicator should be affixed to each product pack for ready identification of irradiated and non-irradiated products.

4.3. Records should be kept in the facility record book which show the nature and kind of the product being treated, its identifying marks if packed or, if not, the shipping details, its bulk density, the type of source or electron machine, the dosimetry, the dosimeters used and details of their calibration, and the date of treatment.

4.4. All products shall be handled, before and after irradiation, according to accepted good manufacturing practices taking into account the particular requirements of the technology of the process.[3] Suitable facilities for refrigerated storage may be required.

* Revised version of the Recommended International Code of Practice for the Operation of Radiation Facilities Used for the Treatment of Foods (CAC/RCP 19-1979).

[1] See Annex A to this Code.

[2] Detailed in "The Manual of Food Irradiation Dosimetry," Tech. Rep. Series No. 178. Int. Atomic Energy Agency, Vienna, 1977.

[3] See Annex B to this Code.

ANNEX A

DOSIMETRY

1. The Overall Average Absorbed Dose

It can be assumed for the purpose of the determination of the wholesomeness of food treated with an overall average dose of 10 kGy or less, that all radiation chemical effects in that particular dose range are proportional to dose.

The overall average dose, $\bar{D}$, is defined by the following integral over the total volume of the goods

$$\bar{D} + \frac{1}{M}\int \rho(x, y, z) \cdot d(x, y, z) \cdot dV$$

where

M = the total mass of the treated sample
ρ = the local density at the point (x, y, z)
d = the local absorbed dose at the point (x, y, z)
$dV = dx\,dy\,dz$ the infinitesimal volume element which in real cases is represented by the volume fractions.

The overall average absorbed dose can be determined directly for homogeneous products or for bulk goods of homogeneous bulk density by distributing an adequate number of dose meters strategically and at random throughout the volume of the goods. From the dose distribution determined in this manner an average can be calculated which is the overall average absorbed dose.

If the shape of the dose distribution curve through the product is well determined the positions of minimum and maximum dose are known. Measurements of the distribution of dose in these two positions in a series of samples of the product can be used to give an estimate of the overall average dose. In some cases the mean value of the average values of the minimum ($\bar{D}_{min}$) and maximum ($\bar{D}_{max}$) dose will be a good estimate of the overall average dose.

i.e. in these cases

$$\text{overall average dose} \approx \frac{\bar{D}_{max} + \bar{D}_{min}}{2}$$

2. Effective and Limiting Dose Values
Some effective treatment e.g. the elimination of harmful microorganisms, or a particular shelf-life extension, or a disinfestation requires a minimum absorbed dose. For other applications too high an absorbed dose may cause undesirable effects or an impairment of the quality of the product.

The design of the facility and the operational parameters have to take into account minimum and maximum dose values required by the process. In some low-dose applications it will be possible within the terms of section 3 on Good Radiation Processing Practice to allow a ratio of maximum to minimum dose of greater than 3.

With regards to the maximum dose value under acceptable wholesomeness considerations and because of the statistical distribution of the dose a mass fraction of product of at least 97.5% should receive an absorbed dose of less than 15 kGy when the overall average dose is 10 kGy.

3. Routine Dosimetry
Measurements of the dose in a reference position can be made occasionally throughout the process. The association between the dose in the reference position and the overall average dose must be known. These measurements should be used to ensure the correct operation of the process. A recognized and calibrated system of dosimetry should be used.

A complete record of all dosimetry measurements including calibration must be kept.

4. Process Control
In the case of a continuous radionuclide facility it will be possible to make automatically a record of transportation speed or dwell time together with indications of source and product positioning. These measurements can be used to provide a continuous control of the process in support of routine dosimetry measurements.

In a batch operated radionuclide facility automatic recording of source exposure time can be made and a record of product movement and placement can be kept to provide a control of the process in support of routine dosimetry measurements.

In a machine facility a continuous record of beam parameters, e.g. voltage, current, scan speed, scan width, pulse repetition and a record of transportation speed through the beam, can be used to provide a continuous control of the process in support of routine dosimetry measurements.

IV. U.S. FOOD AND DRUG ADMINISTRATION

The U.S. Food, Drug and Cosmetic Act defines radiation as a food additive as follows:

Food Additive (Section 201(s))

The term "food additive" means any substance the intended use of which results or may reasonably be expected to result, directly or indirectly, in its becoming a component or otherwise affecting the characteristics of any food (including any substance intended for use in producing, manufacturing, packing, processing, preparing, treating, packaging, transporting or holding food; and including any source of radiation intended for any such use,...)

This definition brings radiation under those sections of the law which regulate the use of food additives. Section 409 of the act states in part,

> A food additive shall with respect to any particular use or intended use of such additives, be deemed to be unsafe ... unless
>
> (1) it and its use or intended use conform to the terms of an exemption which is in effect ... or
>
> (2) there is in effect, and it and its use or intended use are in conformity with a regulation issued under this section prescribing the conditions under which such additive may be safely used.

The Act states that petitions may be submitted to the FDA for regulations. Petitions must contain specified information concerning the additive, including "Full reports of investigations made with respect to the safety for use of such additive, including full information as to the methods and controls used in conducting such investigations." The act provides for action on petitions by the agency, including the determination of the safety of the additive, which is to be in accord with the following:

> Section 409(c)(5): In determining, for the purpose of this section, whether a proposed use of a food additive is safe, the Secretary shall consider among other relevant factors—
>
> (A) the probable consumption of the additive and of any substance formed in or on food because of the use of the additive;
>
> (B) the cumulative effect of such additive in the diet of man or animals, taking into account any chemically or pharmacologically related substance or substances in such diet; and
>
> (C) safety factors which in the opinion of experts qualified by scientific training and experience to evaluate the safety of food additives are generally recognized as appropriate for the use of animal experimentation data.

The Act also provides for the Secretary (of the Department of Health and Human Services) to establish regulations on his or her own initiative.

In part, the Act originally regulated food additives which are chemical substances. It was later amended to define radiation, or more precisely, a radiation source, as an additive, despite the fact that this legal definition does not constitute a scientifically correct identification of a true additive in accord with previous usage of the term. At best, the radiation source "adds" energy to the food. This energy can cause chemical change in the food and the radiolytic products derived from the food could be regarded as additives. Such radiolytic products, however, are, in terms of reasons for the irradiation of a food, usually unintentional additives that are nonfunctional. Arguments have been advanced to treat irradiation as a process instead of an additive in order to conform with common practices employed with other food processes.

If they are irradiated along with the food, food-contacting packaging

Table 14.1

Packaging Materials Approved by the U.S. Food and Drug Administration for Use during the Irradiation of Prepackaged Foods

Material[a]	Maximum dose (kGy)
Paper, kraft (flour only)	0.5
Paper, glassine	10
Paperboard, wax-coated	10
Cellophane, either nitrocellulose coated or vinylidene chloride copolymer coated	10
Polyolefin film	10
Polystyrene film	10
Rubber hydrochloride film	10
Vinylidene chloride–vinyl chloride copolymer film	10
Nylon-11	10
Vegetable parchment	60
Polyethylene film	60
Polyethylene terephthalate film	60
Nylon-6	60
Vinyl chloride–vinyl acetate copolymer film	60
Acrylonitrile copolymers	60

[a] For details regarding film specifications, the use of adjuvants, etc. see U.S. Code of Federal Regulations, CFR Title 21, Part 179, Sub-part C. 179.45 (1985).

materials could undergo changes and possibly result in the transfer of radiolytic products to the food. In order to avoid a consumer health hazard which would originate in this manner, the U.S. FDA has required that only materials for which they have issued regulations be used. Table 14.1 lists the packaging materials which have FDA approval.

The reader may wish to consult the U.S. Code of Federal Regulations and the Federal Register for specific regulations concerning irradiated foods issued by the U.S. Food and Drug Administration.

V. IDENTIFICATION OF IRRADIATED FOODS

From a regulatory point of view it is desirable to have available an objective test procedure to identify a food as having been irradiated. Additionally, it would be desirable to have a means of measuring the applied dose. These capabilities would fit in with established procedures for the use with chemical food additives and other substances such as pesticides which are used with foods.

As a general statement, it can be said that there is no reliable and otherwise satisfactory analytical procedure for the identification of a food as having been irradiated, nor is there any means of establishing the dose employed. While certain changes in foods resulting from irradiation have been identified, there is no specific change which can serve a regulatory need.

A few qualitative identification procedures have been developed. Potatoes irradiated to inhibit sprouting do not sprout within 8 to 10 days when stored at 25°C when treated with the hormones gibberellic acid (300–500 ppm) or kinetin (500 ppm), whereas potatoes treated with chemical antisprout agents do. Certain instances of differing electron spin resonance spectra, which correlate with free-radical content, have been observed. This procedure, however, applies only to low-moisture foods. With extended storage of the food, the free radicals tend to disappear. Indicator compounds such as malonaldehyde have been found in irradiated carbohydrates. In meats a radiation-formed sarcoplasmic protein fraction has been identified.

While such differences can be found between irradiated and nonirradiated foods, they do not provide means of analyses that fulfill regulatory needs. It is generally accepted that regulatory control measures intended to meet the objectives of limiting irradiation to particular foods and to regulate associated quantitative aspects of irradiation will have to be carried out in the irradiation facility.

SOURCES OF ADDITIONAL INFORMATION

"Report of the Working Party on Irradiation of Food." Her Majesty's Stationery Office, London, 1964.

"Technical Basis for Legislation on Irradiated Food." Food and Agriculture Organization of the United Nations, Rome, 1965.

"Wholesomeness of Irradiated Food with Special Reference to Wheat, Potatoes and Onions," Technical Report Series No. 451. World Health Organization, Geneva, 1970.

"Colloquium on the Identification of Irradiated Foodstuffs." Commission of the European Communities, Luxembourg, 1970.

"Identification of Irradiated Food Stuffs," Proceedings of an International Colloquium. Commission of the European Communities, Luxembourg, 1974.

"Wholesomeness of Irradiated Food," Report of a Joint FAO/IAEA/WHO Expert Committee. Technical Report Series 604. World Health Organization, Geneva, 1977.

"International Acceptance of Irradiated Food. Legal Aspects." Int. Atomic Energy Agency, Vienna, 1979.

Ladomery, L. G., and Nocera, F., Technical and legal aspects relating to labelling of irradiated foodstuffs. *Food Irrad. Newsl.* **4,** 32 (1980).

"Recommendations for Evaluating the Safety of Irradiated Foods," Report of the Irradiated Food Committee. Final Report, July, 1980. Bureau of Foods, U.S. Food and Drug Administration, Washington, D.C., 1980.

Policy of Irradiated Foods; Advance Notice of Proposed Procedures for the Regulation of

Irradiated Foods for Human Consumption. Food and Drug Administration. Federal Register **46**(59), 18992-18994 (1981).

"Wholesomeness of Irradiated Food," Report of a Joint FAO/IAEA/WHO Expert Committee. Technical Report Series 659. World Health Organization, Geneva, 1981.

Van Kooij, J. G., International Aspects of Food Irradiation. *In* "Food Irradiation Now." Nijhoff/Junk, The Hague, 1982.

"The Microbiological Safety of Irradiated Food." Codex Alimentarius Commission, Rome, 1983.

U.S. Federal Food, Drug and Cosmetic Act. U.S. Federal Register. U.S. Code of Federal Regulations. All available from U.S. Government Printing Office, Washington, D.C.

CHAPTER 15

Commercial Aspects

I. CONSUMER ACCEPTANCE OF IRRADIATED FOODS

Consumer response to a food is highly subjective in nature. The inputs into this response are multiple and often complex; they are, in addition, individual. With irradiated foods, factors of consumer acceptance include all those normal for any food. In addition, at least with some consumers, irradiated foods are regarded as processed foods and, therefore, a matter of concern because of a general opposition to processed foods. Beyond these neither unique nor unusual aspects of consumer acceptance is a concern about the use of irradiation in treating foods. This concern centers on the safety in consumption of irradiated foods.

In the record of food irradiation development there are instances of consumer or consumer group opposition to irradiated foods. There is likewise record of consumer acceptance of irradiated foods. It is clear that many factors are involved in the determination of the consumer attitude toward the use of radiation in treating foods. Perhaps the overriding aspect is consumers' fear of radiation itself and their lack of understanding of the process of food irradiation.

A 1984 study of consumer reaction to irradiation conducted on a nationwide basis in the United States demonstrated that only one in five consumers had any prior knowledge of food irradiation. The data of this study, however, showed that concern for safety had little to do with awareness of food irradiation. It is clear that acceptance of irradiated foods for some consumers will require action to convince them of the safety of irradiated foods. Doing this can be an important aspect of the commercialization of irradiated foods.

Whether irradiated foods need be identified as such to the consumer by some form of labeling has been a controversial issue. Proponents for labeling have included consumers and government regulatory agencies. Opponents have been the commercial organizations which would be concerned with the sale of the irradiated foods and who see marketing problems in a mandatory labeling requirement. The latter group prefers to have labeling at their option, to be used, if they wish, for promotional purposes.

If some form of labeling is used, just how to do this also has been the subject of controversy. Essentially two approaches have been considered: (1) the use of a scientifically accurate descriptive phrase such as "treated with ionizing radiation" or (2) the use of a coined term such as "Radura," possibly used with an identifiable symbol.*

It would seem that labeling merely to provide identification as an irradiated food is insufficient. Additionally, the consumer must be given other information, including evidence of safety. Since it is impossible to do this with a food label, some form of appropriate consumer education and information transfer to concerned consumers is needed. Figure 15.1 shows the findings of the cited 1984 study on consumer reaction to irradiation, which indicate the positive advantages consumers perceive in food irradiation and which would

* A symbol, shown below, has been devised which has gained international acceptance.

"Irradiated food symbol denoting food–agricultural product-plant in closed package—circle irradiated by penetrating rays." In "Introducing Irradiated Foods to the Producer and Consumer: Peaceful Uses of Atomic Energy" (R. L. Ulmann, ed.), Vol. 12, p. 299. Int. Atomic Energy Agency, Vienna, 1972.

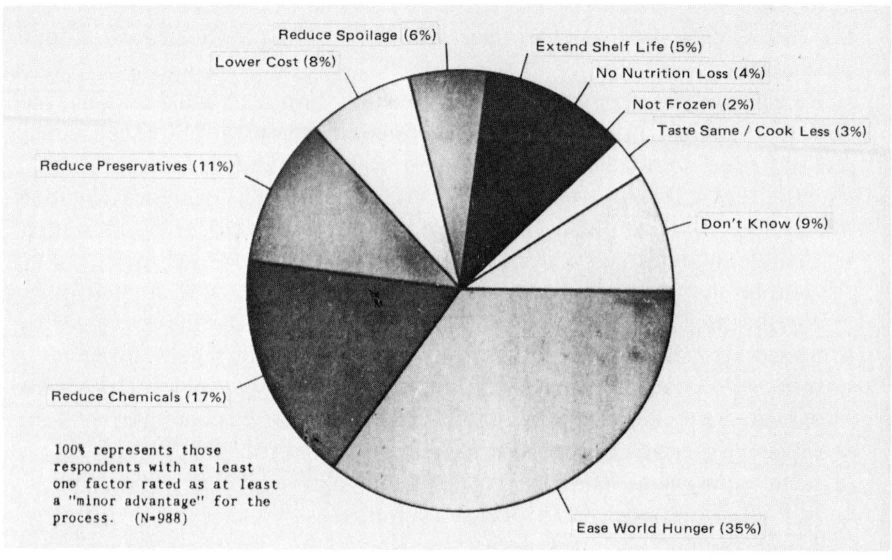

Figure 15.1 Results of a study of U.S. consumer attitude toward food irradiation, indicating the findings regarding certain positive values for food irradiation as seen by consumers. From "Consumer Reaction to the Irradiation Concept." U.S. Dept. of Energy, Albuquerque, New Mexico and Natl. Pork Producers Council, 1984.

justify its use in their view. This study also indicates that consumers have greater concern for chemicals in foods than for food irradiation.

A 1984 Canadian study on the marketability of irradiated fish and seafood yielded similar findings on consumer attitudes, as did a 1973 study in the Netherlands.

Beyond the need to obtain consumer acceptance of the wholesomeness of irradiated foods, there are other considerations regarding consumer acceptance. A review of the potential applications of irradiation shows that most often irradiation does not change foods in ways that are discernible by the consumer. In fact, success in obtaining such a result constitutes a very important aspect of the utility of irradiation. It does, however, obviate claims for irradiated foods being better than nonirradiated foods. In such situations consumer acceptance or preference for irradiated foods would be based on ordinary factors of the market, such as availability, cost, and personal choice. Of these, availability could be very important. For example, radurized seafood available in inland markets as fresh items, rather than frozen or otherwise preserved, would be expected to gain consumer preference, even at some what higher costs.

In some cases radiation-induced changes can be noted by expert panels. Whether such changes are significant in consumer acceptance is not clear. U.S.

Army troops have indicated acceptance of radappertized meats, poultry, and seafood items, as indicated by the data of Table 6.14. (For military foods a rating of 5 or higher on the hedonic scale is acceptable.) Yet expert panels identify a slight "irradiated flavor" in some of these products. Whether acceptance similar to that obtained with army troops will be secured in a civilian market has not been ascertained. In this connection it should be noted that a number of processed foods which have had good consumer acceptance do, in fact, show sensory or other differences from unprocessed counterparts. In view of this, it would appear that similar differences due to radiation processing are not necessarily a hindrance to consumer acceptance. The evaluation of such differences undoubtedly needs to be done in the market.

With certain foods an improvement of some characteristic of a food due to irradiation is discernible to the consumer. Radappertized meats, for example, display better texture than thermally sterilized meats. Such differences can be a basis for greater consumer acceptance of irradiated foods.

A number of test marketings of certain irradiated foods have been conducted in various countries. Additionally, irradiated foods are being marketed commercially in several countries. The results of these activities point to acceptance of irradiated foods by consumers. On the whole, there appears to be a good basis for optimism in anticipating general consumer acceptance of irradiated foods.

II. ECONOMICS OF FOOD IRRADIATION*

A. Introduction

The principal incentive for the commercial use of food irradiation is the opportunity for making a profit. As a part of its activities in securing a profit, business management must know the costs that are involved in producing what it sells. Determination of irradiation costs, therefore is important.

The determination of costs for food irradiation basically is the same as for other food processes. The simplicity of the irradiation process and the availability of knowledge that enables sound engineering, design, and fabrication of irradiation facilities permit the use of straightforward costing

* The treatment of the subject of the economics of food irradiation has been taken largely from the paper Economic Feasibility of Radiation Insect Disinfestation of Foods by W. M. Urbain. *In* "Radiation Disinfestation of Food and Agricultural Products" (J. H. Moy, ed.). Univ. of Hawaii Press, Honolulu, 1985. Proceedings International Conference on Radiation Disinfestation of Food and Agricultural Products, Honolulu, 1985. This use of the cited paper is done with the permission of J. H. Moy, copyright owner.

procedures with no unusual problems.

As is customary, costs may be divided into two kinds: (1) capital or fixed costs and (2) operating or variable costs. Each kind is treated separately here.

B. Capital Costs

Capital costs are associated with the securing of the irradiation facility and other necessary associated facilities. Capital costs are incurred prior to the processing of product. They are recovered by apportioning some fraction of them to each product unit that is processed. How this apportionment is done is determined by the commercial management and, to some degree, by applicable tax laws. The size of the capital costs and the rate of their recovery significantly affect product unit costs.

Capital cost items that have been identified for food irradiation are shown in Table 15.1.

Primary in determining capital costs is the plant size or capacity. Capacity is a function of many factors, some of which interact. Factors which determine capacity are listed in Table 15.2.

The plant capacity relates to the number of units of product needed in some time period in order to meet business requirements. In general the plant production capability should be matched as closely as possible with demand for product. Unused capacity adds to the cost of products irradiated. On the other hand, inadequate capacity can cause lost business.

In order to obtain adequate low product costs the facility must be of a certain minimum size. Above this minimum size, lower product unit costs can be secured, up to a point where increasing plant capacity produces little or no further cost reduction.

A very important variable in sizing a plant is the number of hours it is operated per year. The greater the number of hours, the smaller the plant required to produce a given amount of product. Continuous, round-the-clock, daily operation might seem most effective. Certain circumstances, however, may limit the hours of operation of an irradiation facility. For example, some foods are available for irradiation only on a seasonal basis. Some foods may be so perishable as not to permit storage prior to irradiation. Processing done prior to irradiation may complicate the phasing in of the irradiation step. Transport of product from producing areas to the irradiator may make difficult the efficient scheduling of irradiation. Especially difficult from a cost standpoint are applications calling for a high-volume throughput in a relatively short time period. In such cases the capacity of the facility needs to be appropriately high. If, as a consequence, the facility is used only intermittently, it may be difficult to cover the relatively high capital costs associated with the high capacity.

Table 15.1

Capital Cost Items for Food Irradiation

I. Radiation source
   - Radionuclide or machine
   - Transport to facility
   - Installation
   - First year's decay (radionuclide)

II. Irradiation facility
   - Cell
   - Radionuclide storage and transport system
   - Product conveyor
   - Cell ventilation and cooling
   - Controls
   - Laboratory (dosimetry)

III. Ancillary facilities
   - Receiving
   - Storage (dry)
   - Storage (refrigerated)
   - Storage (frozen)
   - Processing
   - Shipping
   - Maintenance
     1. Shop
     2. Parts storage
   - Office
   - Other

IV. Planning and design engineering

V. Site (land)

VI. Site preparation

VII. Building
   - Construction
   - Modification

VIII. Current working capital (to finance receivables, inventories, prepayments, current operations)

IX. Contingency

Table 15.2

Factors Relating to the Capacity of a Food Irradiation Facility

Product quantity requirement
Hours of operation (per year or other time period)
Product density and dimensions
Dose magnitude and range
Source efficiency
Product transport method

One way to increase usage of a facility is to use it to irradiate more than one kind of product. This can be highly effective but can involve problems of scheduling, of covering a range of different doses, and of cross-contamination among the different products. These kinds of problems require resolution in using a single facility for irradiating a multiplicity of products.

Notwithstanding the problems mentioned, there is real value in operating an irradiator as many hours per year as possible. As long as other costs or other factors do not prevent this, it is an objective to pursue.

For a given product volume the dose largely determines the amount of radiation that must be provided. The larger the dose the greater the amount of radiation needed. In part, the amount of radiation is related to the radiation source size. Hence, capital costs are related to dose. Considering the range of doses (0.05–50 kGy) that possibly can be employed in all the uses of food irradiation, dose is a principal cost variable.

Since dose is specified as energy absorbed per unit weight, the density of the product also needs to be considered with regard to plant capacity. Most foods have densities close to unity, but are not always presented to the irradiator in a way to provide an "effective" or "bulk" density of unity. For example, in a box of oranges there are voids among individual oranges which reduce the effective density. This can reduce the plant capacity as measured by the weight of product irradiated.

Source efficiency plays a major role in determining the size of the radiation source required. The more efficient the source, the smaller it can be. Efficiency may be defined as the percentage of radiation produced by the source that is absorbed by the target material. Source efficiency is related to a number of factors. Because γ rays cannot be focused or otherwise directed, the efficiency of γ-ray radionuclide sources generally is below 40%. Since electron beams, and to some extent X rays also, can be directed, higher efficiencies can be obtained with these sources.

Source design is a major factor in efficiency. For radionuclide sources, the shape of the source material and the configuration of source component assembly greatly affect efficiency.

In order to accomplish the desired result of irradiation a particular minimum dose is required. Since it is impossible to deliver an exact dose, some variation must be accepted. This results in an average dose somewhat above the needed minimum. Generally reduction of the allowable dose range causes reduced source efficiency. On the other hand, a dose larger than the minimum needed to produce the desired effect involves wasted radiation and thereby reduces the efficiency of the operation. The balancing of these opposing effects of dose range is necessary in order to secure the greatest source efficiency that is compatible with other factors.

How the product is moved past the source affects its efficiency, particularly in the case of γ- and X-ray sources. γ and X rays are much more penetrating than electron beams and consequently require greater depths of target material to secure good absorption. Multiple passes of product or "stacking" of product units to obtain greater product depth generally leads to increased absorption. Turning the product between passes can reduce the dose variation secured and in this way improve source efficiency. Because of the limited penetration of electron beams, turning of the product, or its equivalent, with electron beam sources is especially useful for reducing the dose range.

The use of a facility for a multiplicity of products can require delivery of a number of different doses. This can result in reduced efficiency for some doses. Gains obtained in multiple use of a facility may need to be balanced with higher costs due to reduced efficiency.

Machine sources generally have specified capacities. Usage of these kinds of sources below their maximum capacity may be regarded as a kind of inefficiency. It results in higher irradiation costs due to partial use of the installed capacity whose full capital cost must be covered.

Radionuclide sources, on the other hand, have variable capacities related to the amount of radionuclide present. γ-Ray sources can be fitted quite well to capacity requirements. At the start-up of an operation, an initial loading can be made in order to provide an amount of radiation for a selected throughput. As product demand increases, additional radionuclide can be installed, up to the designed maximum capacity. In this way greater efficiency can be secured in the earlier stages of operation.

In the list of capital cost items it can be noted that many are similar to or identical with those for other food-processing facilities. Some, however, are specific to a food irradiation facility. Some of these latter involve selection and specification of particular variables as applicable to the facility usage at hand. All items, nonetheless, are amenable to regular costing procedures. As might be surmised from the listing, the total capital cost of an economically feasible facility must be regarded as fairly high.

C. Operating Costs

Items that have been identified as operating costs are listed in Table 15.3. For a particular accounting period the operating cost total is apportioned over the units of product produced in that period.

With the exception of item 2 (source replenishment), the items of the operating costs listed are similar to those for the operation of any food-processing facility. Item 2 is unique to food irradiation and applies only to radionuclide sources. For ^{60}Co about 13% per year replenishment is needed to

Table 15.3

Operating Cost Items for Food Irradiation

1. Depreciation of facility
2. Source replenishment (radionuclides only)
3. Utilities (power, water, heat, air conditioning, etc.)
4. Labor (manager, office personnel, technicians, engineer, general labor, building maintenance, etc.)
5. Maintenance
6. Supplies
7. Taxes and insurance
8. Interest on borrowed funds
9. Return on equity capital
10. Adjunct requirements

make up for radioactive decay. Items 3 and 4, due to the nature of the irradiation process, should be relatively low in cost.

Item 1 (depreciation) covers the amount of capital investment initially made that is to be recovered as an operating cost in the accounting period covered. From a business operation viewpoint, the magnitude of this item is subject to management determination. Accounting for tax purposes may be on a different basis, as dictated by tax laws. It is possible that different items of the capital cost (Table 15.1) may be handled at different depreciation rates.

Item 9 (return on equity capital) covers the expectation of the investor for the use of his or her money. This item has not always been included in cost estimates for food irradiation. It is different from depreciation cost (item 1). Its magnitude is determined by commercial management requirements. It must be regarded as an important part of irradiation costs. It is likely that in the early usage of food irradiation when there is little actual experience available to guide business management, a relatively high rate of return may be required to serve as compensation for risking capital with something new.

The irradiation of a food is likely to be only one step in its total processing. The use of irradiation conceivably can impose special requirements whose costs must be included in a tabulation of irradiation costs. Such costs are those of item 10 (adjunct requirements). Examples of adjunct costs are (1) special packaging without which irradiation would be ineffective, (2) special product storage facilities, such as may be needed to prevent insect reinfestation, (3) transportation costs for moving the product to and from the irradiator. Depending upon what these adjunct costs are, they can be significant in the total operating costs for irradiation.

D. Economic Feasibility

Considerations discussed so far have been on a general basis and have enabled identification of the various factors which are involved in the costs of irradiation. In a particular situation it should be possible to arrive at the actual cost. These figures, combined with information for those other items that apply such as raw material availability, market potential for the irradiated product, technical feasibility of the radiation treatment, comparative figures for procedures which could be used in place of irradiation, and legal aspects, would enable a decision on commercial viability of an application. It is by dealing with a specific application that arrival at true costs and other business-related factors can be secured. Without this specific knowledge, cost estimates are largely meaningless.

It would be helpful to have available cost figures from actual commercial experience with food irradiation; but food irradiation so far has not had extensive commercial usage. The experience gained in the commercial radiation sterilization of medical supplies can be helpful to a degree, but in several ways this operation does not parallel food irradiation. Information is available on at least one commercial food irradiation application, namely, the irradiation of white potatoes to inhibit sprouting at the Shihoro Agricultural

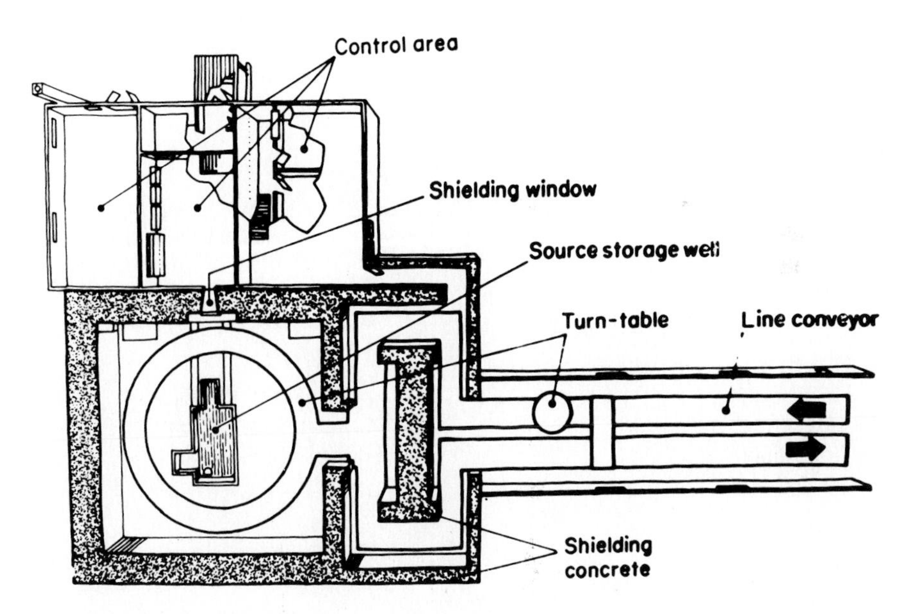

Figure 15.2 Plan of commercial potato irradiator. Hokkaido, Japan. Courtesy of Kawasaki Heavy Industries, Ltd.

Cooperative in Shihoro, Hokkaido, Japan. Commercial operations were started in 1973 and continue today. A description of the facility and details of the costs involved have been made available.

Figure 15.2 shows a floor plan of the irradiator, and Fig. 15.3 is a picture of the irradiation cell interior. Table 15.4 provides descriptive details of the facility and process. Table 15.5 gives the construction cost. (Construction was completed in 1973.) Table 15.6 reports total operating costs for the last 3 years reported along with tonnage of potatoes irradiated. Also shown is the irradiation cost per ton which was slightly greater than $12 (U.S.). The facility was operated only 3 months each year. In the down period labor costs, except for management staff, were not charged. The stated capacity of the plant is 30,000 tons per month, but was not fully secured in any of the years reported.

The Japanese government contributed $866,000 to the construction cost. In the reported figures, this sum was not included in the depreciation item of the operating costs. Its inclusion would have increased the operating cost, possibly of the order of $7 per ton.

Figure 15.3 Interior of commercial potato irradiator. Hokkaido, Japan. Courtesy of Kawasaki Heavy Industries, Ltd.

Table 15.4

Information on Shihoro Potato Irradiator[a]

Item	Detailed information
Radiation source	300 kCi ^{60}Co (pellets)
Target container	0.98 × 1.6 × 1.3 m (1.5 tons of potatoes)
Absorbed dose	0.06–0.15 kGy
Capacity	15 tons/hr
Processing time	2 hr (two-sided irradiation)
Source efficiency	~9.5%

[a] From K. Umeda, Commercial experiences with the Shihoro potato irradiator. *Food Irradiat. Newsl.* 7(3), 19 (1983).

Table 15.5

Construction Cost of Shihoro Potato Irradiator[a]

Item	Cost (U.S. dollars)
Building	$379,900
Equipment	620,000
Source (300 kCi ^{60}Co)	170,000
Other	127,800
TOTAL	$1,297,700

[a] From K. Umeda, Commercial experiences with the Shihoro potato irradiator. *Food Irradiat. Newsl.* 7(3), 19 (1983).

Table 15.6

Operating Costs of Shihoro Potato Irradiator[a]

	Year		
Cost factor	1976–1977	1977–1978	1978–1979
Annual cost	$180,276	$198,680	$177,510
Volume (tons)	14,442	15,435	14,604
Cost (U.S. dollars per ton)	12.48	12.87	12.15

[a] From K. Umeda, Commercial experiences with the Shihoro potato irradiator. *Food Irradiat. Newsl.* 7(3), 19 (1983).

Figure 15.4 shows an estimate of cost versus source capacity for sprout inhibition (a dose of ≤ 0.25 kGy), for radurization (a dose of ≤ 2.5 kGy), and for radappertization (a dose of ≤ 45 kGy). This estimate was made in 1968 and the actual absolute costs (cents per pound) are outdated. The figure does show, however, the relative significance of dose and plant throughput in affecting costs. Dose and throughput are the dominant cost factors. The foregoing discussion of this chapter, however, points out that many other factors are important. Meaningful cost estimates can be secured only through a detailed cost analysis which follows the general approach outlined in this chapter.

A summary statement on costs may be as follows: Operating costs for irradiation generally fall in a range as to be commercially feasible. Fixed costs are moderately high and require fairly large product volumes for their support. A substantial capital investment for an irradiation facility is needed.

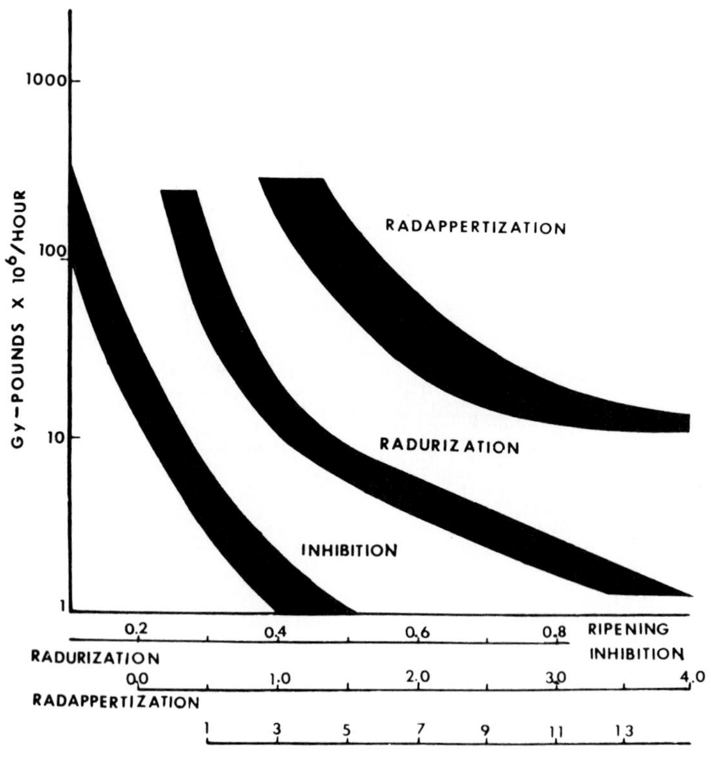

Figure 15.4 Irradiation cost estimates as related to product volume for low-, intermediate-, and high-dose applications. From "The Commercial Prospects for Selected Irradiated Foods," U.S. Dept. of Commerce, Washington, D.C., 1968.

III. IRRADIATION PROCESSING

A. Introduction

Food irradiation facilities have three basic components: (1) the radiation source, (2) a transport system to move the food to and from the source, and (3) a cell structure to confine the radiation so as to provide protection from the radiation for operating personnel. In addition, there can be certain ancillary facilities such as control systems, receiving and shipping facilities, pre- and postirradiation storage areas, laboratory and maintenance facilities, and office space. A typical irradiation plant layout is shown in Fig. 15.5.

Facilities may be built to handle a variety of products with different doses. In such cases it may be necessary to accept reduced efficiency in order to gain the needed flexibility. Consideration of the possibility of cross-contamination among different products passing through the facility may be necessary in some cases, and this may impose limitations on what products may be handled in one facility. In situations where sufficient volume of one type of product is available, facilities can be built specifically for that type of product. Such

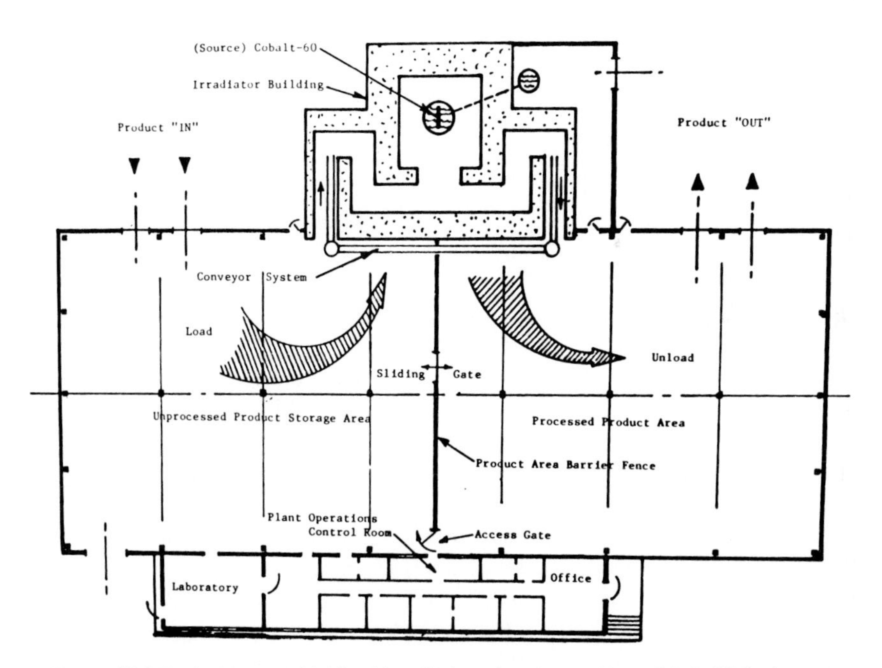

Figure 15.5 Typical commercial food irradiation plant layout. From M. A. Welt. A commercial multipurpose radiation processing facility for Hawaii. *In* "Radiation Disinfestation of Food and Agricultural Products" (J. H. Moy, ed.). Univ. of Hawaii Press, Honolulu, 1985.

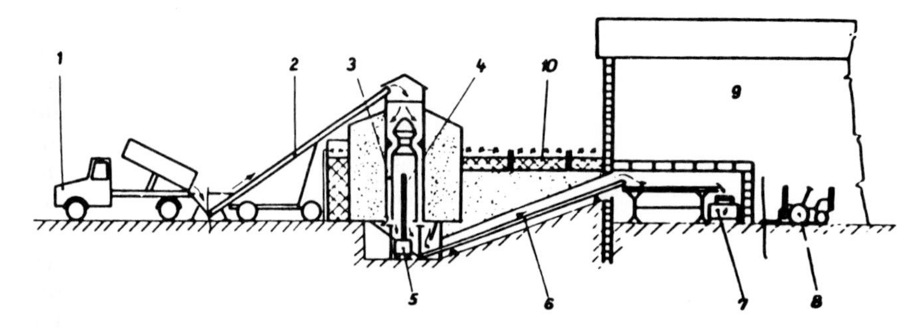

Figure 15.6 Commercial irradiation plant for sprout inhibition of onions. 1, Delivery, truck; 2, incoming conveyor belt; 3, irradiator; 4, radiation source (50,000 Ci ^{60}Co); 5, source container; 6, outgoing conveyor belt; 7, storage crate for irradiated onions; 8, forklift; 9, storage room. From B. Kalman, E. Kedessi, and R. Santa, Introduction of the irradiation technology into the Hungarian food industry. *In* "International Symposium on Food Irradiation Processing, Proceedings." Washington. Int. Atomic Energy Agency, Vienna, 1985.

specialization usually accomplishes greater overall facility efficiency. An irradiator built specifically to irradiate onions is shown schematically in Fig. 15.6. See also Fig. 9.2, which shows an irradiator for irradiating cereal grains.

B. Irradiation Facilities

1. Machine Sources

a. Electron Beam Generators

The principal machine sources have provided electron radiation. X-Ray machines, however, have been available and are likely to become more important. Both electron and X-ray machines utilize electrons, to which energy is given by application of electric fields. The acceleration caused in this way leads to the acquisition of a considerable amount of kinetic energy by the electrons. The machines which accomplish this usually are referred to as accelerators.

A number of different ways have been devised to obtain the needed electric fields for electron acceleration. One such is shown schematically in Fig. 15.7. In this machine, the Dynamitron, incoming alternating current electric power is converted to high-voltage direct current through a series of steps, including conversion to radio frequency (RF) power, raising of voltage, rectification, and series connection of a number of such units to yield a final high voltage, which may be between 400 kV and several million volts. This voltage is applied to an

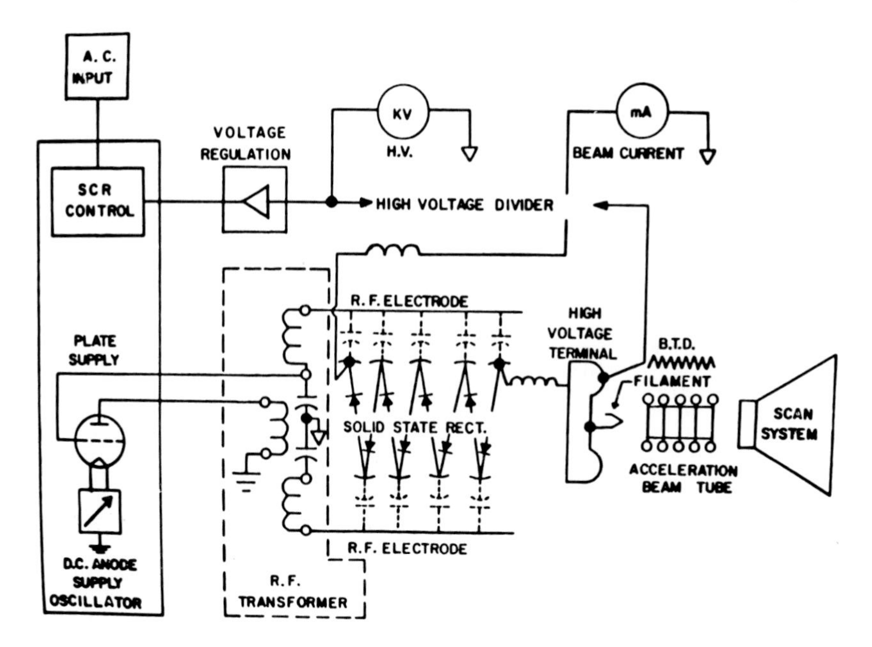

Figure 15.7 Dynamitron electron beam generator—schematic. Courtesy of Radiation Dynamics, Inc., Melville, New York.

evacuated accelerator "beam tube." In this tube electrons emitted by a heated filament are accelerated by the applied voltage and emerge from the tube through a thin metallic window as an intense "pencil beam." Application of a constantly changing magnetic field causes the beam to "scan" the target material in a fan-shaped configuration. A cutaway portrayal of the Dynamitron is shown in Fig. 15.8.

The direct application of a high-voltage field, such as just described, is limited to a maximum acceleration voltage of about 4.5 MV due to electrical breakdown of insulating components at higher voltages. This limitation is overcome in the linear accelerator (linac).

In the linear accelerator, clusters or pulses of electrons acquire energy from the action of an electric field applied in phase as a traveling electromagnetic wave to a series of separate chambers (drift tubes) of the accelerator tube. Each module adds kinetic energy to the electrons and, by providing a sufficient number of such modules, the beam energy can be raised to any desired level. This approach avoids the use of the very high voltage associated with the use of the full electric potential for acceleration. There is virtually no limit, therefore, to the energies that can be given to the electron beam. The electron beam from a linac is essentially monoenergetic and is made up of a series of

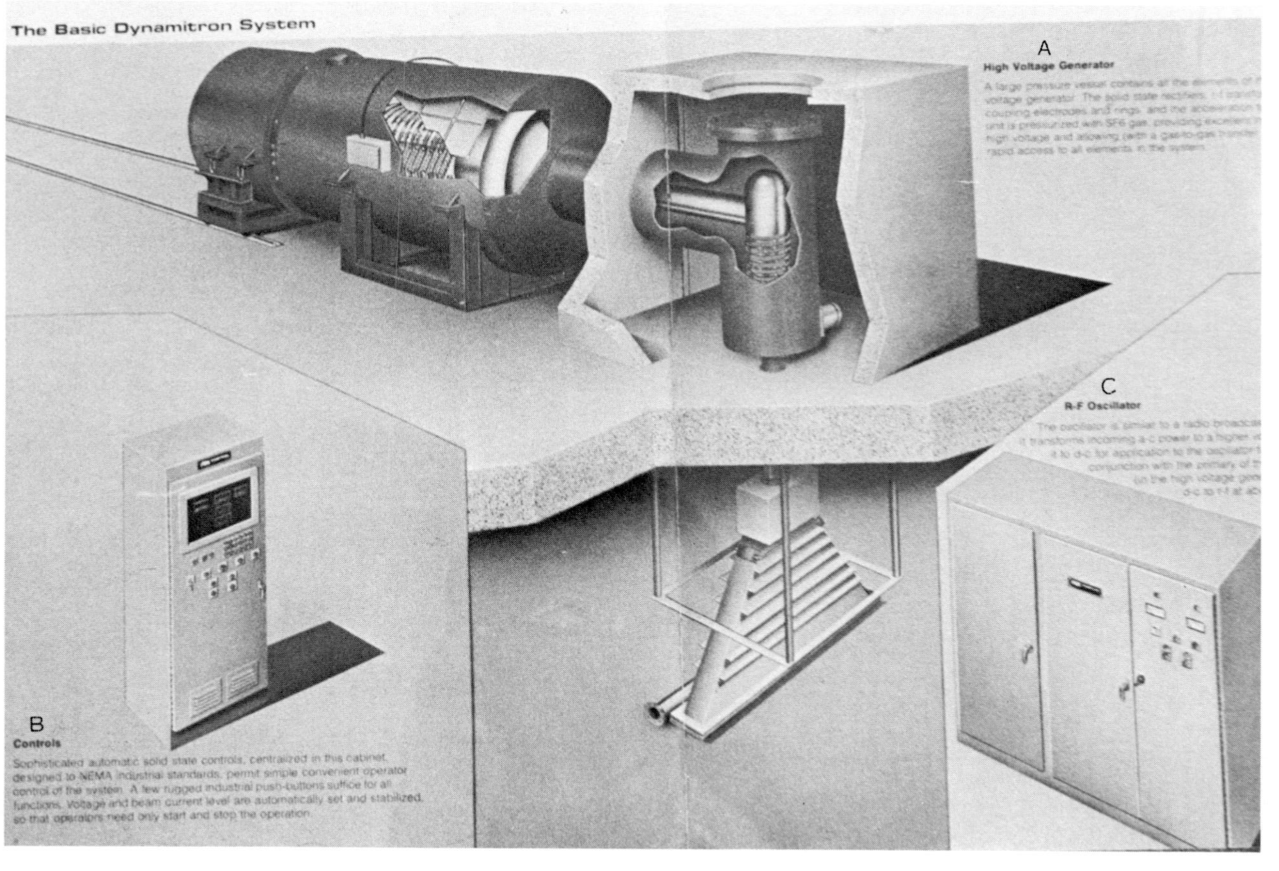
The Basic Dynamitron System
A
High Voltage Generator
B
Controls
Sophisticated automatic solid state controls, centralized in this cabinet, designed to NEMA industrial standards, permit simple convenient operator control of the system. A few rugged industrial push-buttons suffice for all functions. Voltage and beam current level are automatically set and stabilized, so that operators need only start and stop the operation.
C
R-F Oscillator

Figure 15.8

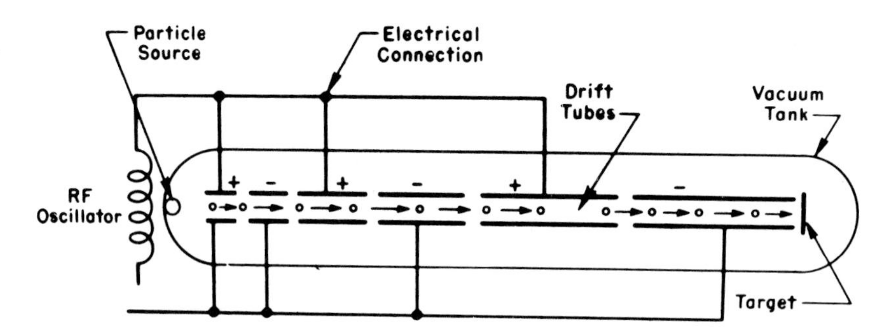

Figure 15.9 Linear accelerator—schematic.

pulses. The beam may be deflected magnetically in order to scan the target material. A schematic diagram of a linear accelerator is shown in Fig. 15.9.

An improved linear accelerator employs a linear array of nonresonant accelerating cavities which operate on the principle of magnetic induction. Higher beam currents and more stable operation are obtained with such a linac.

In order to secure 10-MeV electron beams it is necessary to use a linear accelerator, since other electron accelerators cannot produce beams of this energy. As has been noted (Chapter 1), electron radiation has limited penetration ($\sim$0.5 cm/MeV for water-equivalent material). The practical usable depth limit for 10-MeV electrons in water-equivalent material is 3.9 cm (see Fig. 1.7).

This provides a dose uniformity ratio U of close to 3:1. The spread between maximum and minimum dose can be reduced by two-sided irradiation, as illustrated in Fig. 15.10. In addition, the penetration depth can be doubled to about 8 cm.

Figure 15.8 Dynamitron electron beam generator system. (A) High voltage generator. A large pressure vessel contains all the elements of the high voltage generator: the solid state rectifiers, r-f transformer, coupling electrodes and rings, and the acceleration table. The unit is pressurized with SF_6 gas, providing excellent insulation of high voltage and allowing (with a gas-gas transfer system) rapid access to all elements in the system. (B) Controls. Sophisticated automatic solid state controls, centralized in this cabinet, designed to NEMA industrial standards, permit simple convenient operator control of the system. A few rugged industrial push-buttons suffice for all functions. Voltage and beam current level are automatically set and stabilized so that operators need only start and stop the operation. (C) R-F oscillator. The oscillator is similar to a radio broadcasting transmitter. It transforms incoming a-c power to a higher voltage, then rectifies it to d-c for application to the oscillator table. The table, in conjunction with the primary of the R-F transformer (in the high voltage generator), converts the d-c to r-f at about 100 kHz. Courtesy of Radiation Dynamics, Inc., Melville, New York.

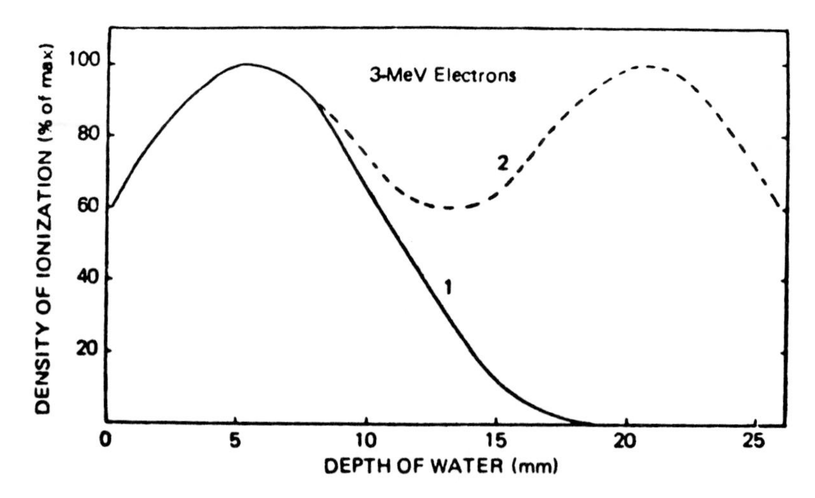

Figure 15.10 Density of ionization at various depths of water irradiated with 3-MeV electrons. Curve 1, irradiated from one side; curve 2, irradiated from the two opposite sides. From "Training Manual on Food Irradiation Technology and Techniques," 2nd Ed. Int. Atomic Energy Agency, Vienna, 1982.

b. X-Ray Generators

When high-energy electrons strike a material, X rays are emitted. The fraction of the electron energy converted to X rays is governed by the initial electron energy and the atomic number of the stopping material, as may be seen from the data of Table 15.7. The overall low efficiency of X-ray production by available methods so far has led only to experimental use of X-ray sources for food irradiation. Nonetheless, a comparison of costs indicates that X-ray irradiation can be competitive with radionuclide γ-ray irradiation, although electron radiation is the least expensive.

Table 15.7
Percentage of Electron Energy Converted to X Rays[a]

	Energy converted to X rays (%)	
Electron energy (MeV)	Iron target (atomic no. 26)	Tungsten target (atomic no. 74)
0.5	1.34	4.77
1.0	2.31	7.63
5.0	8.20	21.17
10.0	14.40	31.78

[a] From M. J. Berger and S. M. Seltzer, NASA Spec. Publ. No. 3012 (1964) as quoted by A. Brynjolfsson and T. Martin, III., *Int. J. Appl. Radiat. Isot.* **22,** 29 (1971).

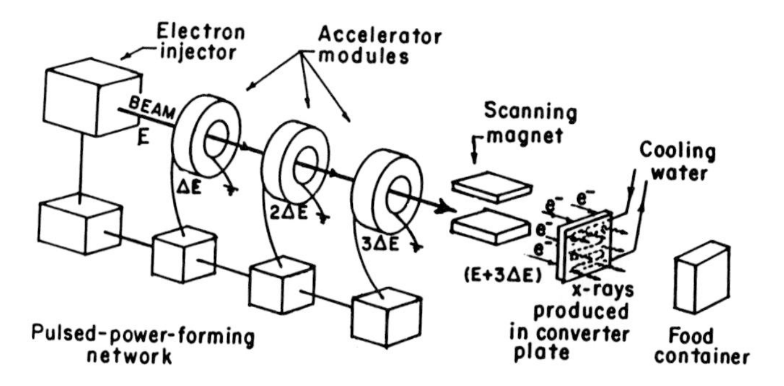

Figure 15.11 X-ray generator employing linear-induction electron accelerator—schematic. From M. C. Lagunas-Solar, *Nuclear Instrum. Methods Physics Res.* **B10/11,** 987 (1985).

A schematic drawing of an X-ray generator employing a 5-MeV induction electron accelerator is shown in Fig. 15.11. The electron beam from the accelerator is converted to X rays at the metallic converter plate. Not all the X radiation travels in the same direction as the impinging electron beam, but this aspect is substantially better than what is obtained with a γ-ray radionuclide source and, therefore, leads to greater source efficiency. Figure 15.12 shows a comparison of the photon energy spectra for X rays from a 5-MeV electron accelerator with γ-rays from ^{60}Co and ^{137}Cs. By interposing 0.63 cm of lead in the X-ray beam, the average energy can be raised from 1.06 to

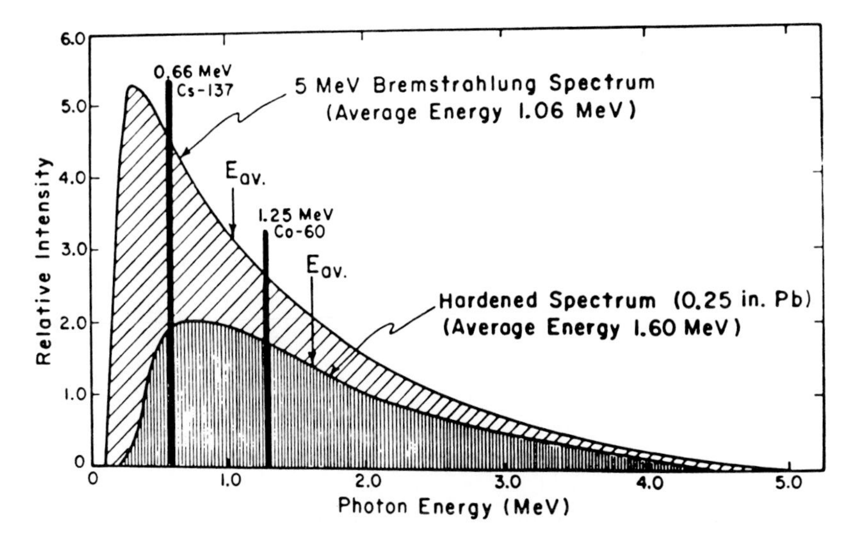

Figure 15.12 Comparison of photon energy spectra of ^{60}Co and ^{137}Cs radionuclide γ radiation with X-ray (bremstrahlung) spectra from 5-MeV electron beam. From M. C. Lagunas-Solar and S. M. Matthews, *Nuclear Instrum. Methods Physics Res.* **B10/11,** 987 (1985).

1.6 MeV. This is done at a loss of about 50% of the total energy, which can be compensated for by increasing the electron beam power. The use of the lead filter aids in reducing the depth–dose variation.

2. Radionuclide Sources

While theoretically there are other man-made radionuclides which could serve as γ-ray sources for food irradiation, only two have emerged as meeting requirements for providing suitable γ rays and also for having availability in amounts and at costs that permit consideration for commercial use. These radionuclides are ^{60}Co and ^{137}Cs. The decay patterns of each are shown in Fig. 15.13. Additional information on these radionuclides is given in Table 15.8.

^{137}Cs is a fission by-product and is available as the chloride. It is obtained by chemical separation from spent nuclear fuel. ^{60}Co is obtained by the activation of ^{59}Co by neutrons according to the following simplified equation:

$$^{59}\mathrm{Co} + n \longrightarrow {}^{60}\mathrm{Co}$$

This is accomplished in a nuclear reactor which provides a slow-neutron flux. The degree and rate of activation are related by the intensity of the neutron flux. The attainment of useful levels of activation ordinarily requires more than 1 year of exposure.

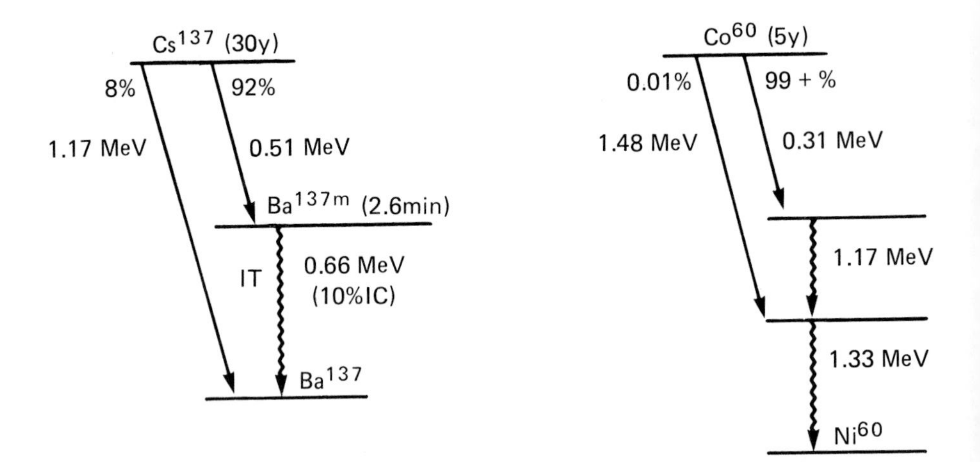

Figure 15.13 Disintegration schemes for ^{60}Co and ^{137}Cs. ↘ β^- Particle; ⌇ γ photon. From "Training Manual on Food Irradiation Technology and Techniques," 1st Ed. Int. Atomic Energy Agency, Vienna, 1970.

Table 15.8

Characteristics of ^{137}Cs and ^{60}Co

Characteristics	^{137}Cs	^{60}Co
Typical source form	CsCl pellets	Metal
Half life (years)	30	5.3
Available Ci/g	25 Maximum	400 Maximum
γ energy (MeV)	0.66	1.17 and 1.33
Power Ci/w	258	68

In order to prevent contamination of the environment, radionuclide sources are encapsulated, usually in stainless steel. Usually a double or triple encapsulation is employed. The physical form varies according to the intended use and includes slugs, pellets, rods, and strips. These primary forms are assembled in various configurations such as cylinders and placques to obtain a

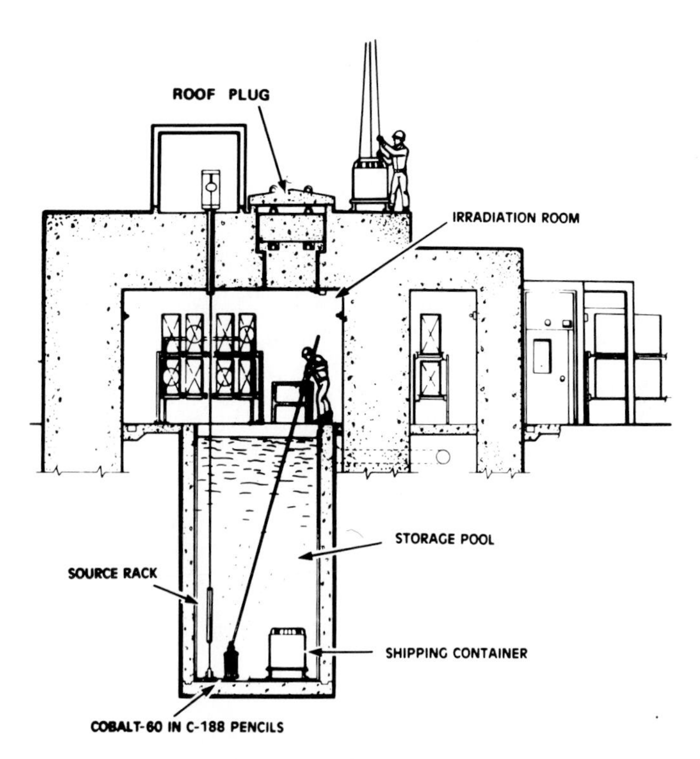

Figure 15.14 Portrayal of radionuclide (^{60}Co) source loading of a γ-irradiation facility. Courtesy Atomic Energy of Canada Ltd., Kanata, Ontario, Canada.

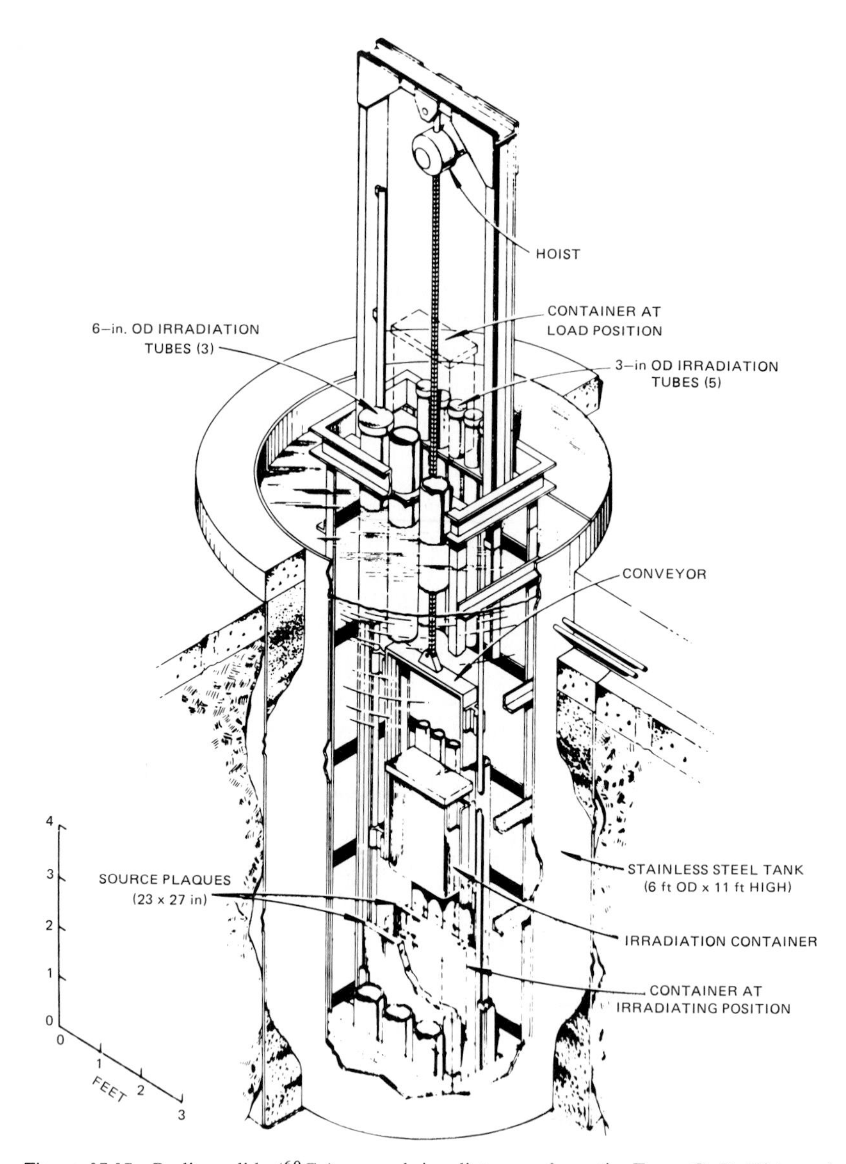

Figure 15.15 Radionuclide (^{60}Co) research irradiator—schematic. From G. R. Dietz and R. H. Lafferty, Jr., *Food Irradiat.* (*Saclay*) **6**(4), A39 (1966).

source for a facility. It is customary to store an assembled source under water when it is not in use. The protection afforded to operating personnel by a sufficient depth of water permits entry into the radiation cell area and enables safe loading and unloading of the radionuclides in the facility. Figure 15.14 portrays the loading of radionuclide source material in an irradiation facility.

A simple research irradiator employing ^{60}Co is shown schematically in Fig. 15.15. A self-contained ^{60}Co source for laboratory use, which employs lead shielding, is shown in Fig. 15.16. A much larger, more complex ^{60}Co research irradiator, which includes a product transport system operated from

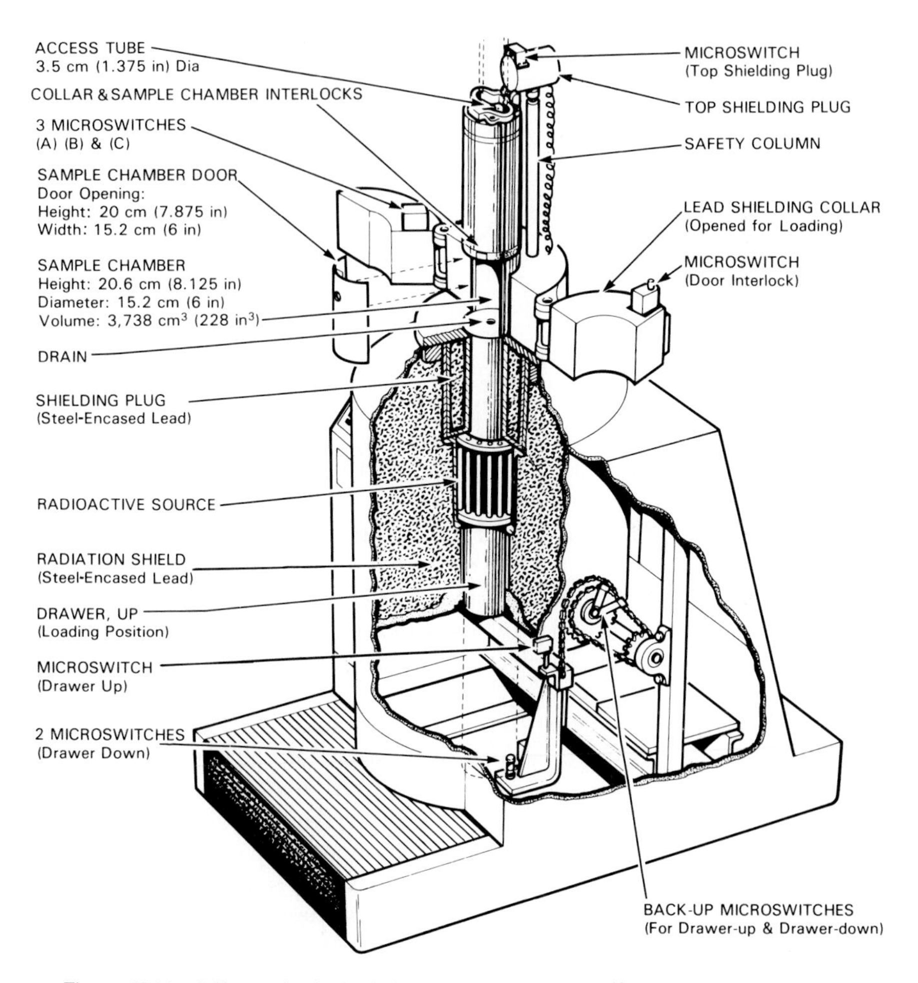

Figure 15.16 Self-contained, lead-shielded radionuclide (^{60}Co) research irradiator—schematic. Courtesy Atomic Energy of Canada Ltd., Kanata, Ontario, Canada.

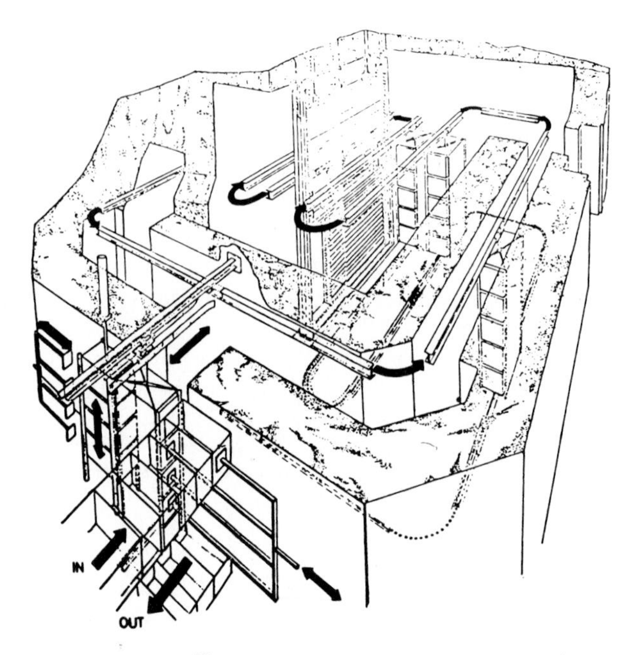

Figure 15.17 Radionuclide (^{60}Co) irradiation pilot plant—schematic. From "Training Manual on Food Irradiation Technology and Techniques," 2nd Ed. Int. Atomic Energy Agency, Vienna, 1982.

outside the radiation cell, is shown in Fig. 15.17. A research and pilot-scale ^{60}Co irradiator is shown in Fig. 15.18.

Irradiators for commercial production may be of two general types, batch or continuous. The Shihoro potato irradiator is a batch unit (see Section II,D). Several more complex commercial irradiators are portrayed in Figs. 15.19 to 15.23. These all use ^{60}Co and provide various product transport systems. A photograph of the loading and unloading area of a commercial irradiator is shown in Fig. 15.24.

The transport system of the commercial irradiator is a highly important part of the facility. Batch-type irradiators have the simplest requirements and can be operated with manual labor. Continuous irradiators move product by mechanized means and can take various forms as indicated by the facilities portrayed in Figs. 15.20 to 15.23. Depending upon the construction of the irradiator, product movement past the source can be more or less complex. Two objectives are sought: facility efficiency and, as near as possible, a value of unity for the dose uniformity ratio U. Principal methods for attaining these objectives involve positioning the target material with respect to source sequentially in a number of locations, including placing two or more product units in depth with respect to the source; turning the product units so as to get,

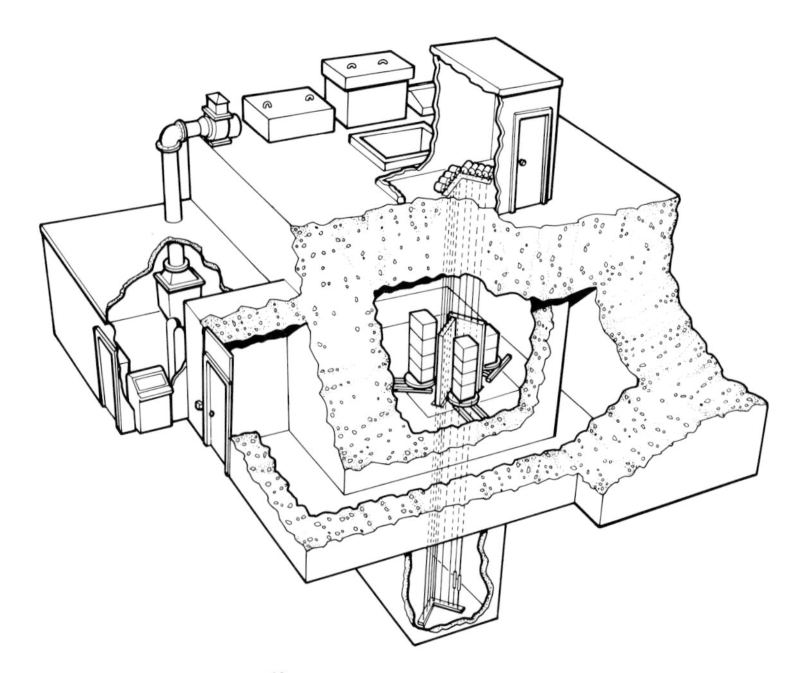

Figure 15.18 Radionuclide (^{60}Co) research and pilot-scale irradiator. Courtesy of Atomic Energy of Canada Ltd., Kanata, Ontario, Canada.

in effect, two-sided irradiation; and moving the product units through different locations both horizontally and vertically within the working area of the source. Usually each position requires a time period without movement in order to allow sufficient absorption of the radiation.

In seeking to maximize efficiency and to obtain as small as possible spread between the minimum and maximum doses, improvements in irradiator operation have been secured. One such type of improvement has been to employ computer control of the positioning of the target material. This has required a more versatile product transport system which does not depend upon one product unit pushing another unit through the system or which does not have a fixed track or overhead rail system.

^{60}Co and ^{137}Cs are not numerically equivalent sources of γ radiation. The γ photon of ^{137}Cs has an energy of 0.66 MeV, whereas ^{60}Co provides two γ photons, one of 1.17 and the other of 1.33 MeV. For this reason ^{60}Co radiation is the more penetrating, which makes it easier to obtain better dose uniformity ratios. The power output of ^{137}Cs is 3.32 kW/MCi; that of ^{60}Co is 14.84. Hence, for the same power output and same efficiency of use, 4.47 times as much ^{137}Cs is required as of ^{60}Co. The half-life of ^{137}Cs is 30 years, that of ^{60}Co 5.3 years. The annual decay rate of ^{137}Cs is 2.28%, that of ^{60}Co 12.39%. The price established by the U.S. government as of this writing for ^{137}Cs is

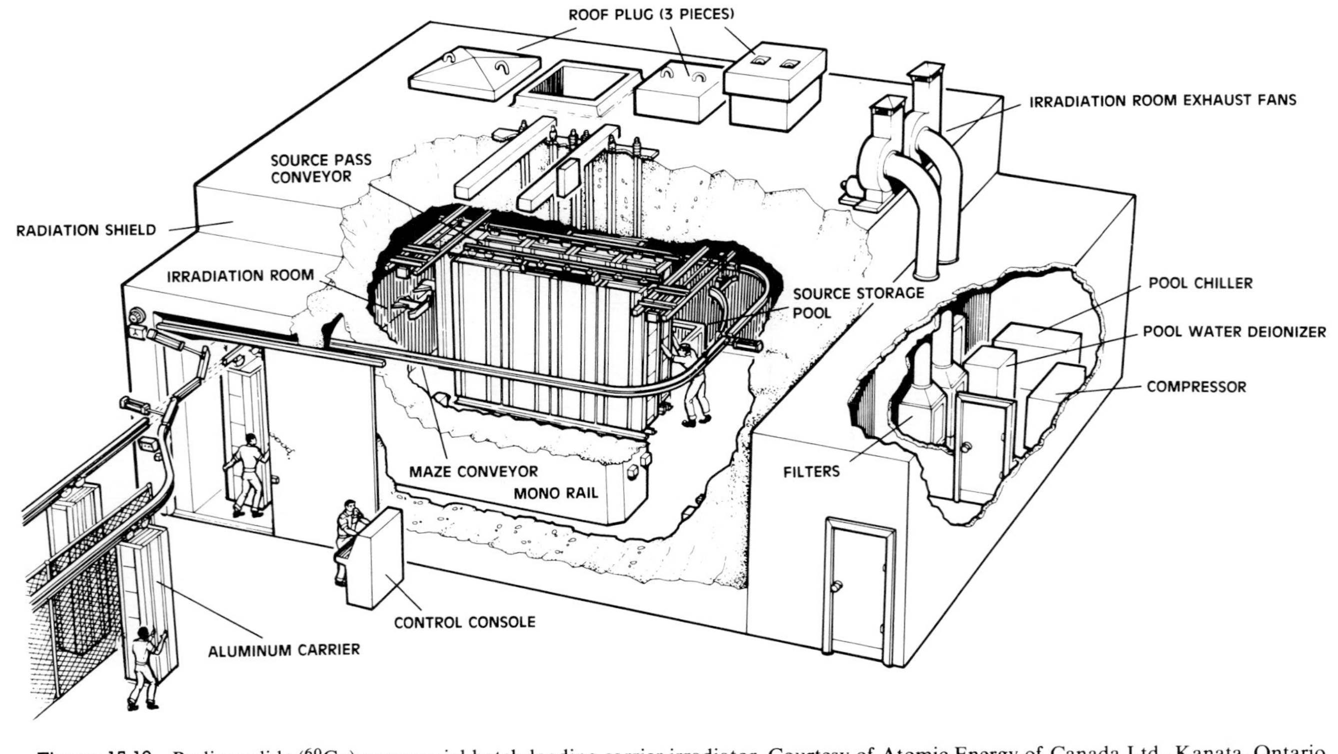

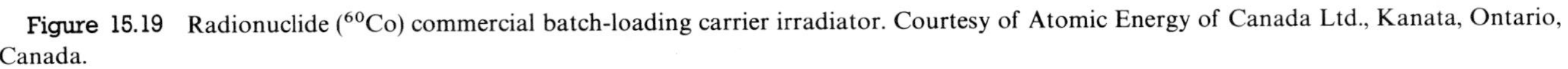
Figure 15.19 Radionuclide (^{60}Co) commercial batch-loading carrier irradiator. Courtesy of Atomic Energy of Canada Ltd., Kanata, Ontario, Canada.

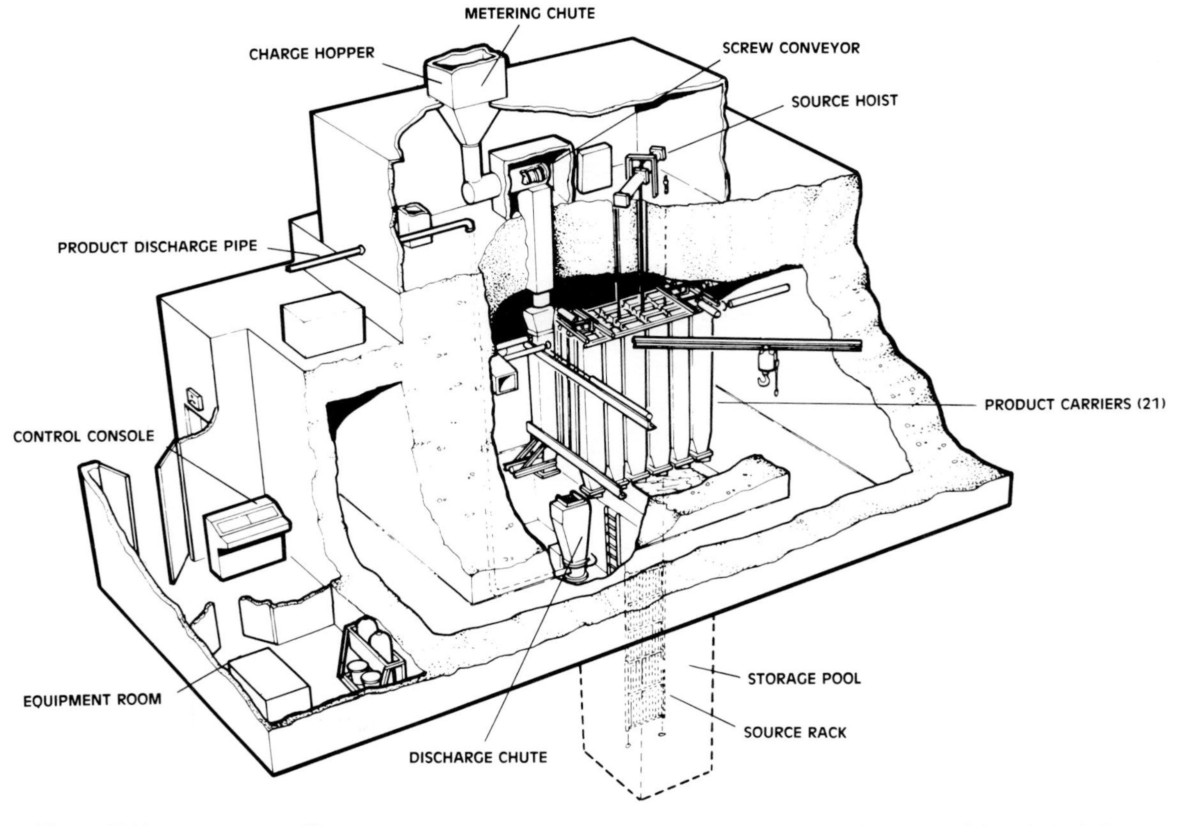

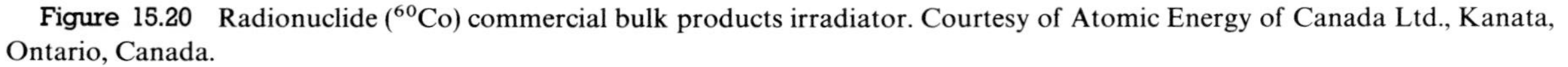

Figure 15.20 Radionuclide (^{60}Co) commercial bulk products irradiator. Courtesy of Atomic Energy of Canada Ltd., Kanata, Ontario, Canada.

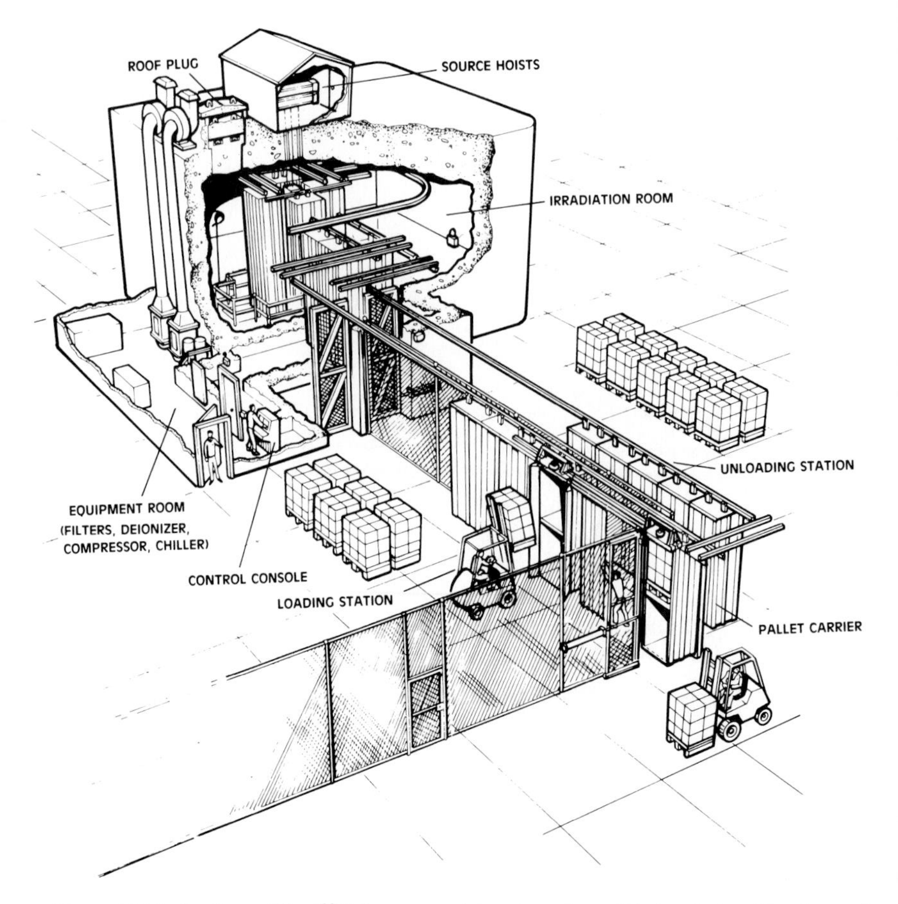

Figure 15.21 Radionuclide (^{60}Co) commercial automatic pallet irradiator. Courtesy of Atomic Energy of Canada Ltd., Kanata, Ontario, Canada.

$0.10 (U.S.) per curie. This compares with recent prices for a curie of ^{60}Co of $1.00 (U.S.). On this basis, ^{137}Cs irradiators of equal capacity are less expensive than those using ^{60}Co.

C. Dose Control

1. Introduction

The objective in irradiating a material is to deposit a specific absorbed dose in it. Consequently, for the operation of an irradiation facility there must be provided dose control measures. There are various methods for measuring

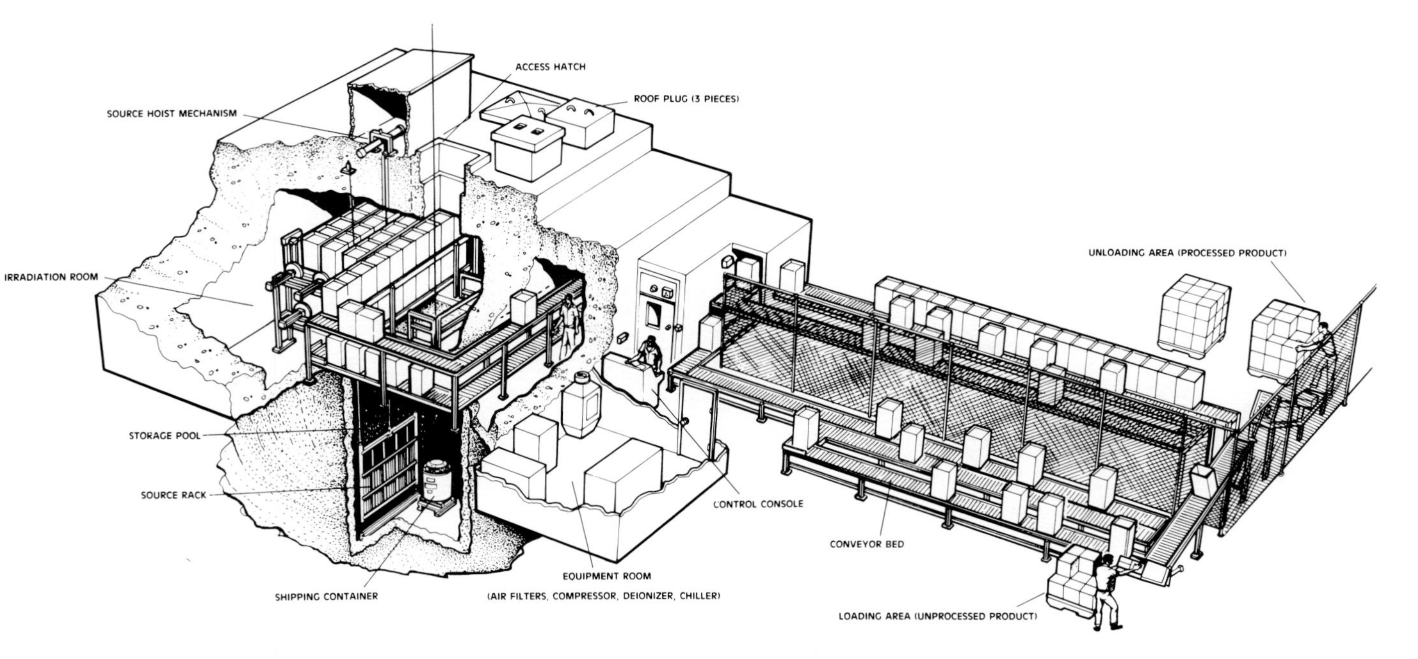

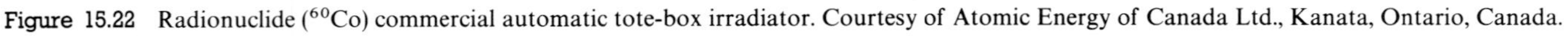

Figure 15.22 Radionuclide (^{60}Co) commercial automatic tote-box irradiator. Courtesy of Atomic Energy of Canada Ltd., Kanata, Ontario, Canada.

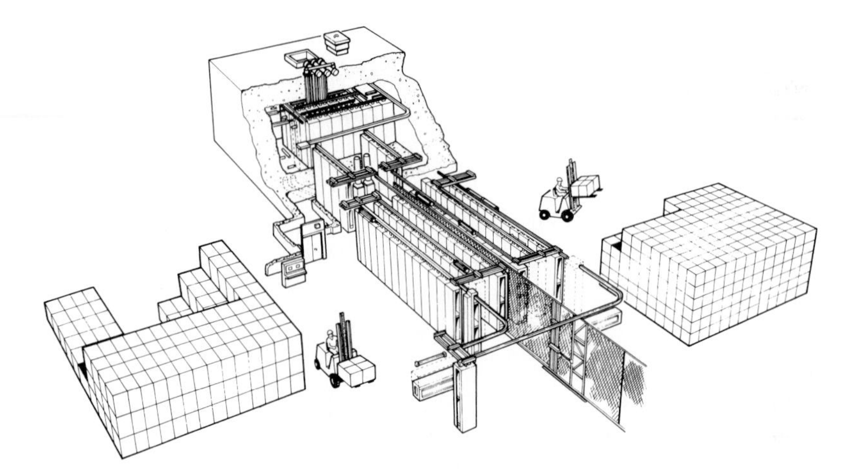

Figure 15.23 Radionuclide (^{60}Co) commercial automatic carrier irradiator. Courtesy of Atomic Energy of Canada Ltd., Kanata, Ontario, Canada.

dose. On a broad basis the methods can be classified as *physical* or *chemical*. In connection with food irradiation they may also be identified as *reference* and *routine* methods. In the latter classification reference dosimetric procedures are employed as primary methods. The routine procedures are those used on a day-to-day basis and generally are chosen partly on the basis of convenience of use.

2. Physical Methods

a. Calorimetry

Most of the absorbed ionizing radiation ($\geq 95\%$) degrades to heat and by measuring the temperature rise in a calibrated system, the dose can be determined. A number of calorimetric dose meters are available.

b. Ionization

The ionization produced by radiation can be employed to measure dose. The electroscope portrayed in Fig. 15.25 illustrates this method. The positive electrode is an insulating rod to which is attached a metal wing or "string." The negative electrode is the wall of the instrument. When the instrument is charged, the wing deflects in proportion to the amount of charge (position A, Fig. 15.25). Ionizing radiation makes the air in the instrument electrically conductive, and this dissipates the initial charge in proportion to the amount of ionization produced, causing a reduction of the deflection of the wing to a

Figure 15.24 Radionuclide (380 kCi ^{60}Co) commercial irradiator. Product loading area. Radurization (2.4 kGy) of strawberries to control *Rhizopus* and *Botrytis*. Courtesy of Radiation Technology Division, Atomic Energy Corporation of South Africa (Pty.) Ltd., Pretoria.

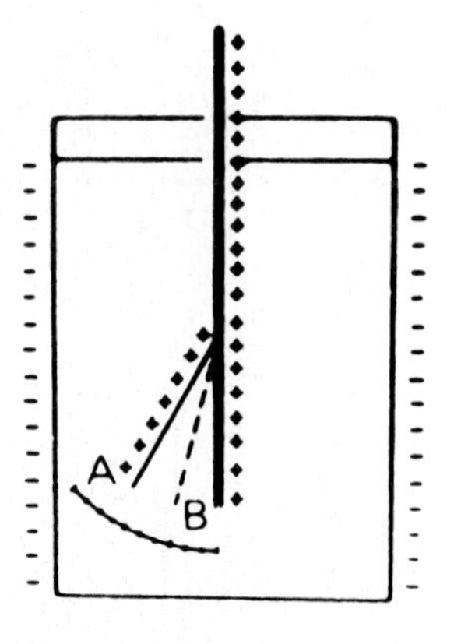

Figure 15.25 Electroscope. A and B indicate the two wing positions referred to in the text discussion of the electroscope. From "Training Manual on Food Irradiation Technology and Techniques," 2nd Ed. Int. Atomic Energy Agency, Vienna, 1982.

new position B. The deflection, properly calibrated, is an index of dose. The electroscope measures the accumulated effect of exposure over a period of time. A commercial model is shown schematically in Fig. 15.26. A commercial model of a "pocket" dosimeter is shown in Fig. 15.27A. The charging device for it is shown in Fig. 15.27B. Such a device may be used for monitoring the exposure of personnel of an irradiation facility.

Electroscopes as just described do not measure ions that recombine before reaching the electrodes. If a voltage is applied across the electrodes which is sufficiently high to discharge all ions formed, the electric current flowing between the electrodes measures the dose rate.

c. Scintillation

An incident photon of ionizing radiation will have a fraction of its energy converted to light in certain liquid and solid materials (e.g., a crystal of thallium-activated NaI). The amount of light produced as measured by a sensitive light meter is a measurement of the radiation impinging on the scintillator material.

3. Chemical Methods

Chemical reactions which result from irradiation can be used as a means of measuring dose. Except for systems with irradiators which have a very high dose rate, such as electron beam generators, chemical dose meters have been preferred for the dosimetry needed for food irradiation.

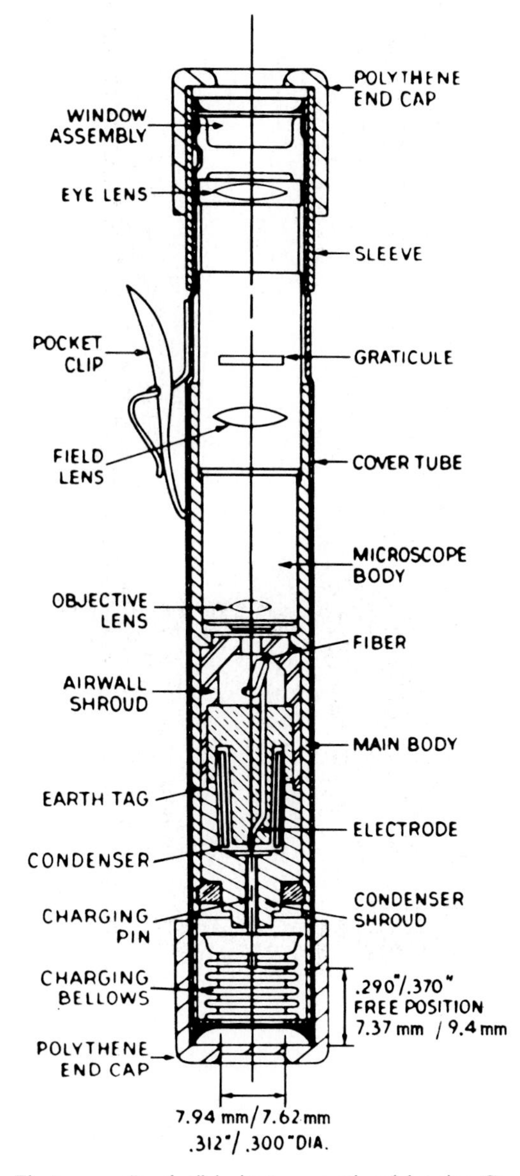

Figure 15.26 Electroscope "pocket" dosimeter—sectional drawing. Courtesy of Baird Corp., Bedford, Massachusetts.

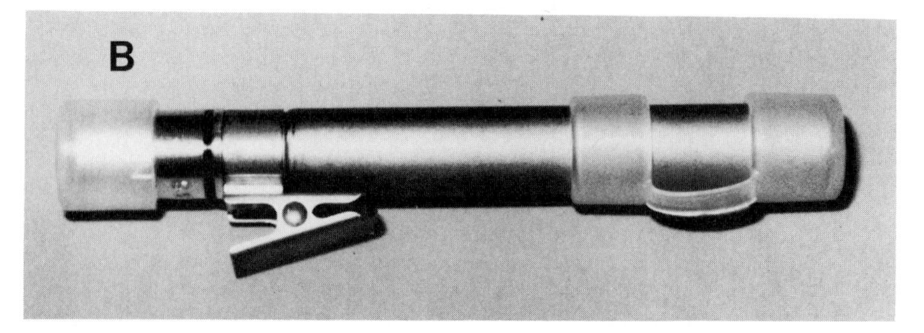

Figure 15.27 Electroscope "pocket" dosimeter (A) and charging unit (B). Courtesy Baird Corp., Bedford, Massachusetts.

a. Reference Dosimetric Systems

The most widely used and accepted reference dosimetric method and recommended for calibration purposes is the ferrous sulfate (Fricke) dose meter. It is based on oxidation of ferrous ions to ferric ions in an acid aqueous solution. The radiation-formed ferric ion concentration is determined spectrophotometrically at 305 nm, the peak wavelength for absorption. The chemical reactions involved in the oxidation are shown below:

$$\text{Radiation} + H_2O = H + OH$$
$$Fe^{2+} + OH = Fe^{3+} + OH^-$$
$$H + O_2 = HO_2$$
$$Fe^{2+} + HO_2 = Fe^{3+} + HO_2^-$$
$$HO_2^- + H^+ = H_2O_2$$
$$Fe^{2+} + H_2O_2 = OH + OH^- + Fe^{3+}$$

In appropriate ranges of dose (40–400 Gy) the oxidation of Fe^{2+} is proportional to dose. The *G* value for Fe^{2+} (number of Fe^{2+} ions oxidized per 100 eV) is 15.6. The accuracy of the measurement is not affected by dose rate between 0.2 and 2×10^7 Gy/sec, nor by irradiation temperatures between 1 and 60°C. The response of the system is essentially independent of radiation energy levels between 0.5 and 16 MeV. Precision within $\pm 1\%$ can be attained with the ferrous sulfate system.

The addition of cupric sulfate to the ferrous sulfate solution reduces the *G* value of Fe^{2+} to 0.66 and extends the usable dose range to 100 to 8000 Gy. A further extension of dose range to above 30 kGy can be secured by modification of the solution.

Details in the use of this dose meter system as given in manuals for dosimetry should be consulted in connection with its use.

The ceric sulfate–cerous sulfate dose meter is somewhat similar to the ferrous sulfate system. Radiation causes the reduction of ceric ions to cerous ions. Electrochemical potentiometry or ultraviolet spectrophotometry is used to determine the dose. The ceric–cerous dose meter is suitable for doses in the range 1–500 Gy and may be used as a reference dosimetric system.

b. Routine Dosimetric Systems

For routine use there are available a number of simple and more convenient chemical dose meters. Among these is the group of polymethylmethacrylate (PMMA) dose meters with commercial names such as Perspex, Lucite, and Plexiglas. These dose meters may be made of clear PMMA or they may be dyed. Irradiation causes a change in absorbance, which is read spectrophotometrically. The absorption wavelength employed varies with the color of the PMMA. The dose limits are governed somewhat by the thickness of the sheet, greater thicknesses (~3 mm) for doses in the range 0.1–25 kGy and lesser thicknesses (~1 mm) for doses in the range 0.3–10 kGy. Manuals of dosimetry giving important details relating to the use of this type of dose meter should be consulted.

Films, solutions, and papers containing triphenylmethane radiochromic leuko dyes such as hexahydroxyl-ethyl-pararosanile cyanide, become colored on irradiation and serve as useful routine dose meters in the range 10–100 kGy. By using a solution of the dye as the core of an optical wave guide, measurement of doses as low as 0.1 Gy can be obtained.

Papers treated with appropriate materials to cause color changes on exposure to radiation can be placed on a product unit to serve as a kind of label to indicate whether that unit has been irradiated or not. These have been termed "go/no-go" dose meters. These "labels" can incorporate other information to assist in record keeping.

A number of other dosimetric systems are available, including thermoluminescent dose meters, solutions of chlorobenzene and water in ethanol, silver-activated phosphate glass dose meters, lyoluminescent dose meters, and measurement of radiation-formed free radicals in crystalline alanine by use of electron spin resonance.

4. General Considerations Regarding the Use of Dose Meters

There are a number of items which need to be considered in the selection and use of dose meters for control of dose in food irradiation processing. Among these are the following:

1. At least initially in establishing a procedure for a given product unit in a particular irradiation facility, dose meters should be placed at a sufficient number of locations to define the spatial dose distribution in the product.
2. Dose meters employed should have dimensions that are relatively small compared with the product unit size receiving irradiation in order that measurements at a number of locations within the unit can be secured.
3. Dose meters employed should be "product equivalent"; that is, they should not differ greatly in density from the product.
4. In order to avoid atypical energy absorption, dose meters should be placed either inside a product unit or surrounded with a sufficient amount of product-equivalent material.
5. The dose meter selected should provide a reproducible adequate response over the dose range used with the product, at the dose rate employed, and with the kind of radiation energy used.
6. The dose meter selected should not be affected significantly by environmental factors that will be encountered, such as temperature, light, and atmosphere.

7. The dose meter selected should have useful stability before and after irradiation, so as to be "practical."

8. The dose meter selected should be convenient to use and not require unusual skill in its use.

IV. PERSONNEL PROTECTION

A. Introduction

Irradiation facilities must be constructed and operated so as to afford protection to operating personnel in particular and to the public in general. All types of sources of radiation must be controlled; radionuclide sources, however, have special problems related to the fact of their continuous production of radiation.

Potential hazards in an irradiation facility, broadly speaking, are of three types: (1) ionizing radiation, (2) toxic substances formed in the atmosphere of the irradiation area of the facility, and (3) contamination of the "pool" water with the radionuclide of the source due to leakage or corrosion of the source capsule (applies only to facilities with radionuclide sources).

B. Radiation Protection

Plant design and construction provide for a "cell" to contain the radiation source. Operators are never inside the cell when the facility is operating. The walls of the cell are between the source and operating personnel when they are outside of the cell and, by absorption, reduce the radiation intensity at the outside surface to a level safe for the operators. Any material which accomplishes the needed reduction of radiation intensity can be used for shielding. In commercial-size plants and for many research facilities the walls usually are made of ordinary concrete. For smaller units, particularly the "self-contained" units, lead may be used. Whatever material is used for shielding, it must completely surround the source, allowing no radiation to pass directly to the exterior in any direction.

The shielding provided must take into account the source strength and the nature of the radiation. Protection with electron radiation must provide for bremstrahlung production as well as the electron radiation. Figures 15.28 and 15.29 indicate the thicknesses of ordinary concrete required to provide adequate shielding for electron radiation and the γ rays of ^{60}Co and ^{137}Cs.

Entry ports to the cell need consideration to prevent leakage of radiation to the outside. Plugs or gates with recessed edges to prevent leakage can be used. Mazes or labyrinths which prevent straight-line escape of radiation can be

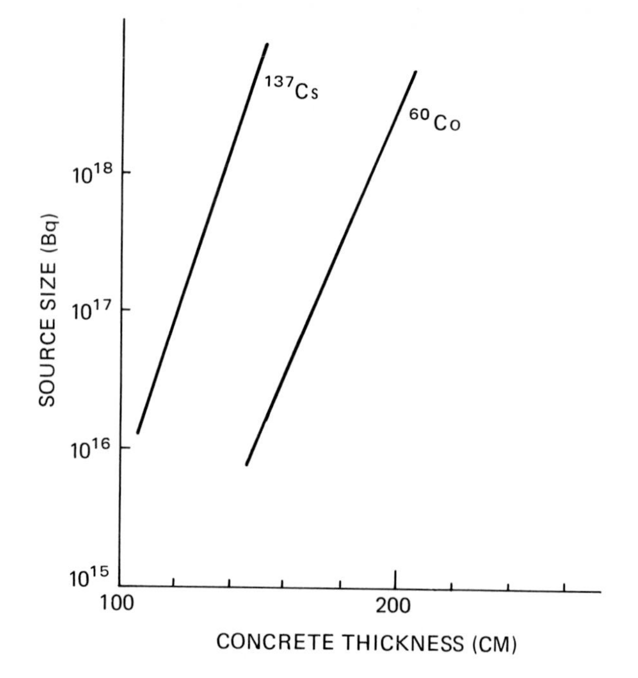

Figure 15.28 Approximate wall thickness of concrete to reduce radiation level at 5 m to 10^{-6} Sv/hr (0.1 m rem) for various sized ^{60}Co and ^{137}Cs gamma sources. Reprinted from T. G. Martin III, Radiation protection and health physics in food irradiation facilities. *In* "Preservation of Food by Ionizing Radiation" (E. S. Josephson and M. S. Peterson, eds.), Vol. I. Copyright CRC Press, Boca Raton, Florida, 1982.

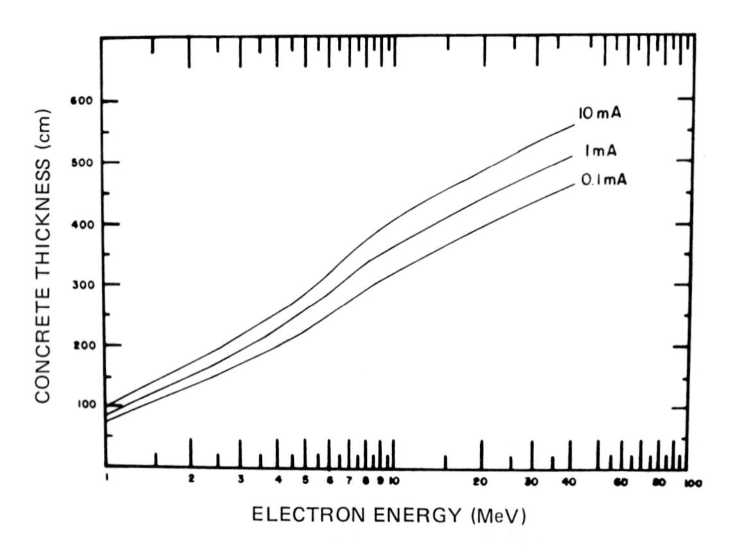

Figure 15.29 Approximate wall thickness of concrete to reduce radiation level at 5 m to 10^{-6} Sv/hr (0.1 m rem) from accelerators producing electrons of varying energies. Upper curve, 10 mA; middle curve, 1 mA; lower curve, 0.1 mA. From A. Brynjolfsson and T. G. Martin III, Bremstrahlung production and shielding of static and linear electron accelerators below 50 MeV. Toxic gas production, required exhaust rates and radiation protection instrumentation. *Int. J. Appl. Radiat. Isot.* **22,** 29 (1971). Copyright (1971) Pergamon, Oxford.

used to provide access to the cell for conveyors or other entry (see Figs. 15.2 and 15.5). Interlocks to prevent accidental entry into the cell by operators or others when the source is operating are required. Appropriate devices located outside the cell should indicate the operating status of the facility.

Radionuclide sources in all but self-contained units ordinarily, when not in use, are stored in "pools," that is, tanks of sufficient depth and filled with water to provide safe shielding for personnel. A mechanism operated from outside the cell to raise the source out of the pool for use is usually employed. Alternatively, the target material may be lowered to the source area, but this is very restrictive in the facility operation and seldom employed.

Water pool shielding is a convenience in handling radionuclide material, especially in transferring it from the shipping container to the source assembly.

C. Atmosphere

No conversion of ^{14}N and ^{16}O to radioactive isotopes occurs with the radiations used in food irradiation. The radiation used, however, converts O_2 to O_3 (ozone), which is toxic. Radiation also produces oxides of nitrogen (NO, NO_2, NO_3, N_2O, N_2O_3, etc.). To remove these substances the irradiation cell area must be well ventilated and the discharged air vented so as to not cause an environmental hazard. After use and when the source is placed in a nonoperating mode, it may be necessary to delay entry into the cell area for a period needed to secure a safe O_3 concentration in the cell air. The maximum permissible safe level of O_3 in air is 0.1 cm^3/m^3.

D. Pool Water Contamination

Because of the possibility that leakage of the capsule in which the radionuclide is contained might occur and lead to contamination of the pool water, a system for filtration of the water is provided. Periodic testing of the water for radioactivity is done to maintain a check on possible contamination.

E. Monitoring of Radiation Exposure of Operating Personnel

Personnel who regularly are concerned with the operation of an irradiation facility are required to wear a radiation exposure monitoring device. Two general types of monitors are used: (1) a photographic film badge, which periodically on a regular basis is collected and processed to be read for the recorded radiation exposure, and (2) a portable electroscope or equivalent device which is used for short-term monitoring and which can be read promptly on location.

Other dose-metering devices may be used at various times in connection with the operation of the irradiation facility.

Table 15.9

Maximum Permissible Ionizing Radiation Exposure of Humans per Year

	Annual exposure limit (Sv)	
Body part	Public	Occupational
Whole body	0.005	0.05
Gonads, red bone marrow	0.005	0.05
Skin, bone, thyroid	0.03	0.3
Hands, forearms, feet, ankles	0.075	0.75
Other single organs	0.015	0.15

There are government regulatory standards for allowable limits of exposure of operating personnel. Those whose work involves possible radiation exposure are allowed greater amounts than the general public. Different limits have been established for the different parts of the body. In food irradiation, the concern, however, is for external exposure of the whole body. The standard for whole-body exposure is the same as that for the more sensitive body parts, namely, gonads and red bone marrow. The maximum exposure permitted for the whole body by the standards of International Commission on Radiological Protection (1969) is 0.05 Sv/year (see Table 15.9). This translates to 1 mSv/week. Generally persons under 18 years of age are excluded from irradiation facilities.

While the standard for maximum exposure of food irradiation facility workers is 1 mSv/week, good management in practice will seek to reduce exposure of personnel to amounts less than this amount, and generally to minimize exposure. Reference may be made to the Codex Alimentarius Code of Practice for food irradiation facilities (see Chapter 14, Section III).

Usually there are particular government regulatory standards regarding operation of facilities employing ionizing radiation and the safety of operating personnel. Usually a radiation facility operates under a license which requires compliance with such standards. In the United States such regulation may be at the federal or state level.

SOURCES OF ADDITIONAL INFORMATION

Consumer Acceptance

Anonymous, "Marketing and Consumer Acceptance of Irradiated Foods. Report of Consultants' Meeting," IAEA TEDOC-290. Int. Atomic Energy Agency, Vienna, 1983.

Anonymous, "Consumer Reaction to the Irradiation Concept. U.S. Dept. of Energy, Albuquerque, New Mexico and Natl. Pork Producers Council, 1984.

Anonymous, "Marketability Testing of Irradiated Fish and Seafood. Concept Development and Testing." Dept. of Fisheries and Oceans, Ottawa, 1984.

Defesche, F., How does the consumer react to irradiated food? *In* "Food Irradiation Now." Nijhoff/Junk, The Hague, 1982.

Urbain, R. W., and Urbain, W. M., Marketing and labelling of radiation insect-disinfested foods. *In* "Radiation Disinfestation of Food and Agricultural Products" (J. H. Moy, ed.). Univ. of Hawaii Press, Honolulu, 1985.

Urbain, W. M., Radiation update. *In* "Feeding the Military Man." U.S. Army Natick Labs., Natick, Massachusetts, 1970.

van der Linde, H. J., Marketing Experience with Radurized Products in South Africa. *In* "Ionizing Energy Treatment of Food, Proceedings of Symposium." ISBN 0 85856 053 4. Sydney. 1983.

van Kooij, J. G., Consumer attitudes toward food irradiation in The Netherlands. *In* "Requirements for the Irradiation of Food on a Commercial Scale." Int. Atomic Energy Agency, Vienna, 1975.

Economics

Anonymous, "Radiation-Pasteurizing Fresh Strawberries and Other Fresh Fruits and Vegetables: Estimates of Costs and Benefits." U.S. Atomic Energy Comm., Washington, D.C., 1965.

Anonymous, "The Commercial Prospects for Selected Irradiated Foods," TID 24058. U.S. Dept. of Commerce, Washington, D.C., 1968.

Baraldi, D., The Fucino food irradiation project and its economic feasibility. Contribution to the savings of energy within the food preservation industry. *Food Irrad. Newsl.* **7**(2), 32 (1983).

Deitch, J., Cost considerations for the establishment and operation of a commercial food irradiation facility. *In* "Radiation Preservation of Food, Proceedings of Symposium, Bombay." Int. Atomic Energy Agency, Vienna, 1973.

Lagunas-Solar, M. C., and Matthews, S. M., "Radionuclide and Electric Accelerator Sources for Food Irradiation." *Rad. Physics Chem.* **25**(4–6), 691 (1985).

Leemhorst, J. G., Economics of irradiator operation as a service facility. *In* "Ionizing Energy Treatment of Foods, Proceedings of Symposium." ISBN 0 85856 053 4. Sydney. 1983.

Sudarsan, P., Techno-economic and commercial feasibility of food irradiation with special reference to developing countries. *Food Irradiat. Newsl.* **7**(2), 19 (1983).

Umeda, K., Commercial experience with the Shihoro potato irradiator. *Food Irradiat. Newsl.* **7**(3), 19 (1983).

Urbain, W. M., Review of factors and conditions influencing the economics of food irradiation applications. *Food Irradiat. Newsl.* **7**(1), 36 (1983).

van der Linde, H. J., Economic considerations for the irradiation preservation of foods in South Africa. *Food Irradiat. Newsl.* **7**(3), 32 (1983).

Irradiation Facilities

Charlesby, A., "Radiation Sources." Pergamon, Oxford, 1964. Distributed in U.S. by Macmillan, New York.

Döllstädt, R., A new onion irradiator. *Food Irradiat. Newsl.* **8**(2), 40 (1984).

Eicholz, G. G., ed. "Radiosotope Engineering." Dekker, New York, 1972.

Fraser, F. M., Gamma radiation processing equipment and associated energy requirements in food irradiation. *In* "Combination Processes in Food Irradiation." Int. Atomic Energy Agency, Vienna, 1981.

Gay, R. G., Design and operation of radiation facilities. *In* "Ionizing Energy Treatment of Foods, Proceedings." ISBN 0 85856 053 4, Sydney. 1983.

Jarrett, R. D., Sr., Isotope (gamma) radiation sources. *In* "Preservation of Food by Ionizing Radiation," (E. S. Josephson and M. S. Peterson, eds.), Vol. I. CRC Press, Boca Raton, Florida, 1982.

McMullen, W. H., and Sloan, D. P., Cesium-137 as a radiation source. *In* "Radiation Disinfestation of Food and Agricultural Products" (J. H. Moy, ed.). Univ. of Hawaii Press, Honolulu, 1985.

Ramler, W. J., Machine sources. *In* "Preservation of Food by Ionizing Radiation," (E. S. Josephson and M. S. Peterson eds.), Vol. I. CRC Press, Boca Raton, Florida, 1982.

Sloan, D. P., and Ahlstrom, S. B., *TPCI* (transportable cesium irradiator): A tool for commodity irradiation research. *In* "Radiation Disinfestation of Food and Agricultural Products" (J. H. Moy, ed.). Univ. of Hawaii Press, Honolulu, 1985.

Welt, M. A., A commercial multipurpose radiation processing facility for Hawaii. *In* "Radiation Disinfestation of Food and Agricultural Products" (J. H. Moy, ed.). Univ. of Hawaii Press, Honolulu, 1985.

Dose Control

Anonymous, "Dosimetry in Agriculture, Industry, Biology and Medicine, Proceedings." Int. Atomic Energy Agency, Vienna, 1973.

Anonymous, "Manual on Radiation Sterilization of Medical and Biological Materials," Tech. Rep. Series No. 149. Int. Atomic Energy Agency, Vienna, 1973.

Anonymous, "Radiosterilization of Medical Products, 1974 Proceedings." Int. Atomic Energy Agency, Vienna, 1975.

Anonymous, "Manual of Food Irradiation Dosimetry," Tech. Rep. Series No. 178. Int. Atomic Energy Agency, Vienna, 1977.

Holm, N. W., and Berry, R. J., eds. "Manual on Radiation Dosimetry." Dekker, New York, 1970.

McLaughlin, W. L., Jarret, R. C., Sr., and Olejnik, T. A., Dosimetry. *In* "Preservation of Food by Ionizing Radiation," (E. S. Josephson and M. S. Peterson, eds.), Vol. I. CRC Press, Boca Raton, Florida, 1982.

Personnel Protection

Aglintsev, K. K., ed. "Applied Dosimetry." Iliffe, London, 1965.

Anonymous, "The Basic Requirements for Personnel Monitoring." Int. Atomic Energy Agency, Vienna, 1965.

Brodsky, A., "Handbook of Radiation Measurement and Protection." CRC Press, Boca Raton, Florida, 1985.

Brynjolffson, A., and Martin, T. G., III., Bremstrahlung production and shielding of static and linear electron accelerators below 50 MeV. Toxic gas production, required exhaust rates and radiation protection instrumentation. *Int. J. Appl. Radiat. Isot.* **22,** 29 (1971).

Henry, H. F., "Fundamentals of Radiation Protection." Wiley-Interscience, New York, 1969.

APPENDIX A

Literature Sources on Food Irradiation

The literature on food irradiation is widely scattered. Much of it, especially reports on original research, is in the regular scientific journals, usually in food or food-related journals. Occasional articles have appeared in trade journals. A number of general review articles on food irradiation and also on certain particular aspects, such as wholesomeness, are available.

The *International Food Information Service*, published monthly in England (Lane End House, Shinfield, Reading RG2 9BB), provides abstracts of papers in publications on a worldwide basis and is a good key to the regular scientific journals, as well as other publications.

There are a few books on food irradiation published by commercial publishing houses and governments. Among those are the following:

R. S. Hannan. "Scientific and Technological Problems Involved in Using Ionizing Radiations for the Preservation of Food, Department of Scientific and Industrial Research Food Investigation, Special Report No. 61. Her Majesty's Stationery Office, London, 1955.

"Radiation Preservation of Food." U.S. Army Quartermaster Corps. U.S. Government Printing Office, Washington, D.C. 1957.

L. E. Brownell. "Radiation Uses in Industry and Science." U.S. Atomic Energy Commission. U.S. Government Printing Office, Washington, D.C., 1961.

L. V. Metlitskii, V. N. Rogachev, and V. G. Krushchev, "Radiation Processing of Food Products," ORNL-11C-14. Translation of Original Russian Oak Ridge National Laboratory, Oak Ridge, Tennessee, 1968.

N. W. Desrosier and H. M. Rosenstock. "Radiation Technology in Food, Agriculture and Biology." AVI, Westport, Connecticut, 1960.

P. S. Elias and A. J. Cohen, Eds., "Radiation Chemistry of Major Food Components." Elsevier, New York. 1977.

E. S. Josephson and M. S. Peterson, eds., "Preservation of Food by Ionizing Radiation," Vols. I, II, III. CRC Press, Boca Raton, Florida, 1982, 1983.

P. S. Elias and A. J. Cohen, eds., "Recent Advances in Food Irradiation." Elsevier, New York. 1983.

Several journals devoted to food irradiation have been published at various times, but in some instances are no longer available as current publications. These journals include the following:

Food Irradiation, Japan
Japanese Research Association for Food Irradiation
1966–

Food Irradiation
Quarterly International Newsletter
European Nuclear Energy Agency (OECD) and the Commissariat à l'Energie Atomique, Saclay, France
1960–1971

Food Irradiation Information
International Project in the Field of Food Irradiation
Karlsruhe, Federal Republic of Germany
1972–1981

Food Irradiation Newsletter
Joint FAO/IAEA Division of Isotope and Radiation Applications of Atomic Energy for Food and Agriculture Development
International Atomic Energy Agency, Vienna
1977–

Isotopes and Radiation Technology
U.S. Atomic Energy Commission,
Washington, D.C.
1963–1972

In the past, and continuing to some degree presently, a great deal of information was recorded in reports which have had only limited distribution. Some were contractor reports to government agencies and were not always distributed through usual government channels. Some were reports of multinational organizations which also had limited distribution. Some were proceedings of meetings or conferences published by public or private groups. In the totality of these reports is a great deal of information which, while very difficult to secure, is very much worth the effort to locate.

The International Atomic Energy Agency (IAEA) in Vienna is the source of a number of important publications on food irradiation. A list of these publications is given below.

"Application of Food Irradiation in Developing Countries" (1966)
"Aspects of the Introduction of Food Irradiation in Developing Countries" (1973)
"Combination Processes in Food Irradiation" (1981)
"Decontamination of Animal Feeds by Irradiation" (1979)
"Disinfestation of Fruit by Irradiation" (1971)
"Elimination of Harmful Organisms from Food and Feed by Irradiation" (1968)
"Enzymological Aspects of Food Irradiation" (1969)
"Factors Influencing the Economical Application of Food Irradiation" (1973)
"Improvement of Food Quality by Irradiation" (1974)

"Microbiological Problems in Food Preservation by Irradiation" (1967)
"Microbiological Specifications and Testing Methods for Irradiated Food" (1970)
"Preservation of Fish by Irradiation" (1970)
"Preservation of Fruit and Vegetables by Radiation" (1968)
"Radiation Control of Salmonellae in Food and Feed Products" (1963)
"Radurization of Scampi, Shrimp and Cod" (1971)
"Requirements for the Irradiation of Food on a Commercial Scale" (1975)
"Training Manual on Food Irradiation Technology and Techniques" (2nd Ed., 1982)

In addition, the IAEA publishes proceedings of symposia and seminars on food irradiation, among them the following:

1. "Food Irradiation,"
   Proceedings of Symposium, Karlsruhe. 1966.
2. "Radiation Preservation of Food,"
   Proceedings of Symposium, Bombay. 1973.
3. "Food Preservation by Irradiation," Vols. I and II.
   Proceedings of Symposium, Wageningen. 1977.
4. International Symposium on Food Irradiation Processing. Proceedings. Washington, D.C. 1985.
5. "Food Irradiation for Developing Countries in Asia and the Pacific," Proceedings of a seminar. IAEA-TECDOC-271. Tokyo. 1981.
6. "Marketing and Consumer Acceptance of Irradiated Foods," Report of Consultants' Meeting, IAEA-TECDOC-290. Vienna. 1982.
7. "Use of Irradiation as a Quarantine Treatment of Agricultural Commodities," IAEA-TECDOC-326. Honolulu. 1983.

IAEA publications may be secured from:

UNIPUB
Box 433
Murray Hill Station
New York, N. Y. 10157

The U.S. Department of Agriculture has established at its National Agricultural Library in Beltsville, Maryland, a national Food Irradiation Information Center. Access to available information is through a computerized data base system.

APPENDIX B

Symbols Used in This Book

A_w	Water activity
β	Beta particle; an electron
Bq	Becquerel
c	Velocity of electromagnetic radiation
Ci	Curie
D	Absorbed dose
$\dot{D}$	Dose rate
D_{10}	Decimal reduction dose
e	Base of natural logarithm
e^-	Electron
e^+	Positron
e_{aq}^-	Aqueous electron
eV	Electron volt
E	Energy
F_0	Thermal death time in minutes at 250°F (122.1°C)
g	Gram
γ	Gamma ray (or radiation)
G value	Number of molecules changed per 100 eV
Gy	Gray
h	Planck's constant
Hz	Hertz, cycles per second
I	Radiation intensity after passing through an absorber
I_0	Intensity of incident radiation
J	Joule
k	Kilo
λ	Wavelength of electromagnetic radiation
m	Milli; meter
m_0	Rest mass of a body
m_v	Mass of a moving body
μ	Linear absorption coefficient
μm	Micrometer
M	Mega

n	Neutron
ν	Frequency of electromagnetic radiation
N	Final or surviving population
N_0	Initial population
p	Proton
Pa	Pascal
r	Roentgen
ρ	Density
R	Roentgen
rad	Radiation absorbed dose
rem	Roentgen equivalent man
rep	Roentgen equivalent physical
Sv	Sievert
U	Dose uniformity ratio
Υ	Mass absorption coefficient
v	Velocity
V	Volt
W	Watt
x	Thickness
Z	Atomic number

APPENDIX C

Glossary of Terms Used in Food Irradiation

Term	Symbol	Definition
Absorbed dose	D	See dose
Absorber	—	Any material in which radiation is deposited and which reduces the intensity of the radiation
Absorption	—	Process by which a material reduces the intensity of radiation
Absorption coefficient, linear	μ	Fraction of energy absorbed per unit length of travel
Absorption coefficient, mass	Υ	Fraction of energy absorbed per gram of material having a surface area of 1 cm^2
Accelerator	—	In food irradiation, a device for producing electron beams by imparting energy to electrons through acceleration
Alpha particle	α	Positively charged particle emitted from a nucleus and composed of two protons and two neutrons; identical in all measured properties with the nucleus of a helium atom
Aqueous electron	e^-_{aq}	Hydrated electron, a radiolytic product of water
Atomic number	Z	Number of protons in an atomic nucleus
Becquerel	Bq	Newer unit to indicate the intensity of radioactivity of a radionuclide; replaces the curie (1 Bq equals one disintegration per second; 3.7×10^{10} Bq = 1 Ci)
Beta particle	β^+ or β^-	A charged elementary particle emitted from nucleus during radioactive decay and having a mass and charge equal in magnitude to those of the electron. A negatively charged β particle is physically identical to the electron
Bremstrahlung	—	Electromagnetic radiation formed when a fast-moving particle interacts with an atomic nucleus and some of the kinetic energy of the particle is converted to radiation

Term	Symbol	Definition
Bulk density	—	Mean density of target material of the irradiation process through which the radiation travels. The densities of air spaces between product units, packaging material, as well as food itself are used to secure the "bulk" density
Cathode ray	—	Stream of electrons emitted by cathode of a gas discharge tube or by a hot filament in a vacuum tube; electron beams generated by accelerators and used in food irradiation
Cerenkov radiation	—	Light emitted when a high-velocity particle leaves one medium and enters another, such that the velocity of light in the second medium is less than that in the first
Curie	Ci	Older unit to indicate the intensity of radioactivity of a radionuclide; replaced by the becquerel (1 Ci = 3.7×10^{10} Bq)
Decimal reduction dose	D_{10}	Radiation dose to reduce a population (e.g., of bacteria) by factor of 10, or one log cycle (10% survivors)
Disinfestation	—	In food irradiation, the inactivation of food-borne insects
Dose	D	Amount of ionizing radiation absorbed by a material; unit for dose is the gray
Dose distribution	—	Variation of dose throughout the target material
Dose equivalent index	H	Index of biological effectiveness of different kinds of ionizing radiation relative to 200-keV X rays. If the dose is in grays, H is in sieverts. This index replaces the previously used relative biological effectiveness
Dose meter	—	Device for measuring dose
$\text{Dose}_{min}-\text{Dose}_{max}$	—	Mean minimum and maximum doses obtained in irradiating a material
Dose rate	$\dot{D}$	Radiation dose per unit time
Dose uniformity ratio	U	Ratio of maximum to minimum absorbed dose in the absorber ($U = D_{max}/D_{min}$)
Dosimetry	—	Process of measuring dose
Electron	e^-	A negatively charged elementary particle that is a constituent of all atoms; has a mass of $\frac{1}{1873}$ that of a proton, or 9.1×10^{-28} g
Electron beam	—	Stream of electrons moving essentially in one direction
Electron volt	eV	Amount of kinetic energy gained by an electron accelerated through an electric potential difference of 1 V (1 eV = 1.6×10^{-19} J). 1 eV absorbed per gram equals 1.6×10^{-16} Gy)

Term	Symbol	Definition
Element	—	Any one of chemical substances that cannot be separated into simpler substances by chemical means; a substance all of whose atoms have the same atomic number
Erg	—	Unit of energy equal to 1 dyne cm; 10^7 ergs = 1 J
Excitation	—	Consequence of the absorption of energy by an atom (or molecule) which moves an orbital electron to a higher than normal energy level while still retaining the electron within the atom. Such atoms are said to be "excited" and while in the excited state can become involved in chemical changes. Physical phenomena, such as fluorescence or heat formation, also may result from excitation
Film badge	—	Radiation monitoring device for personnel made of photographic film enclosed in material transparent to ionizing radiation and opaque to light; amount of darkening of film correlates with amount of radiation exposure.
Free radical	—	An electrically neutral molecule with an unpaired electron in the outer orbit. The · placed in this manner, as in OH·, designates a free radical
G value	—	Number of molecules changed per 100 eV of energy transferred to the system. To convert to SI units (molecules changed J^{-1}), divide *G* value in electron volts by 1.6×10^{-19}
Gamma ray	γ	A high-frequency electromagnetic radiation produced when an unstable atomic nucleus releases energy to gain stability
Genetic effects	—	Physiological effects on a living organism caused by radiation which can be transferred from the exposed individual to its progeny
Gray	Gy	International System of Units (SI) unit for absorbed dose; equal to 1 J/kg; replaces older unit, the rad (1 Gy = 100 rad)
Half-life	—	The time necessary for half any starting amount of the atoms of a radionuclide to decay
Half-thickness	—	Thickness of an absorber which will reduce the radiation intensity to one-half its initial value
Hertz	Hz	For electromagnetic radiation, frequency, or cycles per second
High dose		In food irradiation, doses of 10 kGy or more

Term	Symbol	Definition
Induced radioactivity	—	Radioactivity resulting from certain nuclear reactions in an absorber in which exposure to radiation results in the production of unstable nuclei, which through spontaneous disintegrations give off radiation
Ion	A^+ or A^-	An atom or molecule bearing an electrical charge, either positive or negative, caused by an excess or deficiency of electrons
Ionization	—	Process of adding to or displacing electrons from atoms or molecules, thereby creating ions
Ionizing radiation	—	Radiation having the capability of displacing orbital electrons from an atom and in this way forming ions
Irradiate	—	To apply radiation to a material
Irradiation	—	Process of applying radiation
Isotope	${}^{A}_{Z}X$	Atoms of the same chemical element having the same atomic number but with different atomic weights; those with nuclei having the same number of protons but different numbers of neutrons [X is the symbol of the element, Z the atomic number (number of protons), and A the mass number (sum of number of protons and neutrons); often the Z subscript is dropped, since X adequately identifies the element—example: ${}^{60}Co$ represents the isotope of cobalt having a mass number of 60; a commonly used alternative symbol is X-A, e.g., Co-60]
Low dose	—	In food irradiation, doses less than 1 kGy
Medium dose	—	In food irradiation, doses in the range 1–10 kGy
Neutron	n	An elementary particle having zero charge and a mass approximately that of the proton
Nuclide	—	Any atomic form of an element
Photon	—	A quantum of electromagnetic energy
Positron	e^+	An elementary particle having the same mass as an electron, but positively charged
Proton	${}^{1}_{1}H$ or p	An elementary particle with unit positive charge and unit mass; the nucleus of a hydrogen atom
Quantum	—	Unit quantity of energy, equal to radiation frequency times 6.62×10^{-27} erg sec (one quantum equals one photon)
Radappertization	—	Treatment of food with a dose of ionizing radiation sufficient to reduce the number and/or activity of viable microorganisms

Term	Symbol	Definition
		(viruses excepted) to such a level that very few, if any, are detectable by any recognized bacteriological or mycological testing method applied to the treated food. This treatment must be such that no spoilage or toxicity of microbial origin is detectable no matter how long or under what conditions the food is stored after treatment, provided it is not recontaminated
Radiation	—	Radiant energy; in food irradiation, term is limited to γ rays, X rays, and electron beams
Radiation absorbed dose	rad	Outdated term for absorbed dose (1 rad = 100 ergs of absorbed energy per gram; 1 rad = 10^{-2} Gy)
Radicidation	—	Treatment of food with a dose of ionizing radiation sufficient to reduce the number of viable specific non-spore-forming pathogenic bacteria to such a level that none is detectable in the treated food when it is examined by any recognized bacteriological testing method; term may also be applied to parasites
Radioactivity	—	Spontaneous disintegration of an atomic nucleus (radionuclide) which yields ionizing radiation
Radiolysis	—	Process of chemical changes in a substance caused by radiation
Radiolytic product	—	Substance formed by direct or indirect action of radiation on a material such as a food
Radionuclide	—	Unstable nuclide that decays or disintegrates spontaneously, emitting radiation; replaces older term, radioisotope
Radurization	—	Treatment of food with a dose of ionizing radiation sufficient to enhance its keeping quality by causing a substantial reduction in the numbers of viable specific spoilage microorganisms
Relative biological effectiveness	RBE	Obsolete term now replaced by the dose equivalent index, H
Roentgen	r or R	The dose of γ or X radiation producing ion pairs carrying one electrostatic unit of charge per cubic centimeter of standard air surrounded by air; equal to 88 ergs/g air
Roentgen equivalent man	rem	Obsolete unit of dose equivalent now replaced by the sievert (1 rem = 10^{-2} Sv)
Roentgen equivalent physical	rep	Obsolete term for radiation dose in material other than air; equals to 93 ergs/g

Term	Symbol	Definition
Sievert	Sv	The unit of dose equivalent used for personnel protection purposes; replaces the older unit, the rem. If the dose is in grays, 1 Sv = 100 rem. The sievert is the unit of dose for any ionizing radiation which produces the same biological effect as a unit of absorbed dose of 200 keV X rays. The dose equivalent index (H) indicates the relative biological effectiveness of different types of ionizing radiation. For X rays, γ rays, and electrons of energy greater than 300 keV, $H = 1$
Somatic effects	—	Physiological effects in a living organism caused by radiation; limited to the exposed individual
Source efficiency	—	Percentage of total ionizing energy available from a radiation source that is absorbed by the target material
Source strength	—	Amount of radioactivity in a γ-radionuclide source, expressed as curies or becquerels
Unit prefixes		
pico-	p	10^{-12}
nano-	n	10^{-9}
micro-	μ	10^{-6}
milli-	m	10^{-3}
kilo-	k	10^{3}
mega-	M	10^{6}
X ray	—	A high-frequency electromagnetic radiation produced when high-energy charged particles (e.g., electrons) impinge on a suitable target material
Wholesomeness	—	In connection with irradiated foods, a food is wholesome when the following characterize it: (1) no microorganisms or microbial toxins harmful to humans present, (2) no significant deficiency in nutritional quality relative to the same food not irradiated present, and (3) insignificant amounts, if any, of toxic products formed in the food as a result of irradiation present

Note: The International System of Units (SI) is the worldwide official system as of 1986. In this system, the unit for absorbed dose is the gray (Gy). In accord with this system, the gray is used throughout this book, except when certain material is taken directly from earlier publications and it would be difficult to rework it to use the gray as a unit for dose. As a consequence, the reader will observe some usage of the older now obsolete units, namely, rad and rep.

The rad and the rep are almost the same, the rad being the absorption of 100 ergs/g, and the rep, 93 ergs/g. Generally, in food irradiation this small difference is seldom important. It is suggested, therefore, that the rep and the rad ordinarily be regarded as the same.

Conversion of rad or rep to grays involves simply dividing either by 100, since 100 rad equal 1 Gy. Below is given a table which may be helpful in converting rads (or reps) to grays:

Conversion of Absorbed Dose Units

Rad	Kilorad	Megarad	Gray	Kilogray
rad	krad	Mrad	Gy	kGy
100	0.1	0.0001	1	0.001
1,000	1.0	0.001	10	0.01
10,000	10.0	0.01	100	0.1
100,000	100.0	0.1	1,000	1.0
1,000,000	1,000.0	1.0	10,000	10.0

Index

C

G

H

I

N

O

P

Q

R

S

T

V

W